Structural Details
Manual

Structural Details Manual

David R. Williams

McGraw-Hill

New York   San Francisco   Washington, D.C.   Auckland   Bogotá
Caracas   Lisbon   London   Madrid   Mexico City   Milan
Montreal   New Delhi   San Juan   Singapore
Sydney   Tokyo   Toronto

Library of Congress Cataloging-in-Publication Data

Williams, David R.
   Structural Details Manual / David R. Williams.
      p.   cm.
   Includes index.
   ISBN 0-07-070443-0
    1. Building—Details. I. Title
TH2025.W55   1999
691'.2—dc21   98-43380
                                      CIP

McGraw-Hill
    A Division of The McGraw-Hill Companies

1 2 3 4 5 6 7 8 9 0   DOC/DOC   9 0 3 2 1 0 9 8

ISBN 0-07-070443-0

The sponsoring editor for this book was Larry Hager, the editing supervisor was David E. Fogarty, and the production supervisor was Clare Stanley. It was set in Goudy by North Market Street Graphics.

Reprinted by arrangement with The McGraw-Hill Companies, Inc.

McGraw-Hill books are available at special quantity discounts to use as premiums and sales promotions, or for use in corporate training programs. For more information, please write to the Director of Special Sales, McGraw-Hill, 11 West 19th Street, New York, NY 10011. Or contact your local bookstore.

About the Cover: The cover photograh is a two-story Cenit Bank building in Newport News, Virginia. The architect was Caro, Monroe, Liang Architects in Newport News. The structural engineer was Williams Engineering Associates, P.C., in Virginia Beach Virginia.

Contents

Preface

The purpose of this book is to provide a set of practical *structural* details, commonly referred to as "typical structural details." The book is intended to be used by engineers, architects, contractors, subcontractors, universities, industry, and others involved with building design and construction.

The author is a self-employed registered structural engineer with almost 20 years of practical engineering experience. All details were drawn by the author using computer-aided drafting and design (CADD). The details are arranged in a Construction Specifications Institute (CSI)–related format: foundations (CSI Division 2), concrete (CSI Division 3), masonry (CSI Division 4), steel (CSI Division 5), and wood (CSI Division 6). The appendices contain structural notes and schedules. It is anticipated that in the near future these details, notes, and schedules will be offered in a computer-based (.DXF and/or .DWG file) format.

The book contains a general discussion for the structural systems included within each CSI division. A brief commentary is provided for each detail. Due to spacing considerations, the commentaries are grouped together within each CSI division, as opposed to placing the commentary on the same sheet as the detail. The details are generally taken from actual projects that the author has been involved with in Virginia and the Mid-Atlantic region of the United States. The applicable building code for the state of Virginia is the BOCA National Building Code, published by the Building Officials & Code Administrators International. The model code has a major update every 3 years.

Of course, this book is no substitute for the services of a structural engineer. The field of structural engineering is extremely technical and changes occur very rapidly. It takes a determined effort to understand the latest building code requirements and stay abreast of current design methods. Continuing education in the structural engineering field is absolutely necessary to provide the most appropriate and economical structural solutions to each and every project. Methods which may have worked years ago or even fairly recently, may not be appropriate or economical today. Nonengineers may be competent to design some structural systems; however, *only a registered structural engineer* should be tasked with preparing design calculations, reviewing the drawings, and ensuring that the details are appropriate for the intended application.

Another important consideration is for the design professional to be actively involved during the construction phase. It is one thing to put a design on paper but quite another to see it under construction. Details which may seem great in the office are often found to be inappropriate or too complex when seen in the field. Furthermore, it is often a good idea to get feedback from those actually doing the construction work as to whether the design could be improved upon. However, this frequently leads to design challenges, so be prepared to explain why it was detailed as shown.

There are many people who offered valuable assistance during the preparation of this work. My brother Johnny, a fiction writer in his own right, provided the original inspiration, and he also helped during the editing process. My parents, wife, and daughter were always there, providing support during the long hours while I was working on the book while continuing my full-time consulting practice. I thank my many clients for giving me the opportunity to be of service to them. I thank Bill Booth for his help. I also wish to thank everyone in McGraw-Hill's New York office who helped during the process.

The author's ultimate goal in his own practice is to provide a well-detailed set of drawings, on schedule, which meets the current building code requirements, achieves the owner's needs, and represents the most practical and economical structural solutions available in today's marketplace. It is hoped that this book will help others reach these goals in their own endeavors.

David R. Williams

Structural Details
Manual

Foundations

(CSI Division 2)

Discussion—Foundations

TYPES OF DETAILS INCLUDED

FOUNDATION DETAILS ARE PROVIDED IN THIS CHAPTER FOR THE FOLLOWING:

- ▶ *Shallow Foundation Systems—Column and Wall Footings*
- ▶ *Deep Foundation Systems—Pile Caps and Grade Beams*
- ▶ *Column Base Plates and Anchor Bolts*
- ▶ *Slabs on Grade*
- ▶ *Industrial Building Foundation Sections*
- ▶ *Miscellaneous Foundation Sections and Details*

Shallow Foundation Systems—Column and Wall Footings

Column and wall footings that bear directly on near surface soil or rock strata are considered *shallow foundation systems*. Column and wall footings must be properly sized so that the allowable bearing capacity of the supporting strata is not exceeded. Column and wall loads are calculated for the most critical loading combinations, including full dead load and live load. Live load reduction can be considered where appropriate, such as for columns in multistory buildings. For single-story buildings, wind uplift may be the most critical loading condition. In this case, it is conservative to neglect the live loads and use only a percentage of the anticipated dead loads.

The allowable soil bearing capacity should be determined through a comprehensive geotechnical investigation that includes soil borings and a report. Can this investigation be eliminated for smaller projects? The 1996 BOCA building code allows buildings less than 40 ft or three stories in height to be designed based on a presumptive soil bearing capacity. For many single-story buildings, wind uplift is the major consideration, and the allowable soil bearing capacity is not as critical. However, the model codes provide *minimum requirements*, and the local governing jurisdiction may require a soil investigation and report for all projects.

The scope of the geotechnical investigation will vary depending on the project. For most projects, competitive prices are obtained from several geotechnical engineers. For low-rise buildings, a minimum of three to four soil borings should be provided. A single soil boring would provide very little substantive information with which to determine the foundation design criteria. The boring depth should be left up to the geotechnical engineer; however, 20 ft is a minimum.

Where should soil borings be located? The boring locations can be dictated to the geotechnical engineer; however, due to liability concerns, it is preferable to leave the exact locations up to the geotechnical engineer. If a boring location is dictated and an underground utility is hit and damaged, who is responsible? If this decision is left up to the geotechnical engineer, unforeseen site obstructions can be avoided in the field. It also allows soil borings to be placed in areas of obvious concern to the geotechnical engineer, such as in low areas or filled-in areas.

What information will the geotechnical engineer require from the design professional to complete the geotechnical investigation? The maximum anticipated column and wall loads are important considerations. The building size, description, and structural systems should also be provided.

If the client or owner decides not to provide the geotechnical investigation and report, it is prudent to issue a disclaimer letter to the effect that, for the record, a geotechnical investigation is recommended and only through a geotechnical investigation can the presence of any unsuitable soil layers be determined. These unsuitable layers may result in foundation settlements that may cause cosmetic or serious structural damage to the building. However, unless otherwise directed in writing, an allowable soil bearing capacity of 1500 psf will be assumed.

of the indicator test piles (and load tests), the production pile lengths and driving criteria can be established.

Foundation loads are transmitted to the individual piles through concrete pile caps. Concrete grade beams span between pile groups, allowing more efficient utilization of the piles. Timber piles are normally spaced at least 3'-0" apart. The arrangement of piles within pile groups is such that the piles are relatively evenly loaded under concentric gravity loads. Pile driving tolerances within pile groups should be kept to within several inches. Out of tolerance piles within a group will result in some piles being loaded more than others. For this reason, pile foundation systems are normally somewhat overdesigned.

Pile cap stability is a major consideration. An individual pile below a column presents an unstable situation, unless the pile is exactly centered below the column. Due to pile driving tolerances, the pile may be several inches away from the column centerline, resulting in an unstable condition. There are several options to provide stability. Grade beams can be framed into the pile cap in at least two directions, sized for the potential bending moments that result from the pile eccentricities. Another option is to use a larger pile cap, such as a three-pile cap, which is inherently stable. A two-pile cap can also be used; however, it is unstable in one direction and will require a grade beam in one direction.

The pile cap size, thickness, and reinforcing are normally shown with the typical foundation details. As with footing design, there are a wide variety of design aids and computer programs available to assist with pile cap design. The pile cap mark and top of cap elevation should be shown on the foundation plan. The top of cap elevation for interior caps can be set at nominal depths. At the perimeter, the top of cap elevation must be set such that the cap base is below the frostline.

All of the pile details shown are based on timber piles. The details for other types of piles would be similar.

COLUMN BASE PLATES AND ANCHOR BOLTS

Column base plates are used to distribute the loads of steel columns to the foundation system. The base plate thickness will depend on the actual column load and base plate size, with ⅜" thick plates as a recommended minimum. It is recommended that the minimum plate size be equal to the column size plus 6" (3" all around the column). Steel base plates are shop welded to the steel columns. Directly below the base plate, a ¼" leveling bed of non-shrink grout is recommended. This gives some additional tolerance when leveling the steel frame and also provides a larger area of load distribution to the foundation system.

Due to stability concerns during the steel erection process, a minimum of four anchor bolts should be used. Anchor bolt size and embedment must be determined based on the specific column forces. A minimum embedment of 12 or 15 times the bolt diameter is recommended. For pre-engineered metal buildings, where uplift is a major consideration, a 15" minimum embedment is recommended. Headed anchor bolts are recommended for all columns that have uplift forces. J-type or L-type bolts are not as effective for uplift because the hook will tend to straighten out during loading. A variety of design aids are available to assist with anchor bolt design.

The preferred way to show base plate sizes is to detail them individually or show them in a schedule. Anchor bolt embedments can be called out with the base plate details or shown with the typical foundation details.

SLABS ON GRADE

Many factors have an impact on how to select and specify concrete slabs on grade. Slab thickness is the most important consideration. Other considerations include the slab reinforcing, jointing, concrete strength, flatness, and concrete admixtures.

It is recommended that the minimum slab on grade thickness be 4″. This would normally apply to slabs that do not have wheeled traffic (forklifts, cars, trucks, etc.) and have minimal loading requirements and low durability/performance requirements. For instance, 4″ slabs could normally be used for most office buildings, schools, and other similar building types. Thicker slabs, ranging anywhere from 5 to 12″, should be used for other conditions.

Slabs can be designed as totally unreinforced; however, reinforcing helps improve the performance of the slab and increases the distance between crack control joints. It is recommended that welded wire fabric be used as reinforcing. Welded wire fabric reinforcing helps hold the slab together and prevents crack widths from opening up. The welded wire fabric should be located near the top surface of the slab (not on the bottom). Welded wire fabric sizes vary, with "6×6-W1.4×W1.4" being the smallest size recommended. The "6×6" denotes that the wires are spaced 6″ on center in both directions. The "W1.4" denotes the gauge of the welded wires. For more significant loading requirements, reinforcing bars can be used. Polypropylene fibers have recently been proposed to replace welded wire fabric; however, their structural properties are not equivalent to the welded wire fabric and any cost savings is normally insignificant. Post-tensioned slabs on grade can also be used, but they are very difficult to design and build.

Proper slab jointing is absolutely essential; otherwise, random slab cracking will occur. Due to a variety of factors, concrete slabs shrink during and after the concrete "sets." Several different types of slab joints are recommended. Control joints are intentionally weakened planes in the concrete so that shrinkage cracks occur at these locations instead of at random. Construction joints are used where the slab area is too large for a single pour. There are many different opinions on how far to space slab joints. In general, the slab jointing should be arranged such that the area within a jointed perimeter does not exceed 600 to 900 ft^2. Slab pours are recommended to be laid out such that the shape is approximately rectangular, with a 2 to 1 maximum aspect ratio.

A minimum 28-day concrete slab compressive strength (f'_c) of 3000 psi is recommended for normal performance slabs. For other slabs (such as for warehouses, freezers, wheeled traffic, etc.) a minimum f'_c of 3500 psi to 4000 psi is recommended. This higher strength is generally obtained by using more cement, which also increases the durability characteristics of the slab.

INDUSTRIAL BUILDING FOUNDATION SECTIONS

The construction cost of industrial buildings, such as storage buildings, offices/warehouses, shell buildings, and so forth, is heavily dependent on the foundation system. The buildings themselves, typically pre-engineered metal buildings, are relatively simple and inexpensive. Hence, providing the most economical foundation system will have a significant impact on the overall building cost. Accordingly, a wide variety of industrial building foundation sections are included to address a range of perimeter and interior foundation conditions.

MISCELLANEOUS FOUNDATION SECTIONS AND DETAILS

Foundation sections are also included for a variety of other different types of buildings, including load-bearing masonry, cold-formed metal framing, and wood framing. In general, shallow foundation systems are shown; however, similar details apply to deep foundation systems.

A variety of miscellaneous foundation details are also presented to include equipment pads, strengthening existing footings, masonry screen walls, concrete strap beams, steel jamb channels, steel pipe bollards, concrete ramps, concrete stairs, trench drains, concrete pits, concrete retaining walls, and concrete wall joints.

GENERAL COMMENT

Practices vary from region to region and the details that follow have generally been taken from specific applications. Any member sizes or connections shown were intended for that particular application. Accordingly, the applicability of all details should be investigated and all member sizes and connections should be determined prior to use.

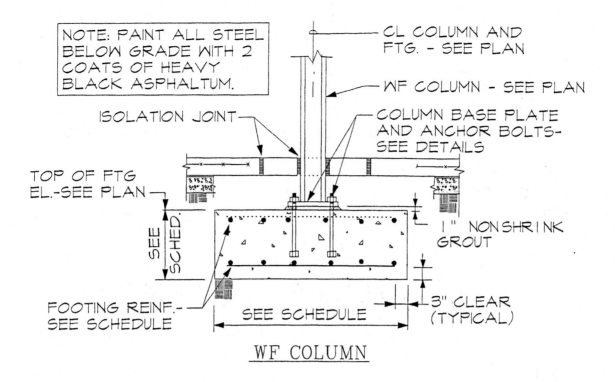

NOTE: PAINT ALL STEEL BELOW GRADE WITH 2 COATS OF HEAVY BLACK ASPHALTUM.

CL COLUMN AND FTG. – SEE PLAN

WF COLUMN – SEE PLAN

COLUMN BASE PLATE AND ANCHOR BOLTS- SEE DETAILS

ISOLATION JOINT

TOP OF FTG EL.-SEE PLAN

SEE SCHED.

1" NONSHRINK GROUT

3" CLEAR (TYPICAL)

FOOTING REINF.- SEE SCHEDULE

SEE SCHEDULE

WF COLUMN

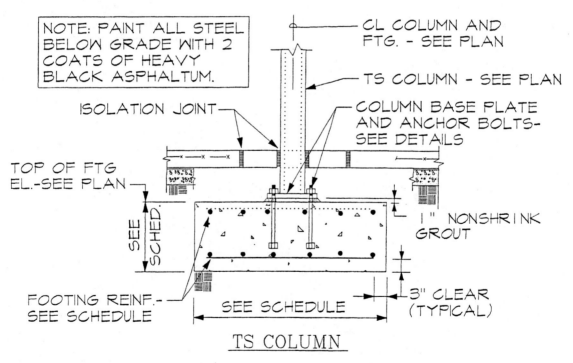

NOTE: PAINT ALL STEEL BELOW GRADE WITH 2 COATS OF HEAVY BLACK ASPHALTUM.

CL COLUMN AND FTG. – SEE PLAN

TS COLUMN – SEE PLAN

COLUMN BASE PLATE AND ANCHOR BOLTS- SEE DETAILS

ISOLATION JOINT

TOP OF FTG EL.-SEE PLAN

SEE SCHED.

1" NONSHRINK GROUT

3" CLEAR (TYPICAL)

FOOTING REINF.- SEE SCHEDULE

SEE SCHEDULE

TS COLUMN

TYP. INTERIOR COL. / FOOTING DETAIL #1
NOT TO SCALE

(DETAIL T2–CFTG1)

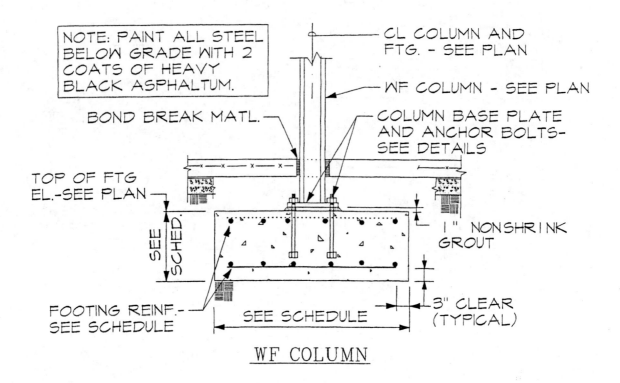

NOTE: PAINT ALL STEEL BELOW GRADE WITH 2 COATS OF HEAVY BLACK ASPHALTUM.

CL COLUMN AND FTG. - SEE PLAN

WF COLUMN - SEE PLAN

BOND BREAK MATL.

COLUMN BASE PLATE AND ANCHOR BOLTS- SEE DETAILS

TOP OF FTG EL.-SEE PLAN

SEE SCHED.

1" NONSHRINK GROUT

FOOTING REINF.- SEE SCHEDULE

SEE SCHEDULE

3" CLEAR (TYPICAL)

WF COLUMN

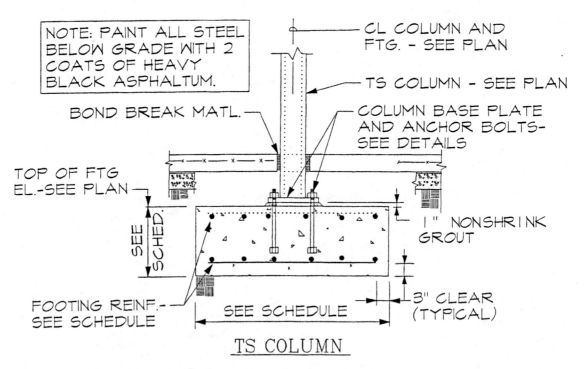

NOTE: PAINT ALL STEEL BELOW GRADE WITH 2 COATS OF HEAVY BLACK ASPHALTUM.

CL COLUMN AND FTG. - SEE PLAN

TS COLUMN - SEE PLAN

BOND BREAK MATL.

COLUMN BASE PLATE AND ANCHOR BOLTS- SEE DETAILS

TOP OF FTG EL.-SEE PLAN

SEE SCHED.

1" NONSHRINK GROUT

FOOTING REINF.- SEE SCHEDULE

SEE SCHEDULE

3" CLEAR (TYPICAL)

TS COLUMN

TYP. INTERIOR COL. / FOOTING DETAIL #2
NOT TO SCALE   (DETAIL T2-CFTG2)

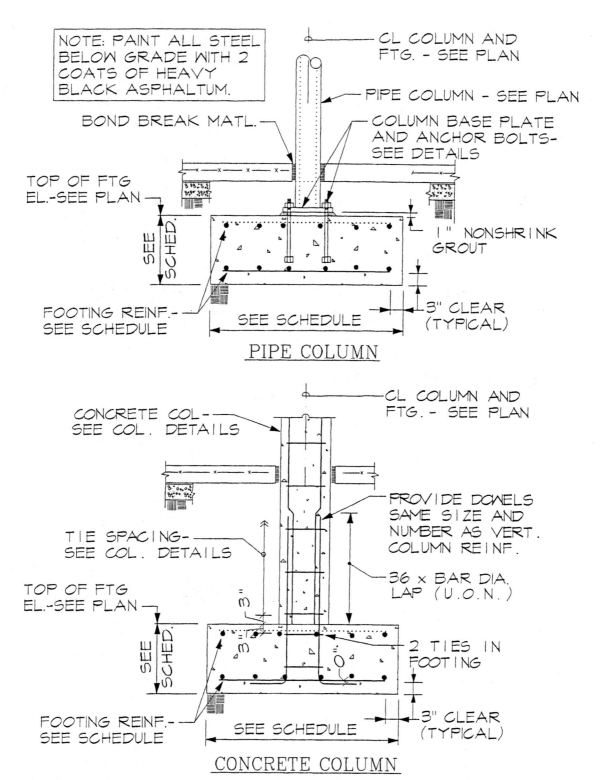

NOTE: PAINT ALL STEEL BELOW GRADE WITH 2 COATS OF HEAVY BLACK ASPHALTUM.

CL COLUMN AND FTG. - SEE PLAN

PIPE COLUMN - SEE PLAN

BOND BREAK MATL.

COLUMN BASE PLATE AND ANCHOR BOLTS- SEE DETAILS

TOP OF FTG EL.-SEE PLAN

SEE SCHED.

1" NONSHRINK GROUT

FOOTING REINF.- SEE SCHEDULE

SEE SCHEDULE

3" CLEAR (TYPICAL)

PIPE COLUMN

CL COLUMN AND FTG. - SEE PLAN

CONCRETE COL.- SEE COL. DETAILS

PROVIDE DOWELS SAME SIZE AND NUMBER AS VERT. COLUMN REINF.

TIE SPACING- SEE COL. DETAILS

36 x BAR DIA. LAP (U.O.N.)

TOP OF FTG EL.-SEE PLAN

3"

SEE SCHED.

3"

2 TIES IN FOOTING

FOOTING REINF.- SEE SCHEDULE

SEE SCHEDULE

3" CLEAR (TYPICAL)

CONCRETE COLUMN

TYP. INTERIOR COL. / FOOTING DETAIL #3
NOT TO SCALE

(DETAIL T2-CFTG3)

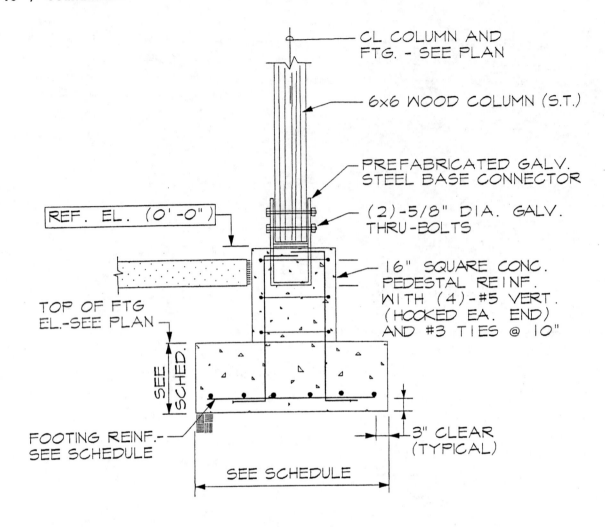

CL COLUMN AND FTG. - SEE PLAN

6x6 WOOD COLUMN (S.T.)

PREFABRICATED GALV. STEEL BASE CONNECTOR

(2)-5/8" DIA. GALV. THRU-BOLTS

REF. EL. (0'-0")

16" SQUARE CONC. PEDESTAL REINF. WITH (4)-#5 VERT. (HOOKED EA. END) AND #3 TIES @ 10"

TOP OF FTG EL.-SEE PLAN

SEE SCHED.

FOOTING REINF.- SEE SCHEDULE

3" CLEAR (TYPICAL)

SEE SCHEDULE

TYP. WOOD POST / FOOTING DETAIL #1

NOT TO SCALE

(DETAIL T2-CFTG4)

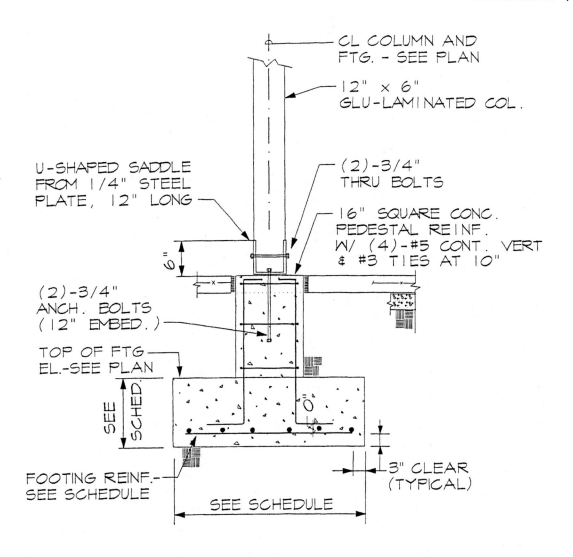

CL COLUMN AND
FTG. - SEE PLAN

12" x 6"
GLU-LAMINATED COL.

U-SHAPED SADDLE
FROM 1/4" STEEL
PLATE, 12" LONG

(2)-3/4"
THRU BOLTS

16" SQUARE CONC.
PEDESTAL REINF.
W/ (4)-#5 CONT. VERT
& #3 TIES AT 10"

6"

(2)-3/4"
ANCH. BOLTS
(12" EMBED.)

TOP OF FTG
EL.-SEE PLAN

SEE SCHED.

FOOTING REINF.-
SEE SCHEDULE

SEE SCHEDULE

3" CLEAR
(TYPICAL)

TYP. WOOD POST / FOOTING DETAIL #2
NOT TO SCALE   (DETAIL T2-CFTG5)

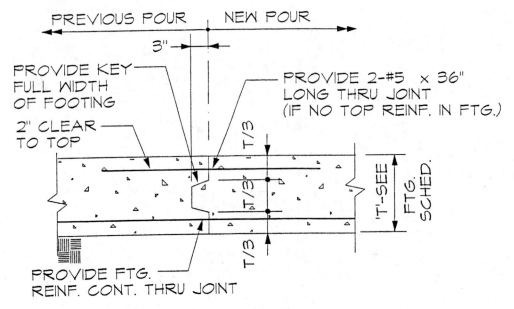

PREVIOUS POUR   NEW POUR

3"

PROVIDE KEY
FULL WIDTH
OF FOOTING

PROVIDE 2-#5 × 36"
LONG THRU JOINT
(IF NO TOP REINF. IN FTG.)

2" CLEAR
TO TOP

T/3
T/3
T/3
T/3

'T'-SEE
FTG.
SCHED.

PROVIDE FTG.
REINF. CONT. THRU JOINT

WALL FOOTING CONSTRUCTION JOINT

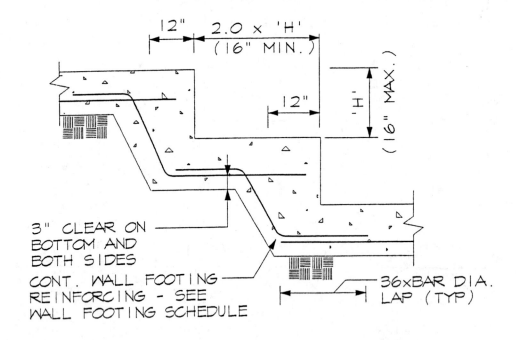

12"   2.0 × 'H'
(16" MIN.)

12"

'H'
(16" MAX.)

3" CLEAR ON
BOTTOM AND
BOTH SIDES

CONT. WALL FOOTING
REINFORCING - SEE
WALL FOOTING SCHEDULE

36×BAR DIA.
LAP (TYP)

STEPPED WALL FOOTING DETAIL

TYPICAL WALL FOOTING DETAILS
NOT TO SCALE   (DETAIL T2-WFTG1)

NOTE: THESE DETAILS DO NOT APPLY IF TOP OF PIPE IS MORE THAN 4" BELOW BOTTOM OF WALL FOOTING.

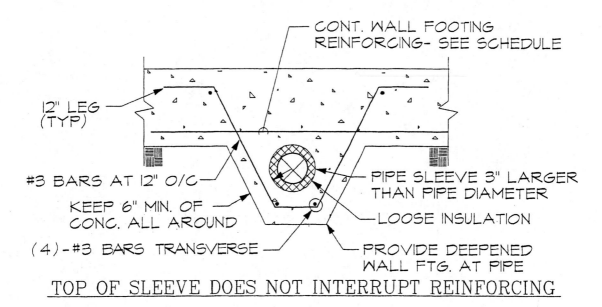

CONT. WALL FOOTING REINFORCING- SEE SCHEDULE

12" LEG (TYP)

#3 BARS AT 12" O/C

KEEP 6" MIN. OF CONC. ALL AROUND

(4)-#3 BARS TRANSVERSE

PIPE SLEEVE 3" LARGER THAN PIPE DIAMETER

LOOSE INSULATION

PROVIDE DEEPENED WALL FTG. AT PIPE

TOP OF SLEEVE DOES NOT INTERRUPT REINFORCING

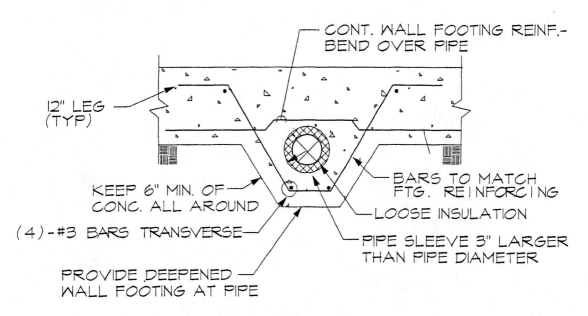

CONT. WALL FOOTING REINF.- BEND OVER PIPE

12" LEG (TYP)

KEEP 6" MIN. OF CONC. ALL AROUND

(4)-#3 BARS TRANSVERSE

PROVIDE DEEPENED WALL FOOTING AT PIPE

BARS TO MATCH FTG. REINFORCING

LOOSE INSULATION

PIPE SLEEVE 3" LARGER THAN PIPE DIAMETER

TOP OF SLEEVE INTERRUPTS BOTTOM REINFORCING

PIPE SLEEVE THRU WALL FTG. DETAILS

NOT TO SCALE   (DETAIL T2-WFTG2)

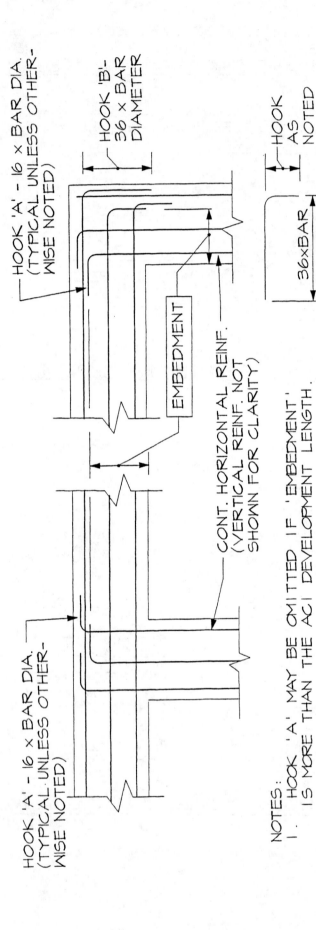

HOOK 'A' - 16 x BAR DIA.
(TYPICAL UNLESS OTHER-
WISE NOTED)

HOOK 'B'-
36 x BAR
DIAMETER

HOOK
AS
NOTED

36xBAR
DIAMETER

OPTIONAL DOWEL

EMBEDMENT

CONT. HORIZONTAL REINF.
(VERTICAL REINF. NOT
SHOWN FOR CLARITY)

HOOK 'A' - 16 x BAR DIA.
(TYPICAL UNLESS OTHER-
WISE NOTED)

NOTES:
1. HOOK 'A' MAY BE OMITTED IF 'EMBEDMENT'
   IS MORE THAN THE ACI DEVELOPMENT LENGTH.

2. DOWELS MAY BE USED IN LIEU OF HOOKING
   CONTINUOUS BARS AS SHOWN.

3. THE DETAIL IS SHOWN FOR A MULTIPLE LINE OF
   REINFORCING. DETAIL IS SIMILAR FOR A SINGLE LINE.

TYP. DETAIL SHOWING CONTINUOUS
REINFORCING AT CORNERS AND INTERSECTIONS

(DETAIL T2-FTGI)

NOT TO SCALE

14

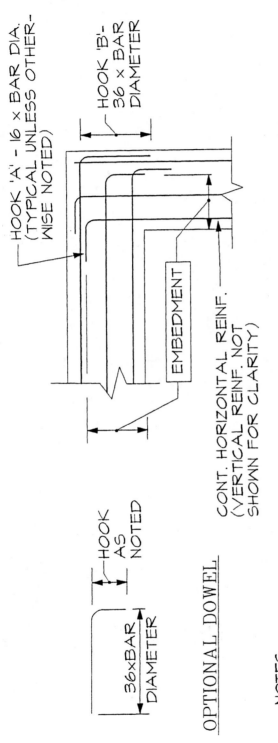

HOOK 'A' - 16 × BAR DIA.
(TYPICAL UNLESS OTHER-
WISE NOTED)

HOOK 'B'-
36 × BAR
DIAMETER

EMBEDMENT

CONT. HORIZONTAL REINF.
(VERTICAL REINF. NOT
SHOWN FOR CLARITY)

36×BAR
DIAMETER

HOOK
AS
NOTED

OPTIONAL DOWEL

NOTES:
1. HOOK 'A' MAY BE OMITTED IF 'EMBEDMENT'
    IS MORE THAN THE ACI DEVELOPMENT LENGTH.

2. DOWELS MAY BE USED IN LIEU OF HOOKING
    CONTINUOUS BARS AS SHOWN.

3. THE DETAIL IS SHOWN FOR A MULTIPLE LINE OF
    REINFORCING. DETAIL IS SIMILAR FOR A SINGLE LINE.

TYP. DETAIL SHOWING
CONTINUOUS REINFORCING AT CORNERS

(DETAIL T2-FTG12)

NOT TO SCALE

15

NOTE: CUT TOPS OF PILE SQUARE TO PILE AXIS AND APPLY 2 COATS OF MATCHING TREATMENT.

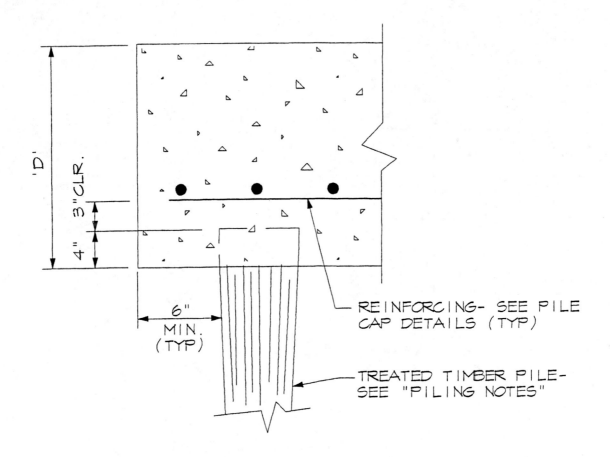

REINFORCING- SEE PILE CAP DETAILS (TYP)

TREATED TIMBER PILE- SEE "PILING NOTES"

TYP. COMPRESSION PILE

TYP. TIMBER PILE EMBEDMENT DETAIL

NOT TO SCALE   (DETAIL T2-PEMB1)

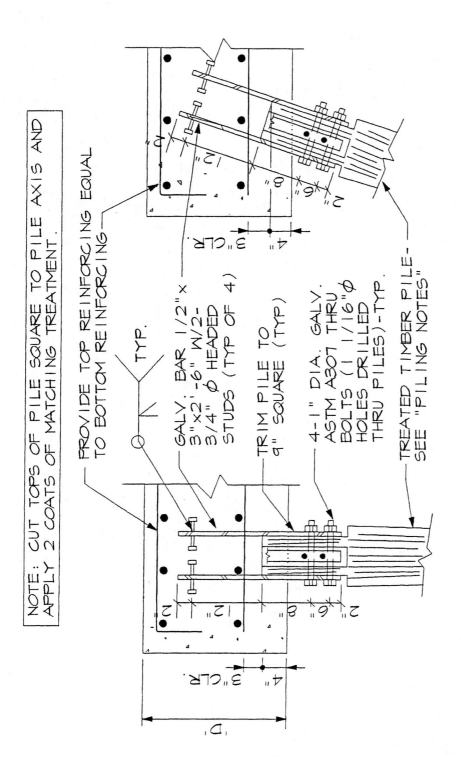

NOTE: CUT TOPS OF PILE SQUARE TO PILE AXIS AND APPLY 2 COATS OF MATCHING TREATMENT.

PROVIDE TOP REINFORCING EQUAL TO BOTTOM REINFORCING

TYP.

GALV. BAR 1/2" × 3"×2'-6" W/2-3/4" ⌀ HEADED STUDS (TYP OF 4)

TRIM PILE TO 9" SQUARE (TYP)

4-1" DIA. GALV. ASTM A307 THRU BOLTS (1 1/16"⌀ HOLES DRILLED THRU PILES)-TYP.

TREATED TIMBER PILE-SEE "PILING NOTES"

BATTERED TENSION PILE

TYP. TENSION PILE

TYP. TIMBER TENSION PILE EMBEDMENT DETAILS

NOT TO SCALE

(DETAIL T2-PEMB2)

17

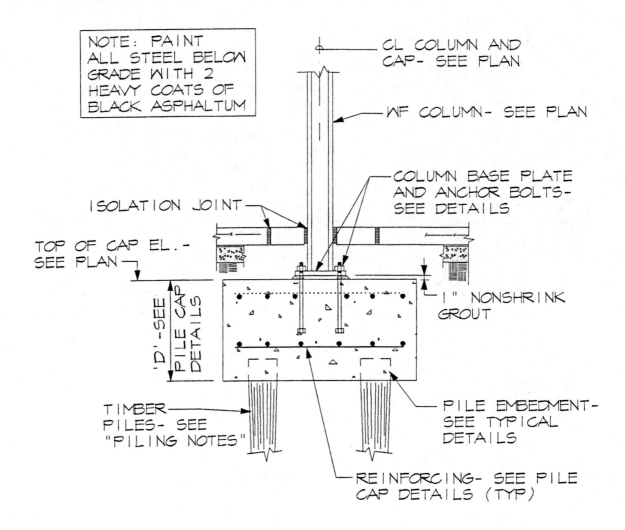

NOTE: PAINT ALL STEEL BELOW GRADE WITH 2 HEAVY COATS OF BLACK ASPHALTUM

CL COLUMN AND CAP- SEE PLAN

WF COLUMN- SEE PLAN

COLUMN BASE PLATE AND ANCHOR BOLTS- SEE DETAILS

ISOLATION JOINT

TOP OF CAP EL.- SEE PLAN

'D'- SEE PILE CAP DETAILS

1" NONSHRINK GROUT

TIMBER PILES- SEE "PILING NOTES"

PILE EMBEDMENT- SEE TYPICAL DETAILS

REINFORCING- SEE PILE CAP DETAILS (TYP)

TYP. COLUMN / PILE CAP DETAIL #1

NOT TO SCALE

(DETAIL T2-PCAP1)

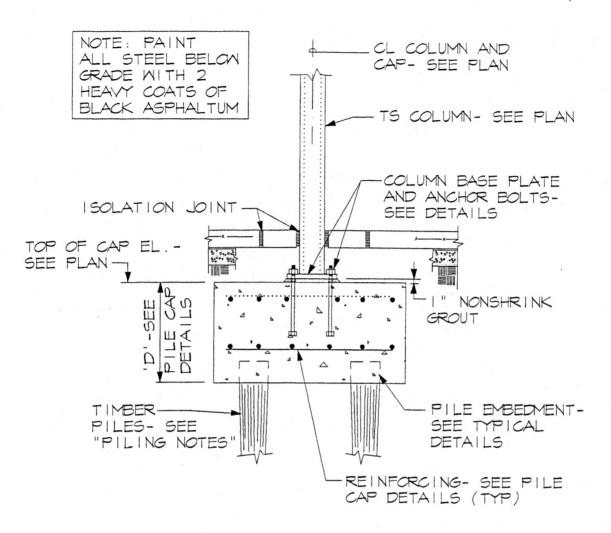

NOTE: PAINT ALL STEEL BELOW GRADE WITH 2 HEAVY COATS OF BLACK ASPHALTUM

CL COLUMN AND CAP- SEE PLAN

TS COLUMN- SEE PLAN

COLUMN BASE PLATE AND ANCHOR BOLTS- SEE DETAILS

ISOLATION JOINT

TOP OF CAP EL.- SEE PLAN

'D'- SEE PILE CAP DETAILS

1" NONSHRINK GROUT

TIMBER PILES- SEE "PILING NOTES"

PILE EMBEDMENT- SEE TYPICAL DETAILS

REINFORCING- SEE PILE CAP DETAILS (TYP)

TYP. COLUMN / PILE CAP DETAIL #2
NOT TO SCALE   (DETAIL T2-PCAP2)

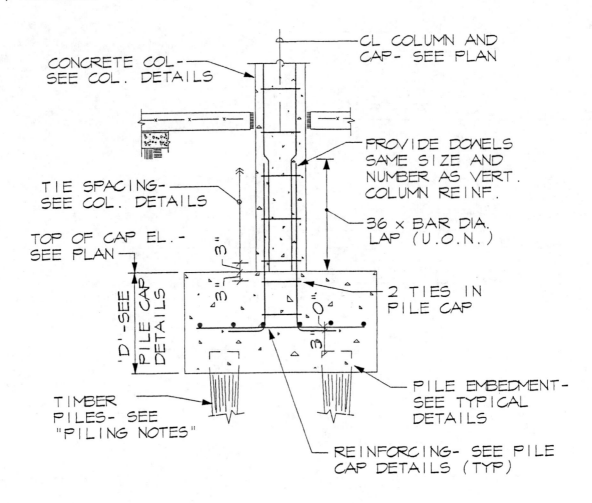

CONCRETE COL-
SEE COL. DETAILS

CL COLUMN AND
CAP- SEE PLAN

PROVIDE DOWELS
SAME SIZE AND
NUMBER AS VERT.
COLUMN REINF.

36 x BAR DIA.
LAP (U.O.N.)

TIE SPACING-
SEE COL. DETAILS

TOP OF CAP EL.-
SEE PLAN

3"

3"

3"

'D'-SEE PILE CAP DETAILS

2 TIES IN
PILE CAP

TIMBER
PILES- SEE
"PILING NOTES"

PILE EMBEDMENT-
SEE TYPICAL
DETAILS

REINFORCING- SEE PILE
CAP DETAILS (TYP)

TYP. COLUMN / PILE CAP DETAIL #3

NOT TO SCALE

(DETAIL T2-PCAP3)

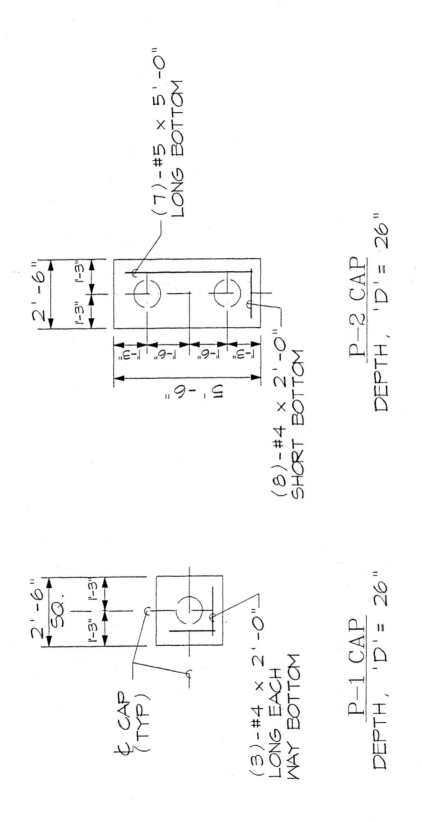

(7)-#5 × 5'-0"
LONG BOTTOM

2'-6"

1'-3"
1'-3"

1'-3" 1'-6" 1'-6" 1'-3"

5'-6"

(8)-#4 × 2'-0"
SHORT BOTTOM

P-2 CAP
DEPTH, 'D' = 26"

2'-6" SQ.

1'-3"
1'-3"

⊄ CAP
(TYP)

(3)-#4 × 2'-0"
LONG EACH
WAY BOTTOM

P-1 CAP
DEPTH, 'D' = 26"

TYPICAL P-1 & P-2 PILE CAP DETAILS
(DETAIL T2-P1&P2)

NOT TO SCALE

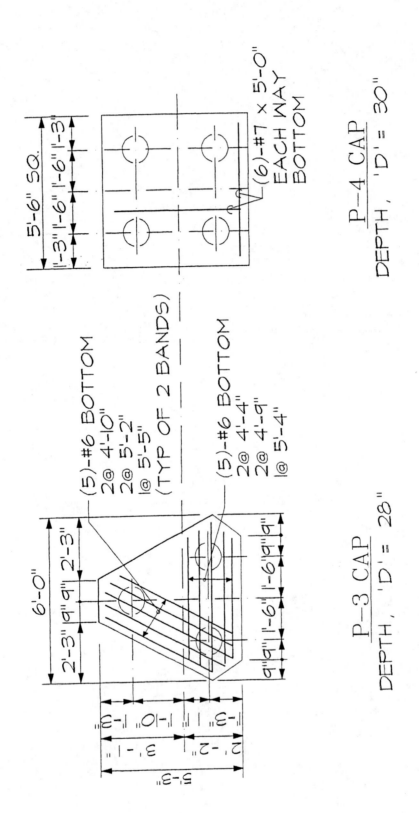

(6)-#7 × 5'-0"
EACH WAY
BOTTOM

5'-6" SQ.

(5)-#6 BOTTOM
2@ 4'-10"
2@ 5'-2"
1@ 5'-5"
(TYP OF 2 BANDS)

(5)-#6 BOTTOM
2@ 4'-4"
2@ 4'-9"
1@ 5'-4"

P-4 CAP
DEPTH, 'D' = 30"

P-3 CAP
DEPTH, 'D' = 28"

6'-0"

5'-3"

TYPICAL P-3 & P-4 PILE CAP DETAILS
(DETAIL T2-P3&P4)

NOT TO SCALE

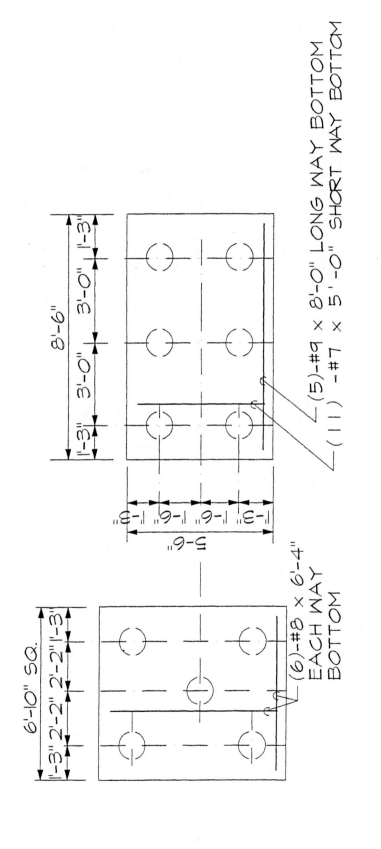

P-6 CAP
DEPTH, 'D' = 36"

P-5 CAP
DEPTH, 'D' = 32"

TYPICAL P-5 & P-6 PILE CAP DETAILS
NOT TO SCALE

(DETAIL T2-P5&P6)

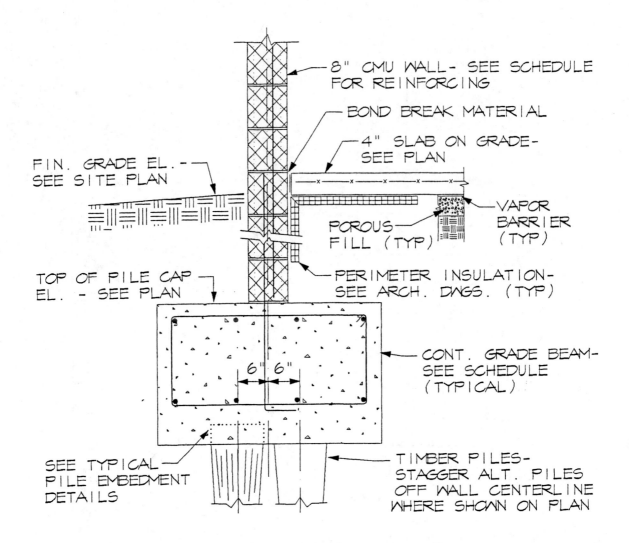

FIN. GRADE EL. - SEE SITE PLAN

8" CMU WALL - SEE SCHEDULE FOR REINFORCING

BOND BREAK MATERIAL

4" SLAB ON GRADE - SEE PLAN

VAPOR BARRIER (TYP)

POROUS FILL (TYP)

TOP OF PILE CAP EL. - SEE PLAN

PERIMETER INSULATION - SEE ARCH. DWGS. (TYP)

6" 6"

CONT. GRADE BEAM - SEE SCHEDULE (TYPICAL)

SEE TYPICAL PILE EMBEDMENT DETAILS

TIMBER PILES - STAGGER ALT. PILES OFF WALL CENTERLINE WHERE SHOWN ON PLAN

PERIMETER FOUNDATION SECTION #1
NOT TO SCALE
(DETAIL T2-PERM1)

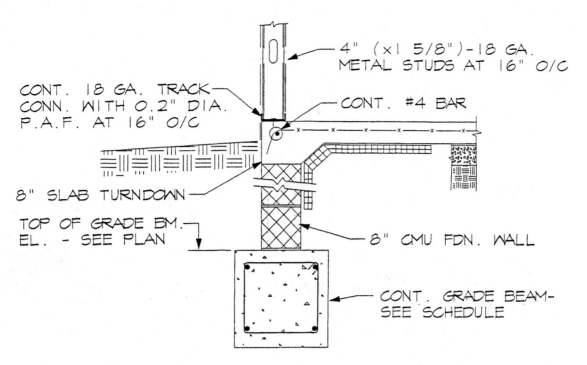

CONT. 18 GA. TRACK
CONN. WITH 0.2" DIA.
P.A.F. AT 16" O/C

4" (×1 5/8")-18 GA.
METAL STUDS AT 16" O/C

CONT. #4 BAR

8" SLAB TURNDOWN

TOP OF GRADE BM.
EL. - SEE PLAN

8" CMU FDN. WALL

CONT. GRADE BEAM-
SEE SCHEDULE

DETAIL AT METAL STUD WALL

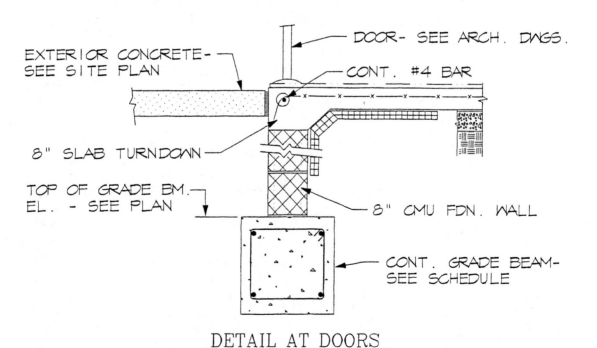

EXTERIOR CONCRETE-
SEE SITE PLAN

DOOR- SEE ARCH. DWGS.

CONT. #4 BAR

8" SLAB TURNDOWN

TOP OF GRADE BM.
EL. - SEE PLAN

8" CMU FDN. WALL

CONT. GRADE BEAM-
SEE SCHEDULE

DETAIL AT DOORS

PERIMETER FOUNDATION SECTION #2
NOT TO SCALE   (DETAIL T2-PERM2)

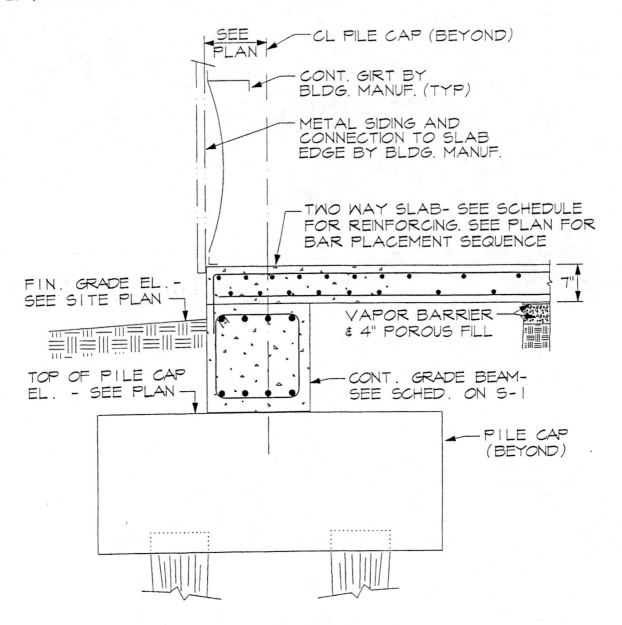

SEE PLAN

CL PILE CAP (BEYOND)

CONT. GIRT BY BLDG. MANUF. (TYP)

METAL SIDING AND CONNECTION TO SLAB EDGE BY BLDG. MANUF.

TWO WAY SLAB- SEE SCHEDULE FOR REINFORCING. SEE PLAN FOR BAR PLACEMENT SEQUENCE

FIN. GRADE EL. - SEE SITE PLAN

7"

VAPOR BARRIER & 4" POROUS FILL

TOP OF PILE CAP EL. - SEE PLAN

CONT. GRADE BEAM- SEE SCHED. ON S-1

PILE CAP (BEYOND)

PERIMETER FOUNDATION SECTION #3
NOT TO SCALE   (DETAIL T2-PERM3)

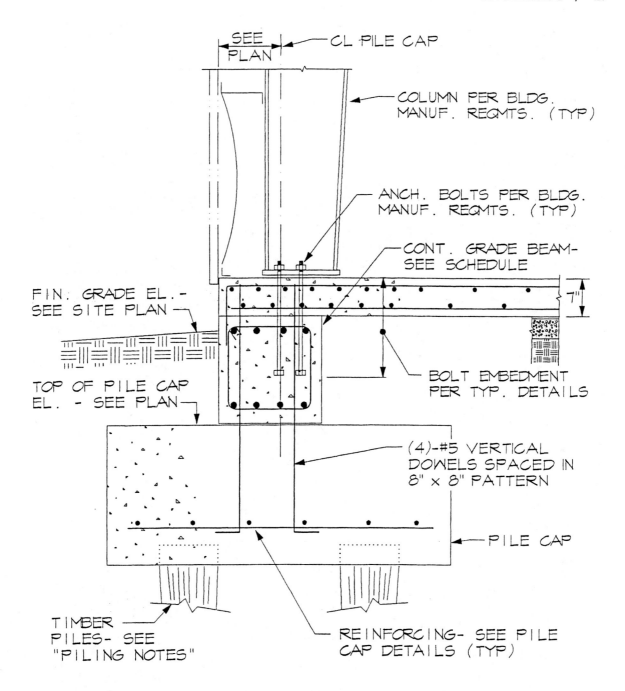

SEE PLAN

CL PILE CAP

COLUMN PER BLDG. MANUF. REQMTS. (TYP)

ANCH. BOLTS PER BLDG. MANUF. REQMTS. (TYP)

CONT. GRADE BEAM- SEE SCHEDULE

FIN. GRADE EL.- SEE SITE PLAN

7"

TOP OF PILE CAP EL. - SEE PLAN

BOLT EMBEDMENT PER TYP. DETAILS

(4)-#5 VERTICAL DOWELS SPACED IN 8" x 8" PATTERN

PILE CAP

TIMBER PILES- SEE "PILING NOTES"

REINFORCING- SEE PILE CAP DETAILS (TYP)

PERIMETER FOUNDATION SECTION #4

NOT TO SCALE

(DETAIL T2-PERM4)

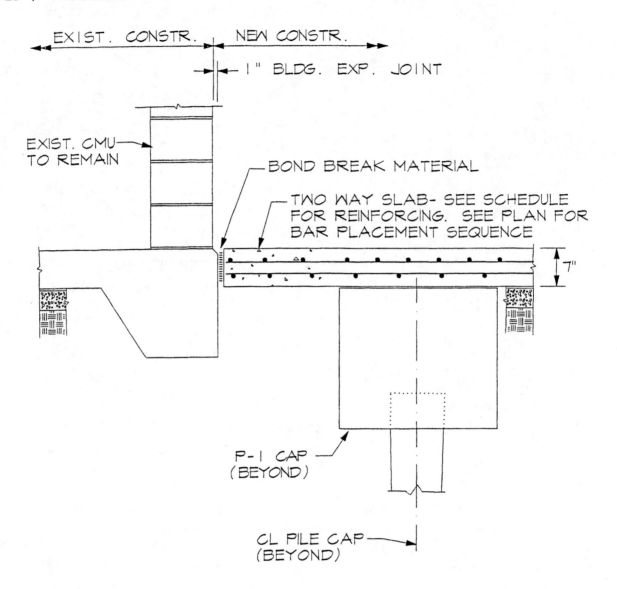

EXIST. CONSTR.

NEW CONSTR.

1" BLDG. EXP. JOINT

EXIST. CMU TO REMAIN

BOND BREAK MATERIAL

TWO WAY SLAB- SEE SCHEDULE FOR REINFORCING. SEE PLAN FOR BAR PLACEMENT SEQUENCE

7"

P-1 CAP (BEYOND)

CL PILE CAP (BEYOND)

PERIMETER FOUNDATION SECTION #5
NOT TO SCALE   (DETAIL T2-PERM5)

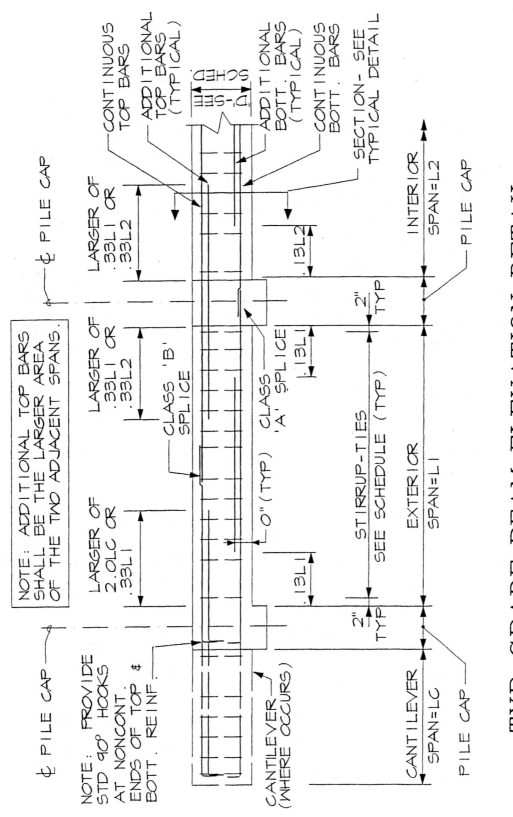

TYP. GRADE BEAM ELEVATION DETAIL

NOT TO SCALE

(DETAIL T2–PGBME)

29

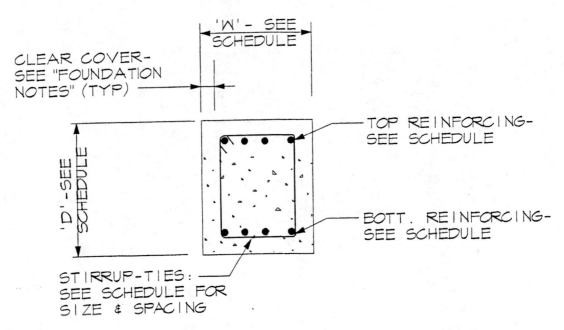

CLEAR COVER-
SEE "FOUNDATION
NOTES" (TYP)

'W' - SEE
SCHEDULE

'D' - SEE
SCHEDULE

TOP REINFORCING-
SEE SCHEDULE

BOTT. REINFORCING-
SEE SCHEDULE

STIRRUP-TIES:
SEE SCHEDULE FOR
SIZE & SPACING

TYPICAL GRADE BEAM SECTION
NOT TO SCALE

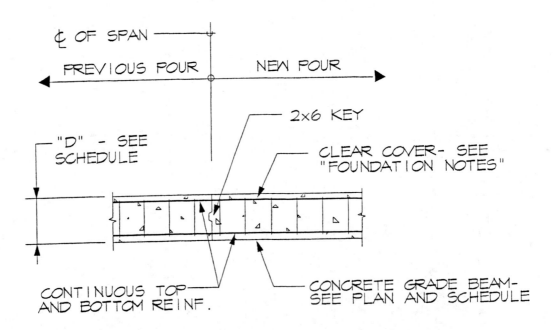

¢ OF SPAN

PREVIOUS POUR

NEW POUR

2x6 KEY

"D" - SEE
SCHEDULE

CLEAR COVER- SEE
"FOUNDATION NOTES"

CONTINUOUS TOP
AND BOTTOM REINF.

CONCRETE GRADE BEAM-
SEE PLAN AND SCHEDULE

GRADE BEAM CONSTR. JOINT DETAIL
NOT TO SCALE
(DETAIL T2-PGBMS)

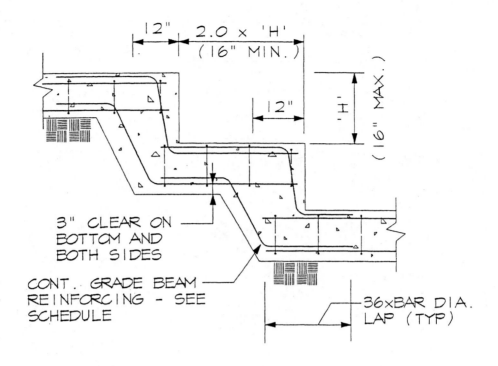

12" 2.0 x 'H'
(16" MIN.)

12"

'H'
(16" MAX.)

3" CLEAR ON
BOTTOM AND
BOTH SIDES

CONT. GRADE BEAM
REINFORCING - SEE
SCHEDULE

36xBAR DIA.
LAP (TYP)

TYP. STEPPED GRADE BEAM DETAIL

NOT TO SCALE   (DETAIL T2-PGBSF)

NOTE: THESE DETAILS DO NOT APPLY IF TOP OF PIPE IS MORE THAN 3" BELOW BOTTOM OF GRADE BEAM.

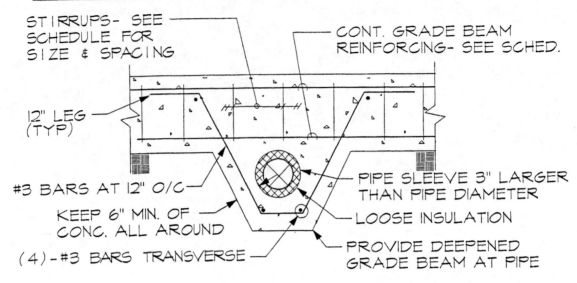

STIRRUPS- SEE SCHEDULE FOR SIZE & SPACING

CONT. GRADE BEAM REINFORCING- SEE SCHED.

12" LEG (TYP)

PIPE SLEEVE 3" LARGER THAN PIPE DIAMETER

#3 BARS AT 12" O/C

LOOSE INSULATION

KEEP 6" MIN. OF CONC. ALL AROUND

PROVIDE DEEPENED GRADE BEAM AT PIPE

(4)-#3 BARS TRANSVERSE

TOP OF SLEEVE DOES NOT INTERRUPT REINFORCING

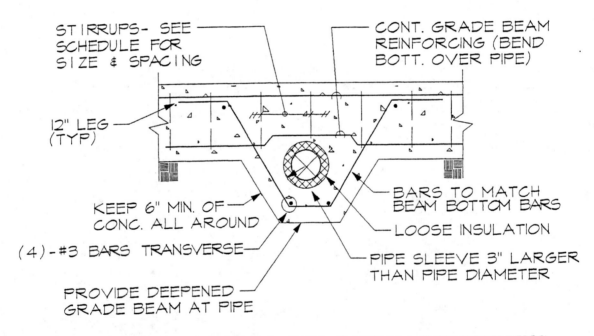

STIRRUPS- SEE SCHEDULE FOR SIZE & SPACING

CONT. GRADE BEAM REINFORCING (BEND BOTT. OVER PIPE)

12" LEG (TYP)

BARS TO MATCH BEAM BOTTOM BARS

KEEP 6" MIN. OF CONC. ALL AROUND

LOOSE INSULATION

(4)-#3 BARS TRANSVERSE

PIPE SLEEVE 3" LARGER THAN PIPE DIAMETER

PROVIDE DEEPENED GRADE BEAM AT PIPE

TOP OF SLEEVE INTERRUPTS BOTTOM REINFORCING

PIPE SLEEVE THRU GRADE BM. DETAILS

NOT TO SCALE

(DETAIL T2-PGBSL)

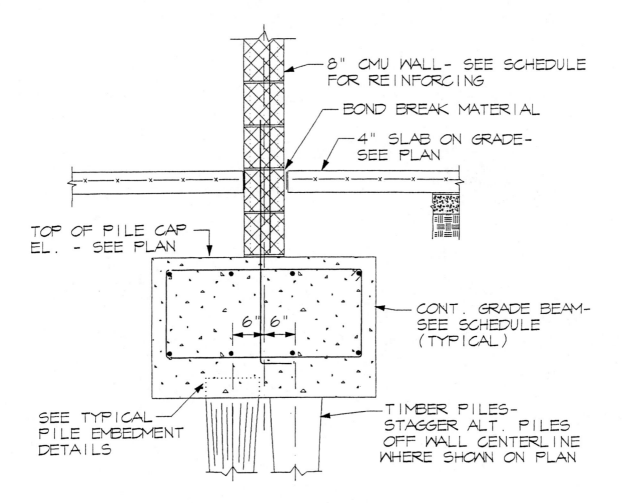

8" CMU WALL- SEE SCHEDULE FOR REINFORCING

BOND BREAK MATERIAL

4" SLAB ON GRADE- SEE PLAN

TOP OF PILE CAP EL. - SEE PLAN

6" 6"

CONT. GRADE BEAM- SEE SCHEDULE (TYPICAL)

SEE TYPICAL PILE EMBEDMENT DETAILS

TIMBER PILES- STAGGER ALT. PILES OFF WALL CENTERLINE WHERE SHOWN ON PLAN

INTERIOR FOUNDATION SECTION #1

NOT TO SCALE   (DETAIL T2-PINT1)

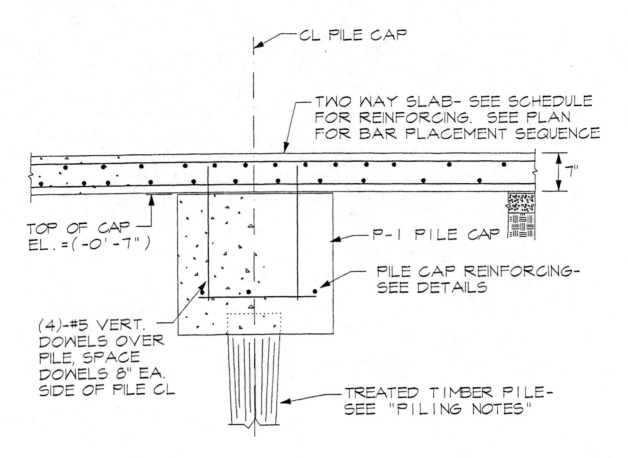

CL PILE CAP

TWO WAY SLAB- SEE SCHEDULE
FOR REINFORCING. SEE PLAN
FOR BAR PLACEMENT SEQUENCE

7"

TOP OF CAP
EL.=(-0'-7")

P-1 PILE CAP

PILE CAP REINFORCING-
SEE DETAILS

(4)-#5 VERT.
DOWELS OVER
PILE, SPACE
DOWELS 8" EA.
SIDE OF PILE CL

TREATED TIMBER PILE-
SEE "PILING NOTES"

INTERIOR FOUNDATION SECTION #2

NOT TO SCALE   (DETAIL T2-PINT2)

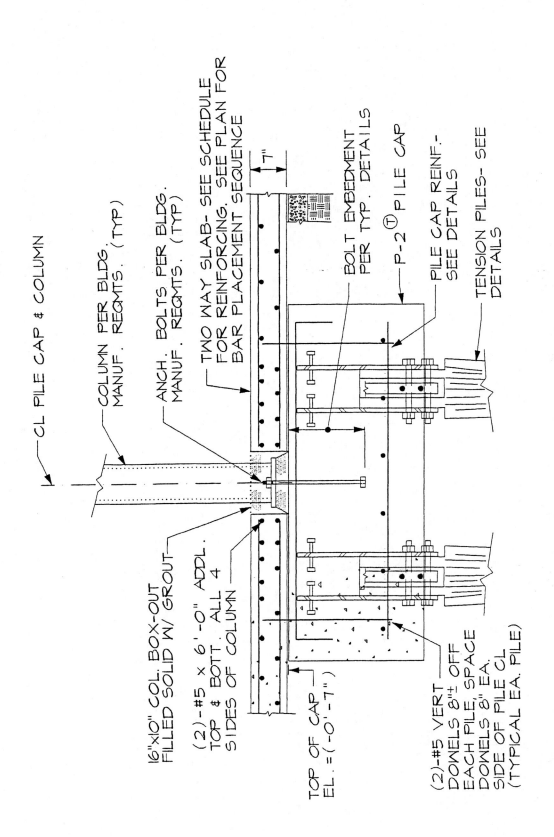

CL PILE CAP & COLUMN

COLUMN PER BLDG. MANUF. REQMTS. (TYP)

ANCH. BOLTS PER BLDG. MANUF. REQMTS. (TYP)

TWO WAY SLAB- SEE SCHEDULE FOR REINFORCING. SEE PLAN FOR BAR PLACEMENT SEQUENCE

7"

BOLT EMBEDMENT PER TYP. DETAILS

P-2 ⓣ PILE CAP

PILE CAP REINF.- SEE DETAILS

TENSION PILES- SEE DETAILS

16"x10" COL. BOX-OUT FILLED SOLID W/ GROUT

(2)-#5 x 6'-0" ADDL. TOP & BOTT. ALL 4 SIDES OF COLUMN

TOP OF CAP EL.=(-0'-7")

(2)-#5 VERT DOWELS 8"± OFF EACH PILE, SPACE DOWELS 8" EA. SIDE OF PILE CL (TYPICAL EA. PILE)

INTERIOR FOUNDATION SECTION #3
(DETAIL T2-PINT3)

NOT TO SCALE

35

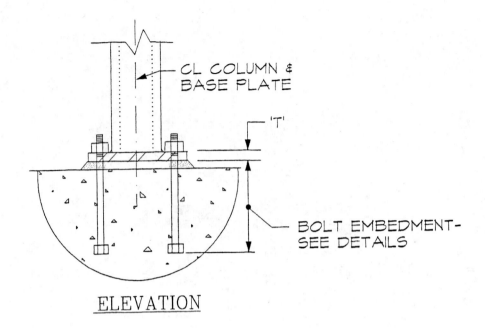

CL COLUMN &
BASE PLATE

'T'

BOLT EMBEDMENT-
SEE DETAILS

ELEVATION

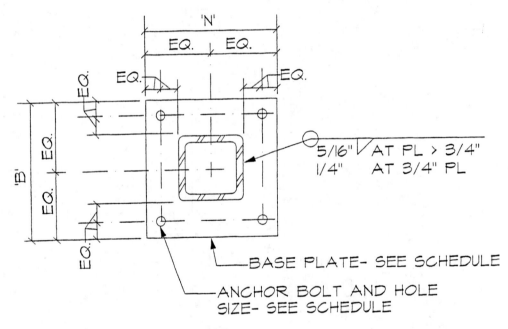

'N'

EQ. | EQ.

EQ. | EQ.

EQ. | EQ.

'B'

EQ.

5/16" AT PL > 3/4"
1/4"   AT 3/4" PL

BASE PLATE- SEE SCHEDULE

ANCHOR BOLT AND HOLE
SIZE- SEE SCHEDULE

PLAN

COLUMN BASE PLATE AND
ANCHOR BOLT LAYOUT DETAILS

NOT TO SCALE   (DETAIL T2-BP1)

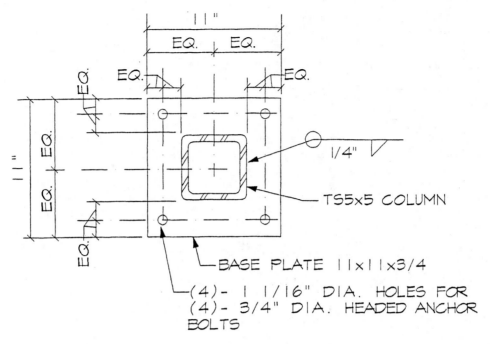

TS5x5 COLUMN

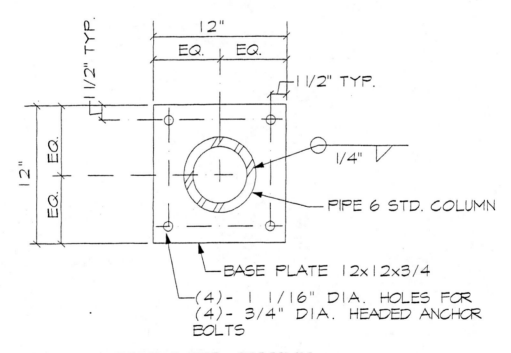

PIPE 6 STD. COLUMN

COLUMN BASE PLATE DETAIL #2

NOT TO SCALE   (DETAIL T2-BP2)

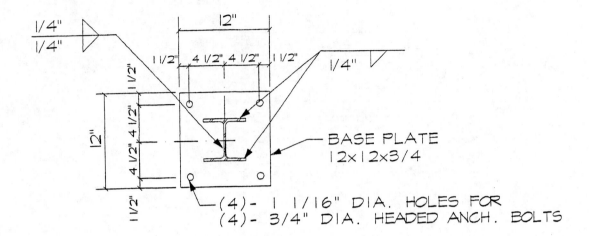

1/4"
1/4"

12"

1 1/2" 4 1/2" 4 1/2" 1 1/2"

1/4"

1 1/2"

12"

4 1/2" 4 1/2"

1 1/2"

BASE PLATE
12x12x3/4

(4)- 1 1/16" DIA. HOLES FOR
(4)- 3/4" DIA. HEADED ANCH. BOLTS

W6 COLUMN

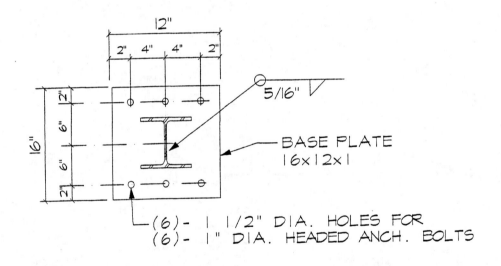

12"

2" 4" 4" 2"

5/16"

16"

2"
6"
6"
2"

BASE PLATE
16x12x1

(6)- 1 1/2" DIA. HOLES FOR
(6)- 1" DIA. HEADED ANCH. BOLTS

W8 COLUMN

COLUMN BASE PLATE DETAIL #3

NOT TO SCALE   (DETAIL T2-BP3)

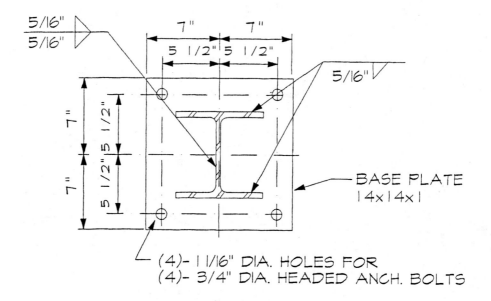

W10 COLUMN

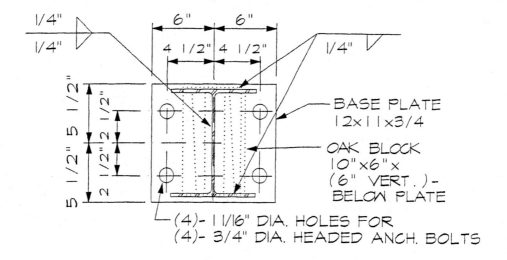

W10 COLUMN

COLUMN BASE PLATE DETAIL #4

NOT TO SCALE   (DETAIL T2-BP4)

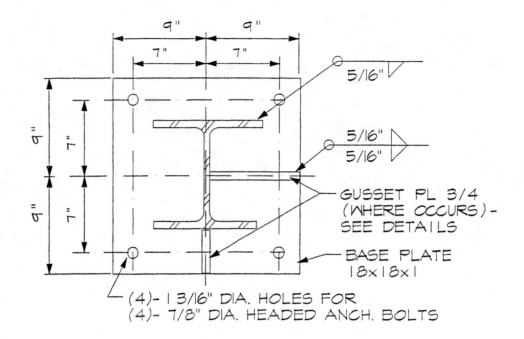

W12 COLUMN AT BRACED BAYS

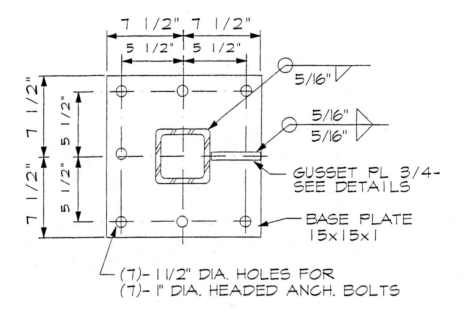

TS COLUMN AT BRACED BAYS

COLUMN BASE PLATE DETAIL #5

NOT TO SCALE   (DETAIL T2–BP5)

NOTE: J-TYPE AND L-TYPE ANCHOR BOLTS ARE PROHIBITED FROM USE.

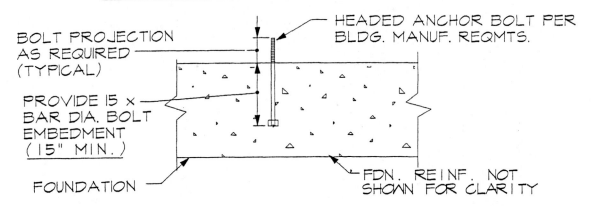

BOLT PROJECTION AS REQUIRED (TYPICAL)

HEADED ANCHOR BOLT PER BLDG. MANUF. REQMTS.

PROVIDE 15 x BAR DIA. BOLT EMBEDMENT (15" MIN.)

FOUNDATION

FDN. REINF. NOT SHOWN FOR CLARITY

AT PRE-ENGINEERED METAL BUILDING COLUMNS

NOTE: J-TYPE AND L-TYPE ANCHOR BOLTS ARE PROHIBITED FROM USE.

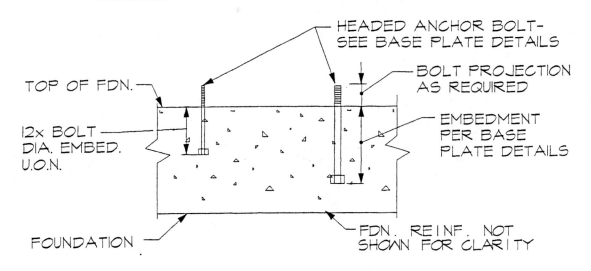

HEADED ANCHOR BOLT- SEE BASE PLATE DETAILS

BOLT PROJECTION AS REQUIRED

TOP OF FDN.

12x BOLT DIA. EMBED. U.O.N.

EMBEDMENT PER BASE PLATE DETAILS

FOUNDATION

FDN. REINF. NOT SHOWN FOR CLARITY

TYPE 1 BOLTS   TYPE 2 BOLTS

TYPICAL ANCHOR BOLT DETAIL #1

NOT TO SCALE   (DETAIL T2-AB1)

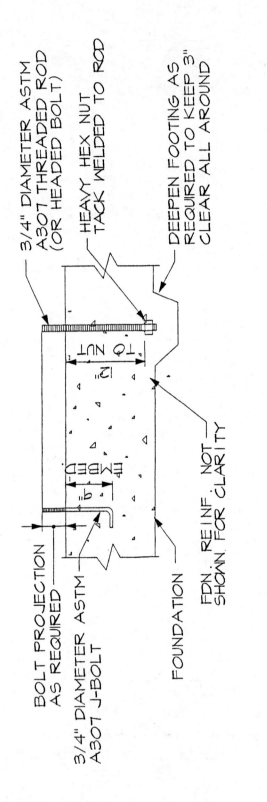

BOLT PROJECTION AS REQUIRED

3/4" DIAMETER ASTM A307 J-BOLT

FOUNDATION

FDN. REINF. NOT SHOWN FOR CLARITY

3/4" DIAMETER ASTM A307 THREADED ROD (OR HEADED BOLT)

HEAVY HEX NUT TACK WELDED TO ROD

DEEPEN FOOTING AS REQUIRED TO KEEP 3" CLEAR ALL AROUND

9" EMBED.

12" TO NUT

NON-WIND UPLIFT LOADING

WIND UPLIFT LOADING

TYPICAL ANCHOR BOLT DETAIL #2
(DETAIL T2-AB2)

NOT TO SCALE

NOTE: J-TYPE AND L-TYPE ANCHOR
BOLTS ARE PROHIBITED FROM USE.

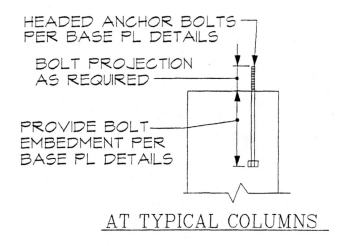

HEADED ANCHOR BOLTS
PER BASE PL DETAILS

BOLT PROJECTION
AS REQUIRED

PROVIDE BOLT
EMBEDMENT PER
BASE PL DETAILS

AT TYPICAL COLUMNS

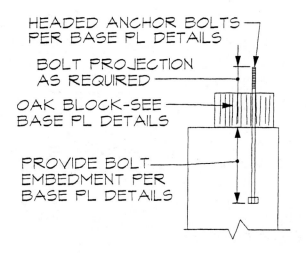

HEADED ANCHOR BOLTS
PER BASE PL DETAILS

BOLT PROJECTION
AS REQUIRED

OAK BLOCK-SEE
BASE PL DETAILS

PROVIDE BOLT
EMBEDMENT PER
BASE PL DETAILS

AT FREEZER BUILDING COLUMNS

TYPICAL ANCHOR BOLT DETAIL #3

NOT TO SCALE   (DETAIL T2-AB3)

NOTE: J-TYPE AND L-TYPE ANCHOR BOLTS ARE PROHIBITED FROM USE.

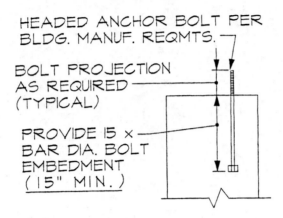

HEADED ANCHOR BOLT PER BLDG. MANUF. REQMTS.

BOLT PROJECTION AS REQUIRED (TYPICAL)

PROVIDE 15 x BAR DIA. BOLT EMBEDMENT (15" MIN.)

PRE-ENGINEERED BLDG. COLUMNS

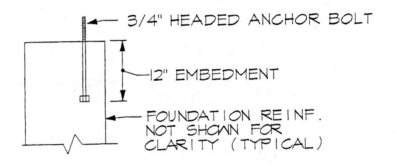

3/4" HEADED ANCHOR BOLT

12" EMBEDMENT

FOUNDATION REINF. NOT SHOWN FOR CLARITY (TYPICAL)

TS CANOPY COLUMNS

TYPICAL ANCHOR BOLT DETAIL #4
NOT TO SCALE   (DETAIL T2-AB4)

NOTE: AT CONTRACTOR'S OPTION, PROVIDE A SAWCUT, TOOLED, OR PREMOLDED INSERT SLAB JOINT. REMOVE INSERT STRIP AFTER CONCRETE HAS SET. PROVIDE 1/8" RADIUS EDGES FOR TOOLED AND INSERT JOINTS.

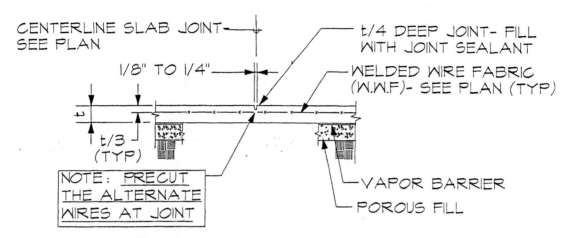

CENTERLINE SLAB JOINT— SEE PLAN

1/8" TO 1/4"

t

t/3 (TYP)

NOTE: PRECUT THE ALTERNATE WIRES AT JOINT

t/4 DEEP JOINT- FILL WITH JOINT SEALANT

WELDED WIRE FABRIC (W.W.F)- SEE PLAN (TYP)

VAPOR BARRIER

POROUS FILL

SAWCUT JOINTS SHALL BE MADE AS SOON AS THE CONCRETE HAS CURED SUCH THAT THE BLADE DOES NOT DISLODGE AGGREGATE AND THE CUT EDGES DO NOT CRUMBLE. DO NOT WAIT MORE THAT 8 HOURS AFTER CONCRETE HAS 'SET'.

SLAB CONTRACTION JOINT DETAIL

NOT TO SCALE

(DETAIL T2–SCJ)

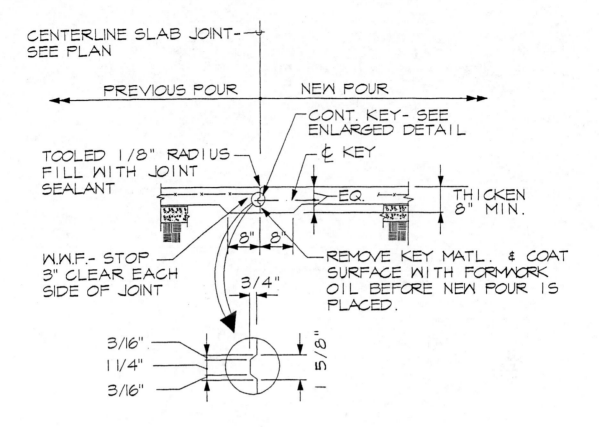

CENTERLINE SLAB JOINT-
SEE PLAN

PREVIOUS POUR   NEW POUR

CONT. KEY- SEE
ENLARGED DETAIL

TOOLED 1/8" RADIUS
FILL WITH JOINT
SEALANT

℄ KEY

EQ.

THICKEN
8" MIN.

W.W.F.- STOP
3" CLEAR EACH
SIDE OF JOINT

8"   8"

REMOVE KEY MATL. & COAT
SURFACE WITH FORMWORK
OIL BEFORE NEW POUR IS
PLACED.

3/4"

3/16"
1 1/4"
3/16"

1 5/8"

ENLARGED KEY DETAIL

KEYED CONSTRUCTION JOINT DETAIL #1
NOT TO SCALE   (DETAIL T2–SKCJ)

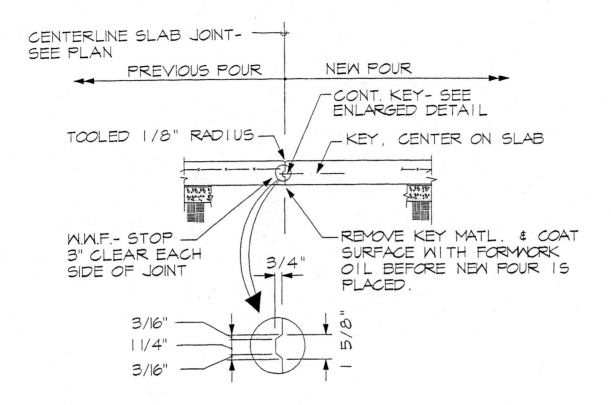

CENTERLINE SLAB JOINT-
SEE PLAN

PREVIOUS POUR   NEW POUR

CONT. KEY- SEE
ENLARGED DETAIL

TOOLED 1/8" RADIUS

KEY, CENTER ON SLAB

W.W.F.- STOP
3" CLEAR EACH
SIDE OF JOINT

REMOVE KEY MATL. & COAT
SURFACE WITH FORMWORK
OIL BEFORE NEW POUR IS
PLACED.

3/4"

3/16"
1 1/4"
3/16"

1 5/8"

ENLARGED KEY DETAIL

KEYED CONSTRUCTION JOINT DETAIL #2

NOT TO SCALE   (DETAIL T2-SKCJ2)

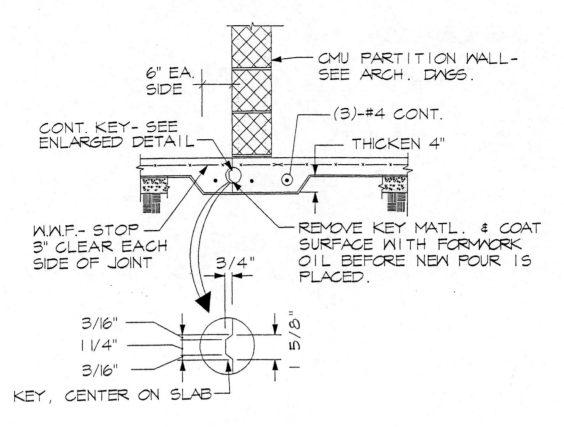

CMU PARTITION WALL-
SEE ARCH. DWGS.

6" EA. SIDE

(3)-#4 CONT.

CONT. KEY- SEE
ENLARGED DETAIL

THICKEN 4"

W.W.F.- STOP
3" CLEAR EACH
SIDE OF JOINT

REMOVE KEY MATL. & COAT
SURFACE WITH FORMWORK
OIL BEFORE NEW POUR IS
PLACED.

3/4"

3/16"

1 1/4"

1 5/8"

3/16"

KEY, CENTER ON SLAB

ENLARGED KEY DETAIL

KEYED CONSTRUCTION JOINT DETAIL #3

NOT TO SCALE   (DETAIL T2-SKCJ3)

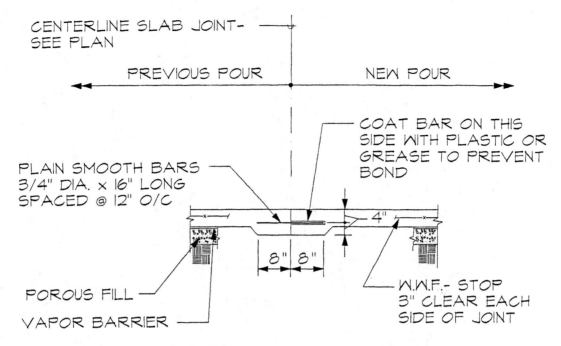

CENTERLINE SLAB JOINT-
SEE PLAN

PREVIOUS POUR   NEW POUR

COAT BAR ON THIS
SIDE WITH PLASTIC OR
GREASE TO PREVENT
BOND

PLAIN SMOOTH BARS
3/4" DIA. x 16" LONG
SPACED @ 12" O/C

4"

8" 8"

POROUS FILL

VAPOR BARRIER

W.W.F.- STOP
3" CLEAR EACH
SIDE OF JOINT

NOTES:
1. FILL JOINT WITH SEALANT - SEE
   "SLAB ON GRADE NOTES".
2. COAT VERTICAL SURFACE WITH OIL
   OR CURING COMPOUND BEFORE
   PLACING NEW POUR.

DOWELED CONSTRUCTION JOINT DETAIL

NOT TO SCALE   (DETAIL T2-SDCJ)

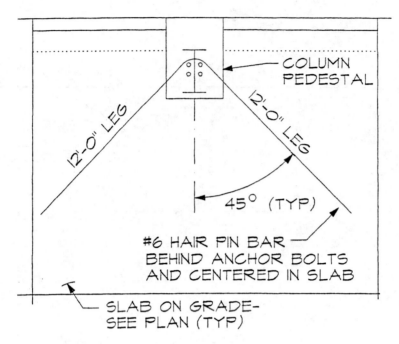

COLUMN PEDESTAL

12'-0" LEG

12'-0" LEG

45° (TYP)

#6 HAIR PIN BAR
BEHIND ANCHOR BOLTS
AND CENTERED IN SLAB

SLAB ON GRADE-
SEE PLAN (TYP)

<u>AT INDIVIDUAL COLUMN PEDESTALS</u>

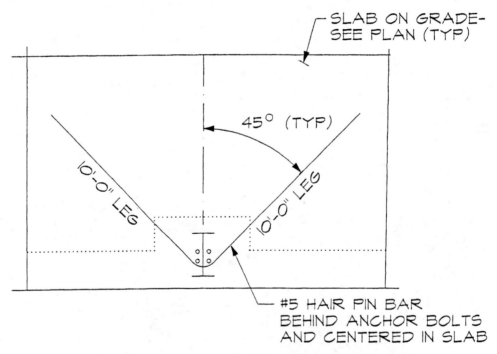

SLAB ON GRADE-
SEE PLAN (TYP)

45° (TYP)

10'-0" LEG

10'-0" LEG

#5 HAIR PIN BAR
BEHIND ANCHOR BOLTS
AND CENTERED IN SLAB

<u>AT MONOLITHIC SLAB FOUNDATIONS</u>

TYPICAL HAIR PIN BAR DETAILS

NOT TO SCALE   (DETAIL T2-SHPIN)

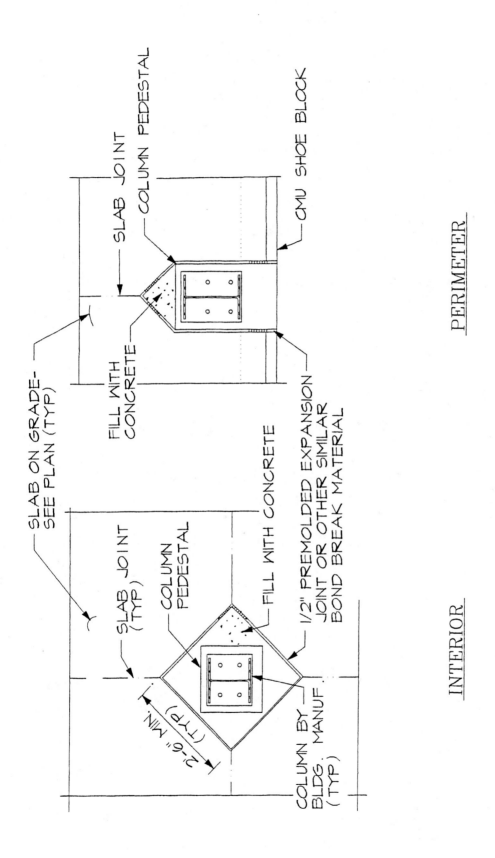

TYP. COLUMN ISOLATION JOINT DETAIL #1
(DETAIL T2-SISO1)

NOT TO SCALE

PERIMETER

INTERIOR

Labels in drawing:
- SLAB JOINT
- COLUMN PEDESTAL
- CMU SHOE BLOCK
- SLAB ON GRADE- SEE PLAN (TYP)
- FILL WITH CONCRETE
- SLAB JOINT (TYP)
- COLUMN PEDESTAL
- FILL WITH CONCRETE
- 1/2" PREMOLDED EXPANSION JOINT OR OTHER SIMILAR BOND BREAK MATERIAL
- COLUMN BY BLDG. MANUF. (TYP)
- 2'-6" MIN. (TYP)

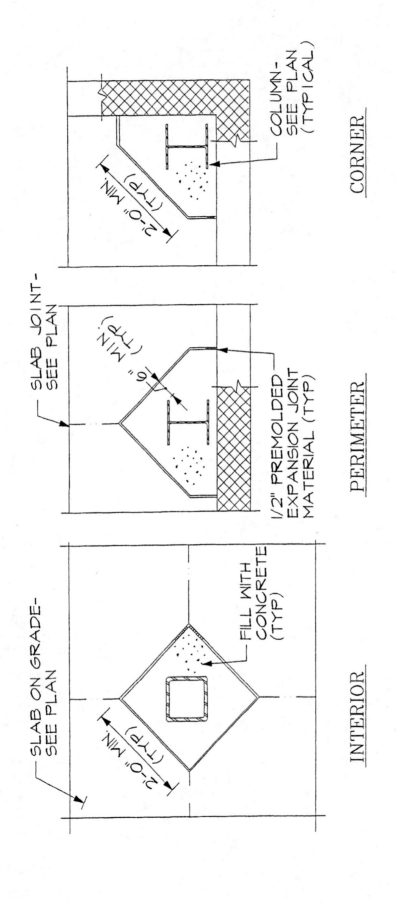

TYP. COLUMN ISOLATION JOINT DETAIL #2
(DETAIL T2-SIS02)

NOT TO SCALE

INTERIOR   PERIMETER   CORNER

SLAB ON GRADE-
SEE PLAN

2'-0" MIN (TYP)

FILL WITH
CONCRETE
(TYP)

SLAB JOINT-
SEE PLAN

6" MIN (TYP)

1/2" PREMOLDED
EXPANSION JOINT
MATERIAL (TYP)

2'-0" MIN
(TYP)

COLUMN-
SEE PLAN
(TYPICAL)

52

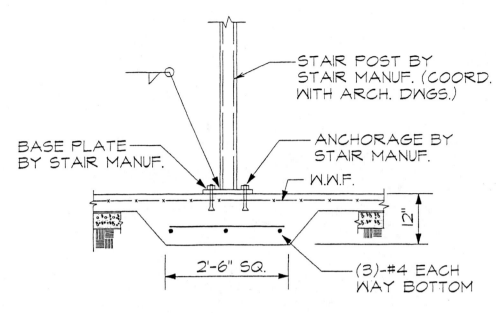

STAIR POST BY
STAIR MANUF. (COORD.
WITH ARCH. DWGS.)

BASE PLATE
BY STAIR MANUF.

ANCHORAGE BY
STAIR MANUF.

W.W.F.

2"

2'-6" SQ.

(3)-#4 EACH
WAY BOTTOM

THICKENED SLAB AT STAIR POSTS

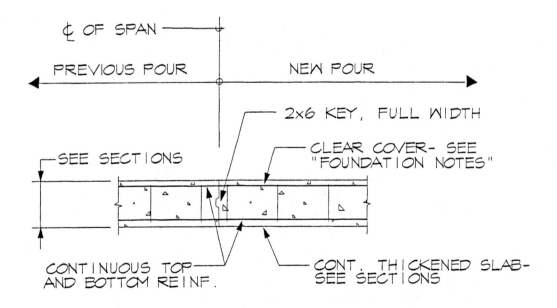

⊄ OF SPAN

PREVIOUS POUR

NEW POUR

2x6 KEY, FULL WIDTH

CLEAR COVER- SEE
"FOUNDATION NOTES"

SEE SECTIONS

CONTINUOUS TOP
AND BOTTOM REINF.

CONT. THICKENED SLAB-
SEE SECTIONS

THICKENED SLAB EDGE CONSTRUCTION JOINT

THICKENED SLAB DETAIL #1

NOT TO SCALE   (DETAIL T2-STS1)

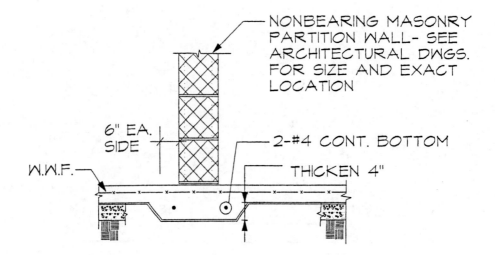

NONBEARING MASONRY
PARTITION WALL- SEE
ARCHITECTURAL DWGS.
FOR SIZE AND EXACT
LOCATION

6" EA. SIDE

2-#4 CONT. BOTTOM

W.W.F.

THICKEN 4"

THICKENED SLAB AT PARTITION WALL

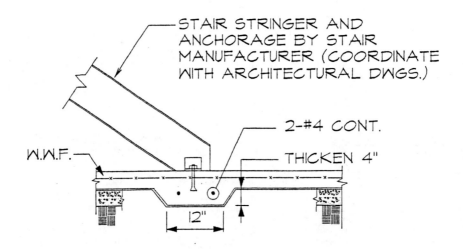

STAIR STRINGER AND
ANCHORAGE BY STAIR
MANUFACTURER (COORDINATE
WITH ARCHITECTURAL DWGS.)

2-#4 CONT.

W.W.F.

THICKEN 4"

12"

THICKENED SLAB AT STAIR STRINGERS

THICKENED SLAB DETAIL #2
NOT TO SCALE   (DETAIL T2–STS2)

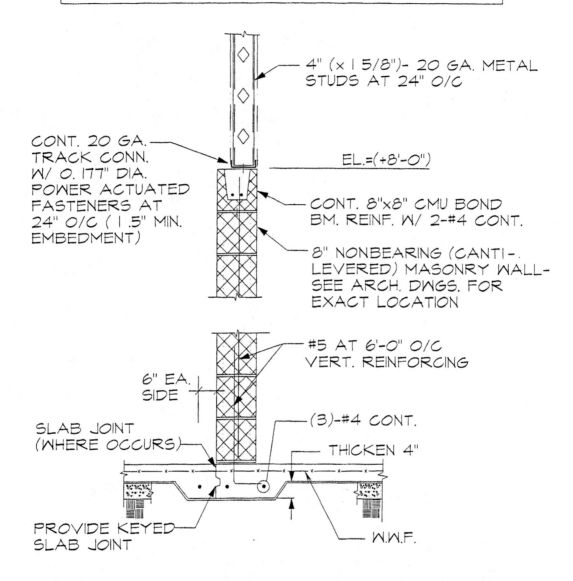

NOTE:
STUD MANUFACTURER TO CONNECT TOP
OF WALL TO ROOF CONSTRUCTION BASED ON AN
INTERIOR WIND LOAD OF 5 PSF.

4" (× 1 5/8")- 20 GA. METAL
STUDS AT 24" O/C

CONT. 20 GA.
TRACK CONN.
W/ 0.177" DIA.
POWER ACTUATED
FASTENERS AT
24" O/C (1.5" MIN.
EMBEDMENT)

EL.=(+8'-0")

CONT. 8"×8" CMU BOND
BM. REINF. W/ 2-#4 CONT.

8" NONBEARING (CANTI-.
LEVERED) MASONRY WALL-
SEE ARCH. DWGS. FOR
EXACT LOCATION

#5 AT 6'-0" O/C
VERT. REINFORCING

6" EA.
SIDE

SLAB JOINT
(WHERE OCCURS)

(3)-#4 CONT.

THICKEN 4"

PROVIDE KEYED
SLAB JOINT

W.W.F.

THICKENED SLAB AT INTERIOR CANTILEVERED WALL

THICKENED SLAB DETAIL #3

NOT TO SCALE   (DETAIL T2-STS3)

NOTE: SEE ARCHITECTURAL DRAWINGS FOR THE
EXACT SIZE AND EXTENT OF SLAB DEPRESSION.

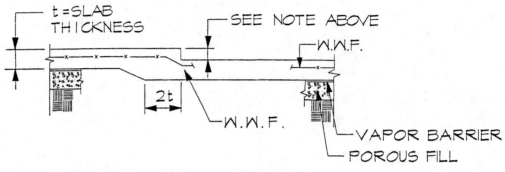

SLAB DEPRESSION DETAIL

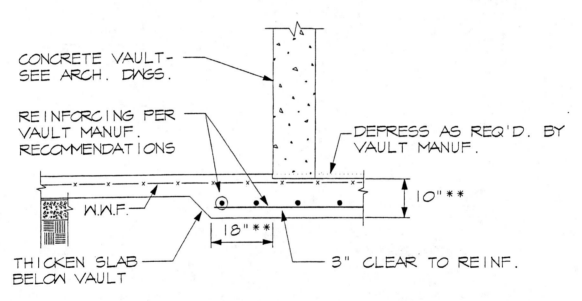

NOTE: ** DENOTES TO COORDINATE PER THE
VAULT MANUF. RECOMMENDATIONS.

SLAB DETAIL AT BANK VAULT

MISCELLANEOUS SLAB DETAIL #1

NOT TO SCALE   (DETAIL T2-SMIS1)

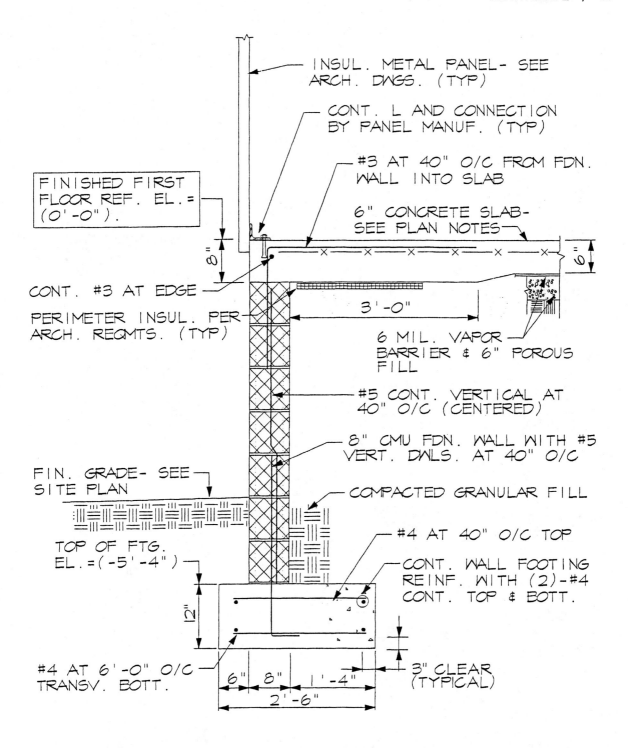

INSUL. METAL PANEL- SEE ARCH. DWGS. (TYP)

CONT. L AND CONNECTION BY PANEL MANUF. (TYP)

#3 AT 40" O/C FROM FDN. WALL INTO SLAB

6" CONCRETE SLAB- SEE PLAN NOTES

FINISHED FIRST FLOOR REF. EL.= (0'-0").

CONT. #3 AT EDGE

PERIMETER INSUL. PER ARCH. REQMTS. (TYP)

6 MIL. VAPOR BARRIER & 6" POROUS FILL

#5 CONT. VERTICAL AT 40" O/C (CENTERED)

8" CMU FDN. WALL WITH #5 VERT. DWLS. AT 40" O/C

FIN. GRADE- SEE SITE PLAN

COMPACTED GRANULAR FILL

TOP OF FTG. EL.=(-5'-4")

#4 AT 40" O/C TOP

CONT. WALL FOOTING REINF. WITH (2)-#4 CONT. TOP & BOTT.

#4 AT 6'-0" O/C TRANSV. BOTT.

3" CLEAR (TYPICAL)

12"

6"   8"   1'-4"

2'-6"

8"

3'-0"

6"

PERIMETER FOUNDATION SECTION #1

NOT TO SCALE   (DETAIL T2-IBPS1)

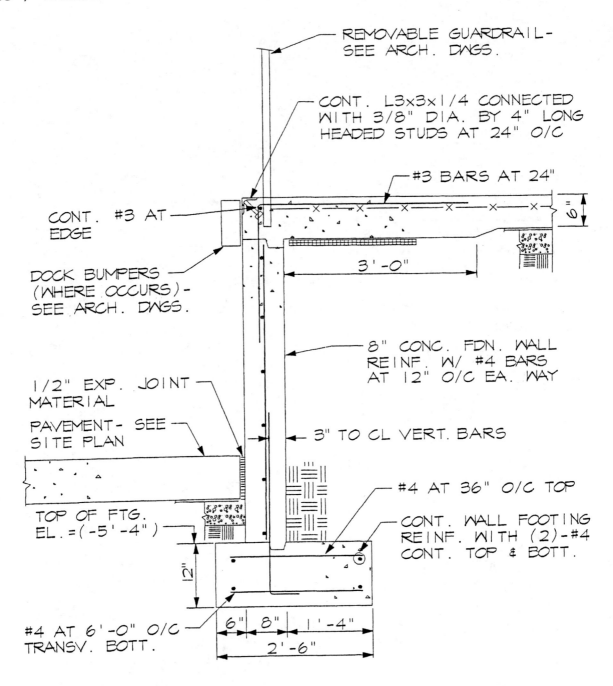

REMOVABLE GUARDRAIL-
SEE ARCH. DWGS.

CONT. L3x3x1/4 CONNECTED
WITH 3/8" DIA. BY 4" LONG
HEADED STUDS AT 24" O/C

#3 BARS AT 24"

6"

CONT. #3 AT
EDGE

DOCK BUMPERS
(WHERE OCCURS)-
SEE ARCH. DWGS.

3'-0"

8" CONC. FDN. WALL
REINF. W/ #4 BARS
AT 12" O/C EA. WAY

1/2" EXP. JOINT
MATERIAL

PAVEMENT- SEE
SITE PLAN

3" TO CL VERT. BARS

#4 AT 36" O/C TOP

CONT. WALL FOOTING
REINF. WITH (2)-#4
CONT. TOP & BOTT.

TOP OF FTG.
EL.=(-5'-4")

12"

#4 AT 6'-0" O/C
TRANSV. BOTT.

6" 8" 1'-4"

2'-6"

PERIMETER FOUNDATION SECTION #2

NOT TO SCALE   (DETAIL T2-IBPS2)

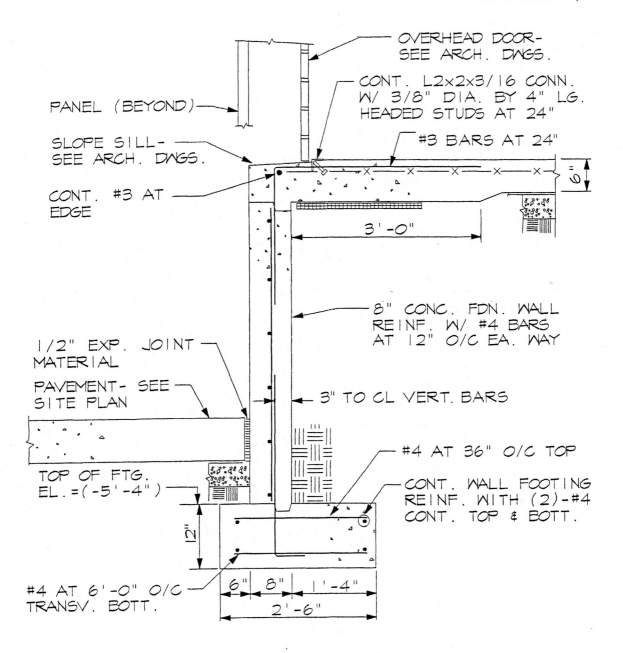

PANEL (BEYOND)

OVERHEAD DOOR-
SEE ARCH. DWGS.

CONT. L2x2x3/16 CONN.
W/ 3/8" DIA. BY 4" LG.
HEADED STUDS AT 24"

SLOPE SILL-
SEE ARCH. DWGS.

#3 BARS AT 24"

CONT. #3 AT
EDGE

6"

3'-0"

8" CONC. FDN. WALL
REINF. W/ #4 BARS
AT 12" O/C EA. WAY

1/2" EXP. JOINT
MATERIAL

PAVEMENT- SEE
SITE PLAN

3" TO CL VERT. BARS

#4 AT 36" O/C TOP

TOP OF FTG.
EL.=(-5'-4")

CONT. WALL FOOTING
REINF. WITH (2)-#4
CONT. TOP & BOTT.

12"

#4 AT 6'-0" O/C
TRANSV. BOTT.

6" 8" 1'-4"

2'-6"

PERIMETER FOUNDATION SECTION #3

NOT TO SCALE   (DETAIL T2-IBPS3)

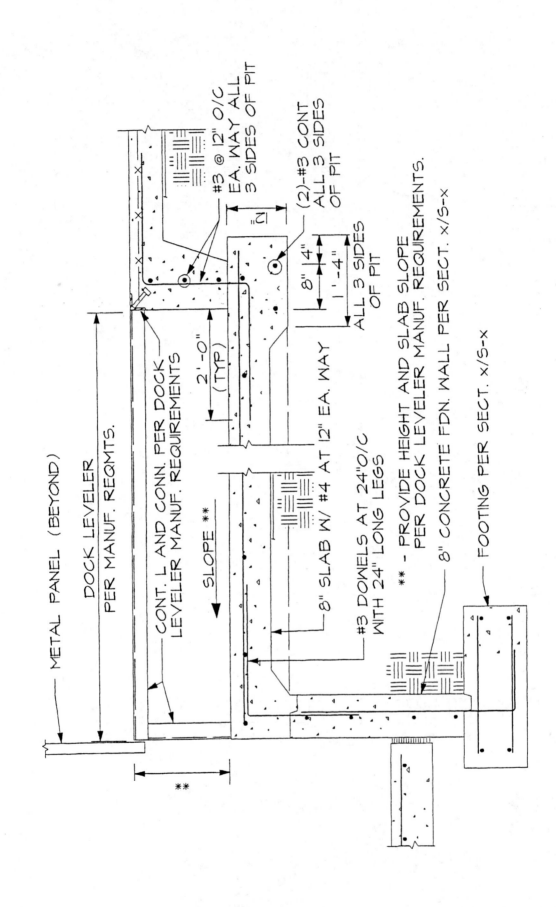

PERIMETER FOUNDATION SECTION #4

(DETAIL T2-IBPS4)

NOT TO SCALE

Labels within the figure:

METAL PANEL (BEYOND)

DOCK LEVELER PER MANUF. REQMTS.

CONT. L AND CONN. PER DOCK LEVELER MANUF. REQUIREMENTS.

SLOPE **

2'-0" (TYP)

8" SLAB W/ #4 AT 12" EA. WAY

#3 DOWELS AT 24"O/C WITH 24" LONG LEGS

** - PROVIDE HEIGHT AND SLAB SLOPE PER DOCK LEVELER MANUF. REQUIREMENTS.

8" CONCRETE FDN. WALL PER SECT. x/S-x

FOOTING PER SECT. x/S-x

#3 @ 12" O/C EA. WAY ALL 3 SIDES OF PIT

(2)-#3 CONT ALL 3 SIDES OF PIT

8" 4"

1'-4"

ALL 3 SIDES OF PIT

12"

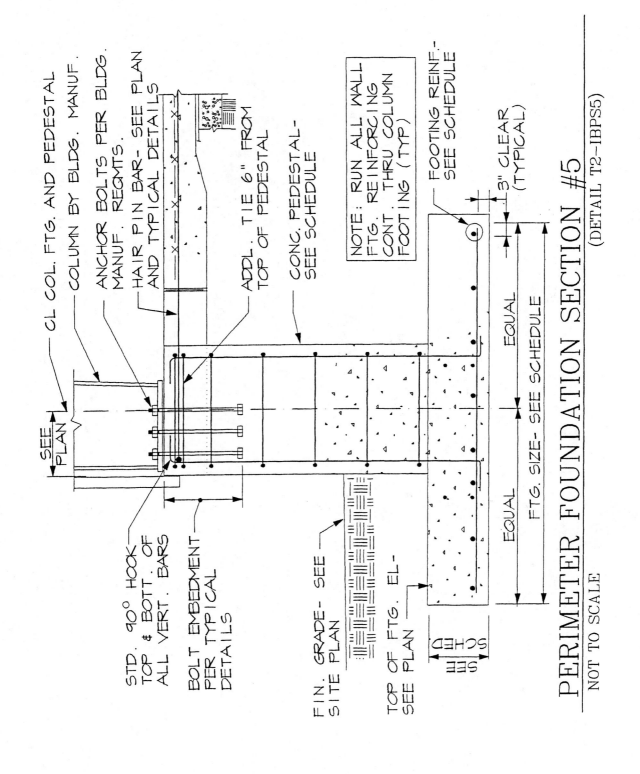

CL COL FTG. AND PEDESTAL

COLUMN BY BLDG. MANUF.

ANCHOR BOLTS PER BLDG. MANUF. REQMTS.

HAIR PIN BAR- SEE PLAN AND TYPICAL DETAILS

ADDL. TIE 6" FROM TOP OF PEDESTAL

CONC. PEDESTAL- SEE SCHEDULE

NOTE: RUN ALL WALL FTG. REINFORCING CONT. THRU COLUMN FOOTING (TYP)

FOOTING REINF- SEE SCHEDULE

3" CLEAR (TYPICAL)

EQUAL

FTG. SIZE- SEE SCHEDULE

EQUAL

SEE PLAN

STD. 90° HOOK TOP & BOTT. OF ALL VERT. BARS

BOLT EMBEDMENT PER TYPICAL DETAILS

FIN. GRADE- SEE SITE PLAN

TOP OF FTG.- SEE PLAN

SEE SCHED

PERIMETER FOUNDATION SECTION #5

NOT TO SCALE

(DETAIL T2-1BPS5)

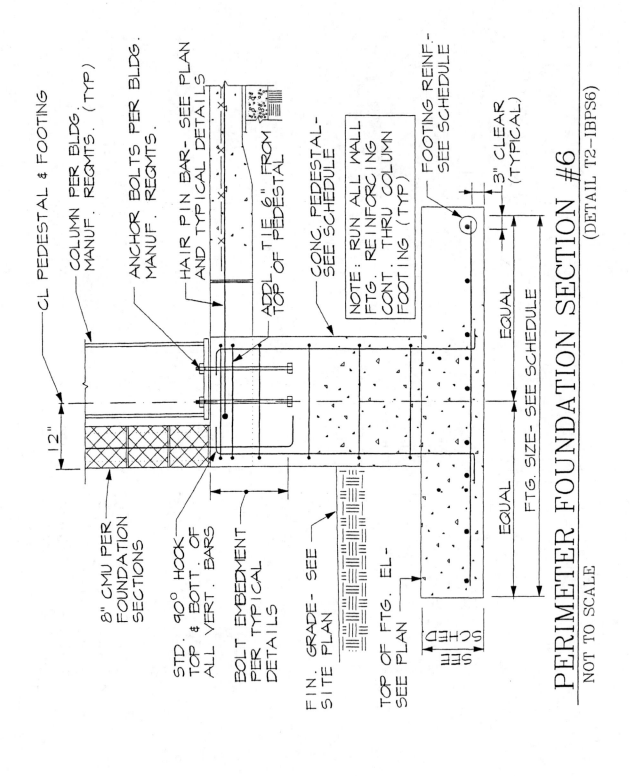

CL PEDESTAL & FOOTING

COLUMN PER BLDG. MANUF. REQMTS. (TYP)

ANCHOR BOLTS PER BLDG. MANUF. REQMTS.

HAIR PIN BAR- SEE PLAN AND TYPICAL DETAILS

ADDL. TIE 6" FROM TOP OF PEDESTAL

CONC. PEDESTAL- SEE SCHEDULE

NOTE: RUN ALL WALL FTG. REINFORCING CONT. THRU COLUMN FOOTING (TYP)

FOOTING REINF.- SEE SCHEDULE

3" CLEAR (TYPICAL)

8" CMU PER FOUNDATION SECTIONS

STD. 90° HOOK TOP & BOTT. OF ALL VERT. BARS

BOLT EMBEDMENT PER TYPICAL DETAILS

FIN. GRADE- SEE SITE PLAN

TOP OF FTG. EL- SEE PLAN

12"

EQUAL

EQUAL

FTG. SIZE- SEE SCHEDULE

SEE SCHED

PERIMETER FOUNDATION SECTION #6

NOT TO SCALE

(DETAIL T2-IBPS6)

62

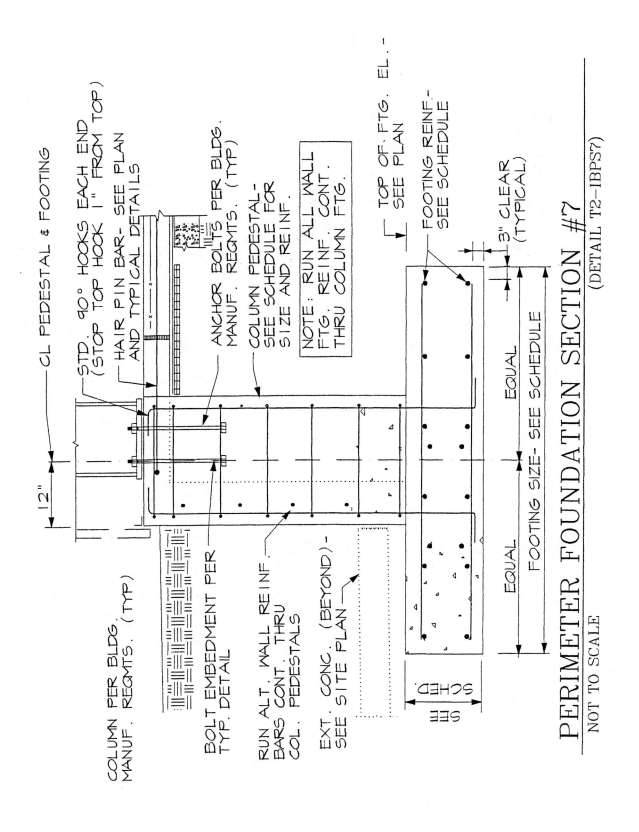

PERIMETER FOUNDATION SECTION #7

NOT TO SCALE

(DETAIL T2-IBPS7)

CL PEDESTAL & FOOTING

STD. 90° HOOKS EACH END
(STOP TOP HOOK 1" FROM TOP)

HAIR PIN BAR- SEE PLAN
AND TYPICAL DETAILS

ANCHOR BOLTS PER BLDG.
MANUF. REQMTS. (TYP)

COLUMN PEDESTAL-
SEE SCHEDULE FOR
SIZE AND REINF.

NOTE: RUN ALL WALL
FTG. REINF. CONT.
THRU COLUMN FTG.

TOP OF FTG. EL. -
SEE PLAN

FOOTING REINF-
SEE SCHEDULE

3" CLEAR
(TYPICAL)

EQUAL

FOOTING SIZE- SEE SCHEDULE

EQUAL

12"

COLUMN PER BLDG.
MANUF. REQMTS. (TYP)

BOLT EMBEDMENT PER
TYP. DETAIL

RUN ALT. WALL REINF.
BARS CONT. THRU
COL. PEDESTALS

EXT. CONC. (BEYOND)-
SEE SITE PLAN

SEE
SCHED.

63

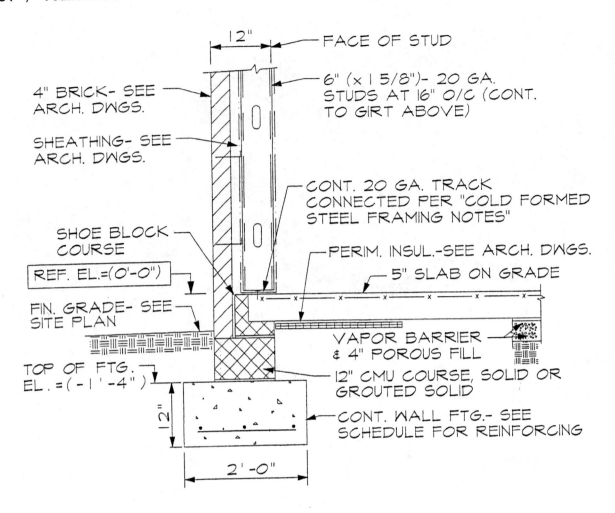

4" BRICK- SEE ARCH. DWGS.

SHEATHING- SEE ARCH. DWGS.

SHOE BLOCK COURSE

REF. EL.=(0'-0")

FIN. GRADE- SEE SITE PLAN

TOP OF FTG. EL.=(-1'-4")

12"

FACE OF STUD

6" (× 1 5/8")- 20 GA. STUDS AT 16" O/C (CONT. TO GIRT ABOVE)

CONT. 20 GA. TRACK CONNECTED PER "COLD FORMED STEEL FRAMING NOTES"

PERIM. INSUL.-SEE ARCH. DWGS.

5" SLAB ON GRADE

VAPOR BARRIER & 4" POROUS FILL

12" CMU COURSE, SOLID OR GROUTED SOLID

CONT. WALL FTG.- SEE SCHEDULE FOR REINFORCING

12"

2'-0"

PERIMETER FOUNDATION SECTION #8

NOT TO SCALE   (DETAIL T2—IBPS8)

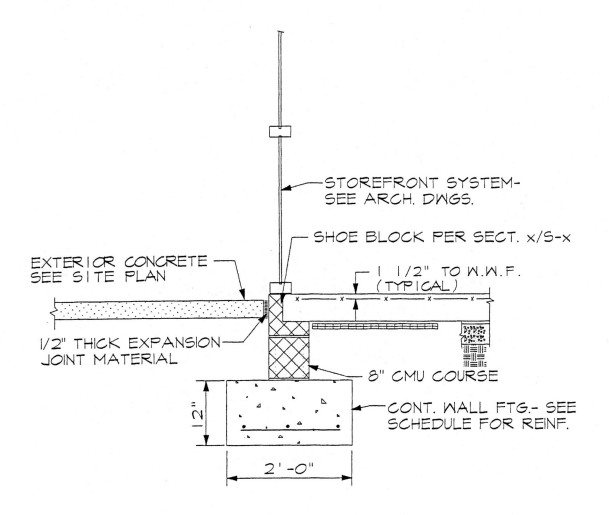

STOREFRONT SYSTEM-
SEE ARCH. DWGS.

SHOE BLOCK PER SECT. x/S-x

1 1/2" TO W.W.F.
(TYPICAL)

EXTERIOR CONCRETE
SEE SITE PLAN

1/2" THICK EXPANSION
JOINT MATERIAL

8" CMU COURSE

12"

CONT. WALL FTG.- SEE
SCHEDULE FOR REINF.

2'-0"

PERIMETER FOUNDATION SECTION #9

NOT TO SCALE   (DETAIL T2-IBPS9)

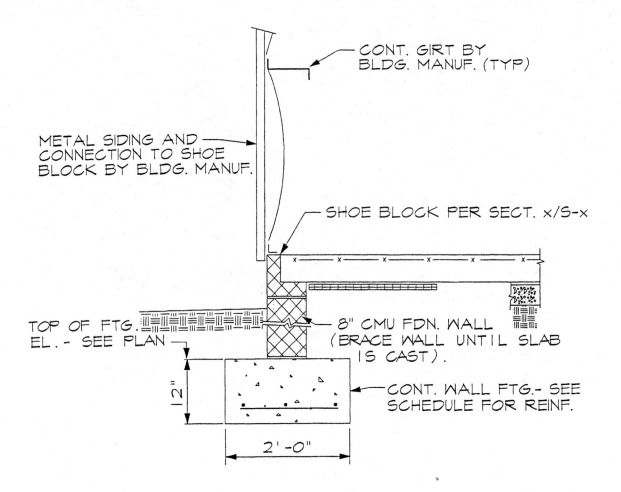

CONT. GIRT BY
BLDG. MANUF. (TYP)

METAL SIDING AND
CONNECTION TO SHOE
BLOCK BY BLDG. MANUF.

SHOE BLOCK PER SECT. x/S-x

TOP OF FTG.
EL.- SEE PLAN

8" CMU FDN. WALL
(BRACE WALL UNTIL SLAB
IS CAST).

12"

CONT. WALL FTG.- SEE
SCHEDULE FOR REINF.

2'-0"

PERIMETER FOUNDATION SECTION #10
NOT TO SCALE   (DETAIL T2–IBP10)

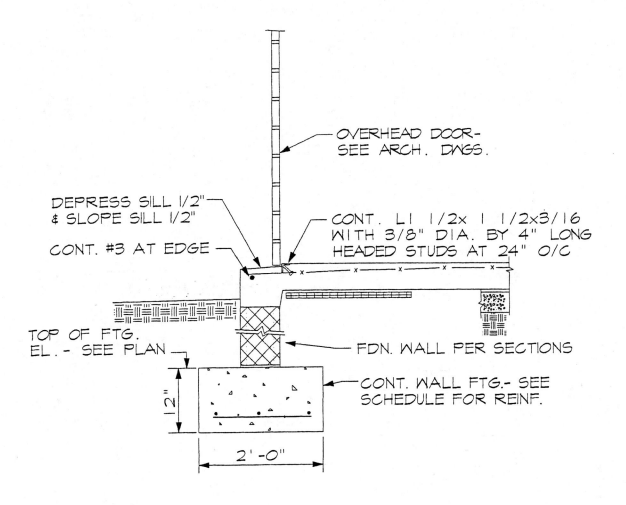

OVERHEAD DOOR-
SEE ARCH. DWGS.

DEPRESS SILL 1/2"
& SLOPE SILL 1/2"

CONT. #3 AT EDGE

CONT. L1 1/2x 1 1/2x3/16
WITH 3/8" DIA. BY 4" LONG
HEADED STUDS AT 24" O/C

TOP OF FTG.
EL.- SEE PLAN

FDN. WALL PER SECTIONS

CONT. WALL FTG.- SEE
SCHEDULE FOR REINF.

12"

2'-0"

PERIMETER FOUNDATION SECTION #11
NOT TO SCALE   (DETAIL T2–IBP11)

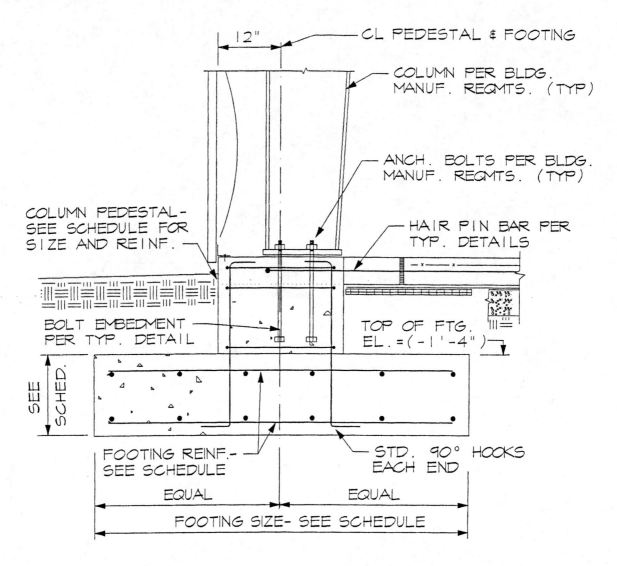

12"

CL PEDESTAL & FOOTING

COLUMN PER BLDG.
MANUF. REQMTS. (TYP)

ANCH. BOLTS PER BLDG.
MANUF. REQMTS. (TYP)

COLUMN PEDESTAL-
SEE SCHEDULE FOR
SIZE AND REINF.

HAIR PIN BAR PER
TYP. DETAILS

BOLT EMBEDMENT
PER TYP. DETAIL

TOP OF FTG.
EL.=(-1'-4")

SEE SCHED.

FOOTING REINF.-
SEE SCHEDULE

STD. 90° HOOKS
EACH END

EQUAL

EQUAL

FOOTING SIZE- SEE SCHEDULE

PERIMETER FOUNDATION SECTION #12

NOT TO SCALE   (DETAIL T2-IBP12)

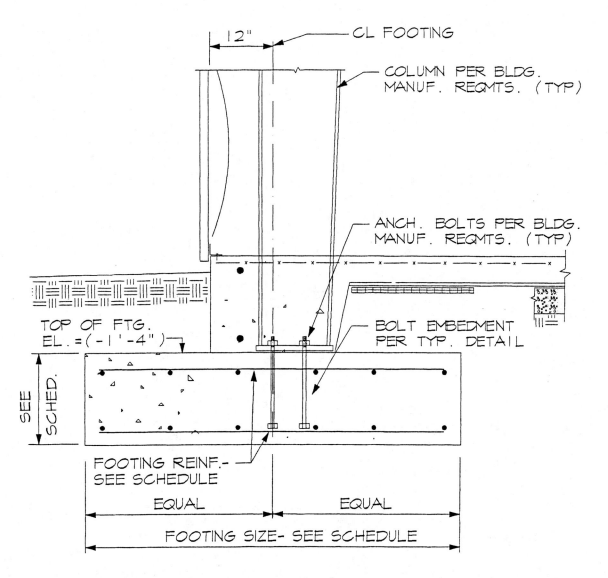

CL FOOTING

12"

COLUMN PER BLDG.
MANUF. REQMTS. (TYP)

ANCH. BOLTS PER BLDG.
MANUF. REQMTS. (TYP)

TOP OF FTG.
EL.=(-1'-4")

BOLT EMBEDMENT
PER TYP. DETAIL

SEE SCHED.

FOOTING REINF.-
SEE SCHEDULE

EQUAL

EQUAL

FOOTING SIZE- SEE SCHEDULE

PERIMETER FOUNDATION SECTION #12A

NOT TO SCALE   (DETAIL T2-IB12A)

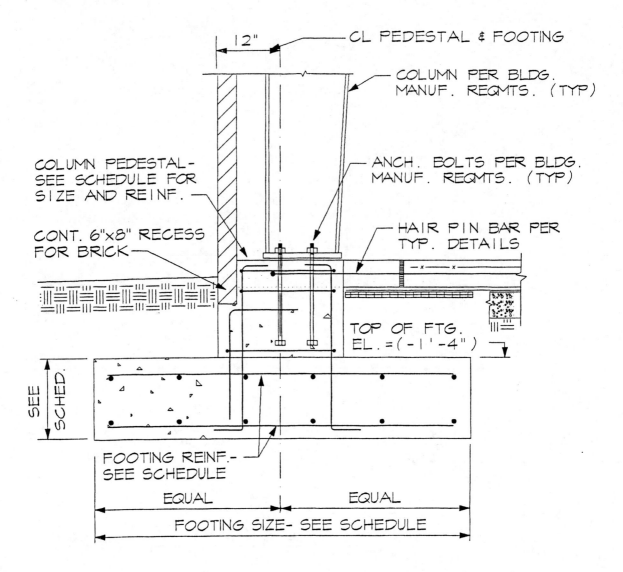

CL PEDESTAL & FOOTING

COLUMN PER BLDG. MANUF. REQMTS. (TYP)

COLUMN PEDESTAL- SEE SCHEDULE FOR SIZE AND REINF.

ANCH. BOLTS PER BLDG. MANUF. REQMTS. (TYP)

CONT. 6"x8" RECESS FOR BRICK

HAIR PIN BAR PER TYP. DETAILS

TOP OF FTG. EL.=(-1'-4")

SEE SCHED.

FOOTING REINF.- SEE SCHEDULE

EQUAL   EQUAL

FOOTING SIZE- SEE SCHEDULE

12"

PERIMETER FOUNDATION SECTION #13
NOT TO SCALE   (DETAIL T2–IBP13)

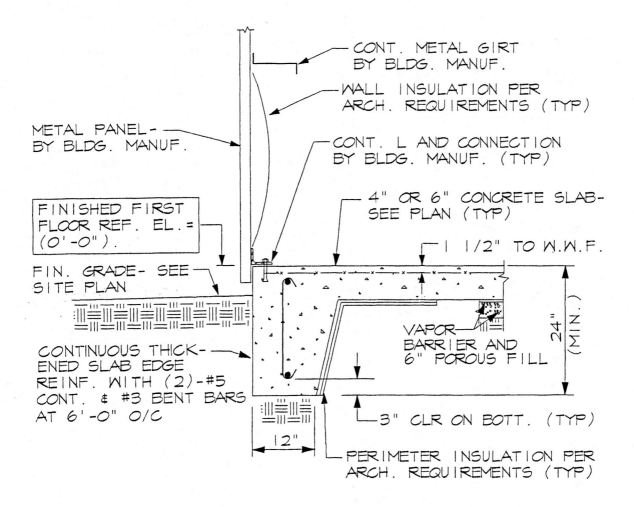

CONT. METAL GIRT
BY BLDG. MANUF.

WALL INSULATION PER
ARCH. REQUIREMENTS (TYP)

METAL PANEL-
BY BLDG. MANUF.

CONT. L AND CONNECTION
BY BLDG. MANUF. (TYP)

4" OR 6" CONCRETE SLAB-
SEE PLAN (TYP)

FINISHED FIRST
FLOOR REF. EL.=
(0'-0").

1 1/2" TO W.W.F.

FIN. GRADE- SEE
SITE PLAN

VAPOR
BARRIER AND
6" POROUS FILL

24" (MIN.)

CONTINUOUS THICK-
ENED SLAB EDGE
REINF. WITH (2)-#5
CONT. & #3 BENT BARS
AT 6'-0" O/C

3" CLR ON BOTT. (TYP)

12"

PERIMETER INSULATION PER
ARCH. REQUIREMENTS (TYP)

PERIMETER FOUNDATION SECTION #14
NOT TO SCALE   (DETAIL T2-IBP14)

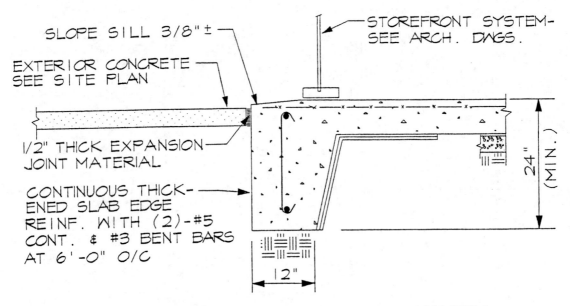

SLOPE SILL 3/8"±

EXTERIOR CONCRETE
SEE SITE PLAN

STOREFRONT SYSTEM-
SEE ARCH. DWGS.

1/2" THICK EXPANSION
JOINT MATERIAL

CONTINUOUS THICK-
ENED SLAB EDGE
REINF. WITH (2)-#5
CONT. & #3 BENT BARS
AT 6'-0" O/C

24" (MIN.)

12"

SLAB EDGE AT STOREFRONT SYSTEM

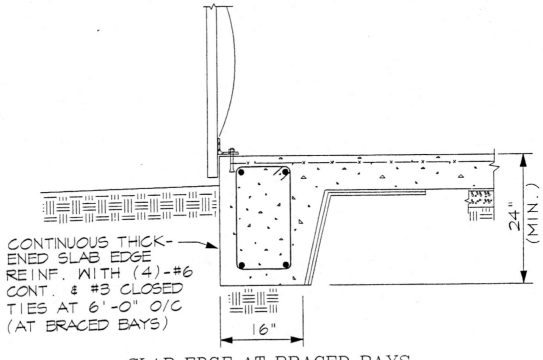

CONTINUOUS THICK-
ENED SLAB EDGE
REINF. WITH (4)-#6
CONT. & #3 CLOSED
TIES AT 6'-0" O/C
(AT BRACED BAYS)

24" (MIN.)

16"

SLAB EDGE AT BRACED BAYS

PERIMETER FOUNDATION SECTION #15
NOT TO SCALE   (DETAIL T2-IBP15)

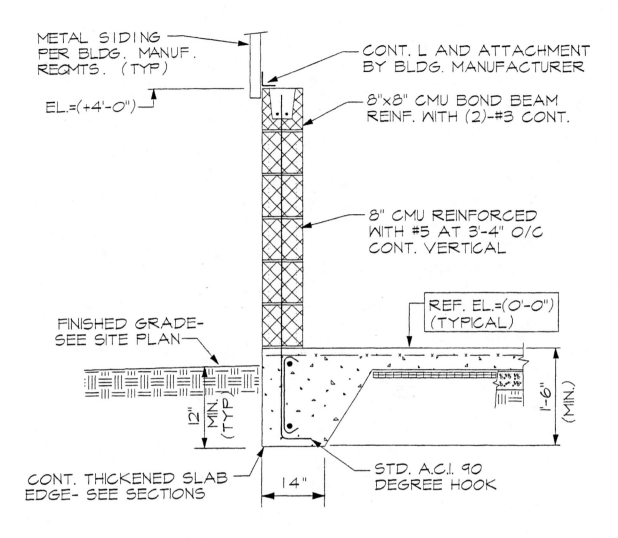

METAL SIDING
PER BLDG. MANUF.
REQMTS. (TYP)

EL.=(+4'-0")

CONT. L AND ATTACHMENT
BY BLDG. MANUFACTURER

8"x8" CMU BOND BEAM
REINF. WITH (2)-#3 CONT.

8" CMU REINFORCED
WITH #5 AT 3'-4" O/C
CONT. VERTICAL

REF. EL.=(0'-0")
(TYPICAL)

FINISHED GRADE-
SEE SITE PLAN

12" MIN. (TYP)

1'-6" (MIN.)

CONT. THICKENED SLAB
EDGE- SEE SECTIONS

14"

STD. A.C.I. 90
DEGREE HOOK

PERIMETER FOUNDATION SECTION #16
NOT TO SCALE   (DETAIL T2–IBP16)

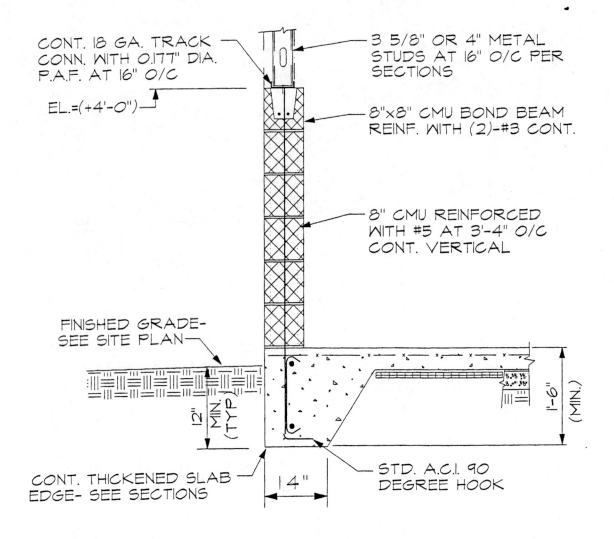

CONT. 18 GA. TRACK
CONN. WITH 0.177" DIA.
P.A.F. AT 16" O/C

EL.=(+4'-0")

3 5/8" OR 4" METAL
STUDS AT 16" O/C PER
SECTIONS

8"x8" CMU BOND BEAM
REINF. WITH (2)-#3 CONT.

8" CMU REINFORCED
WITH #5 AT 3'-4" O/C
CONT. VERTICAL

FINISHED GRADE-
SEE SITE PLAN

12" MIN. (TYP.)

1'-6" (MIN.)

CONT. THICKENED SLAB
EDGE- SEE SECTIONS

14"

STD. A.C.I. 90
DEGREE HOOK

PERIMETER FOUNDATION SECTION #17
NOT TO SCALE   (DETAIL T2-IBP17)

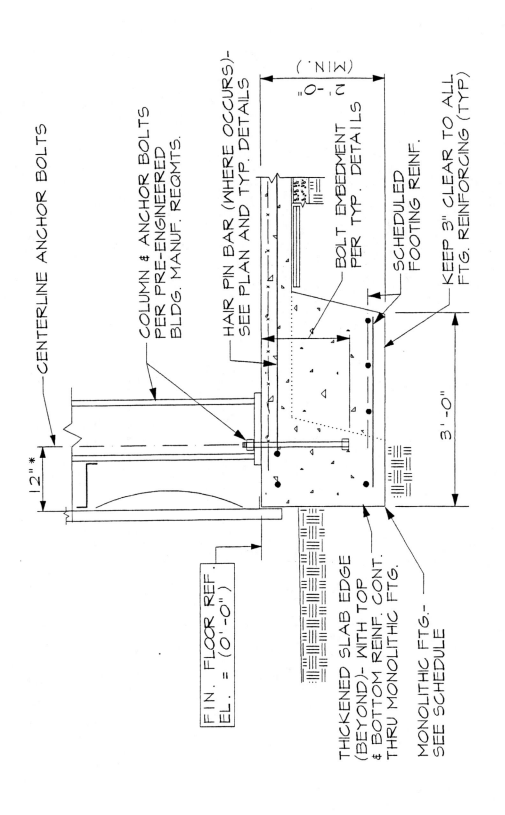

CENTERLINE ANCHOR BOLTS

COLUMN & ANCHOR BOLTS PER PRE-ENGINEERED BLDG. MANUF. REQMTS.

HAIR PIN BAR (WHERE OCCURS)- SEE PLAN AND TYP. DETAILS

BOLT EMBEDMENT PER TYP. DETAILS

SCHEDULED FOOTING REINF.

KEEP 3" CLEAR TO ALL FTG. REINFORCING (TYP)

2'-0" (MIN.)

3'-0"

12"*

FIN. FLOOR REF. EL. = (0'-0")

THICKENED SLAB EDGE (BEYOND)- WITH TOP & BOTTOM REINF. CONT. THRU MONOLITHIC FTG.

MONOLITHIC FTG.- SEE SCHEDULE

PERIMETER FOUNDATION SECTION #18
(DETAIL T2-IBP18)

NOT TO SCALE

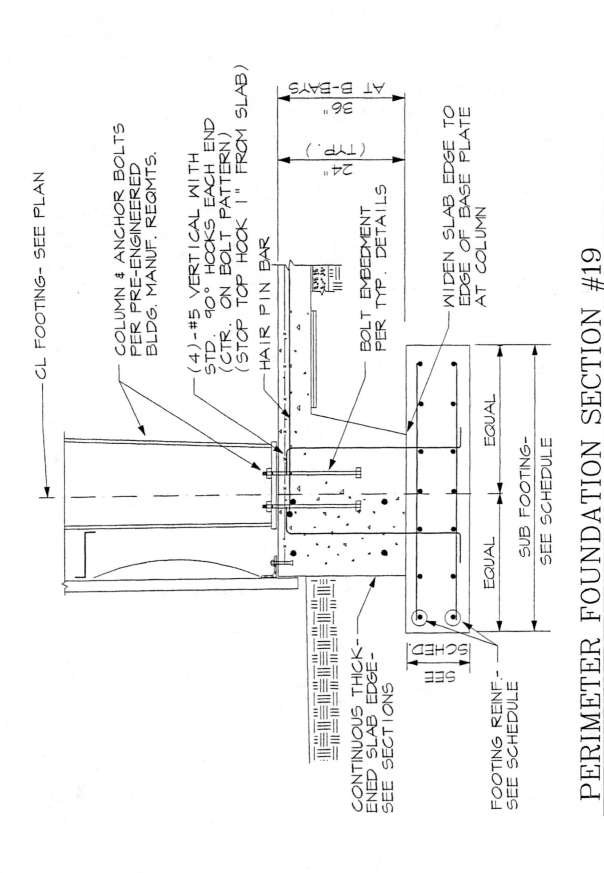

CL FOOTING- SEE PLAN

COLUMN & ANCHOR BOLTS PER PRE-ENGINEERED BLDG. MANUF. REQMTS.

(4)-#5 VERTICAL WITH STD. 90° HOOKS EACH END (CTR. ON BOLT PATTERN) (STOP TOP HOOK 1" FROM SLAB)

HAIR PIN BAR

BOLT EMBEDMENT PER TYP. DETAILS

WIDEN SLAB EDGE TO EDGE OF BASE PLATE AT COLUMN

AT B-BAYS

36"

24" (TYP.)

EQUAL

EQUAL

SUB FOOTING- SEE SCHEDULE

CONTINUOUS THICK- ENED SLAB EDGE- SEE SECTIONS

SEE SCHED.

FOOTING REINF.- SEE SCHEDULE

PERIMETER FOUNDATION SECTION #19

(DETAIL T2-IBP19)

NOT TO SCALE

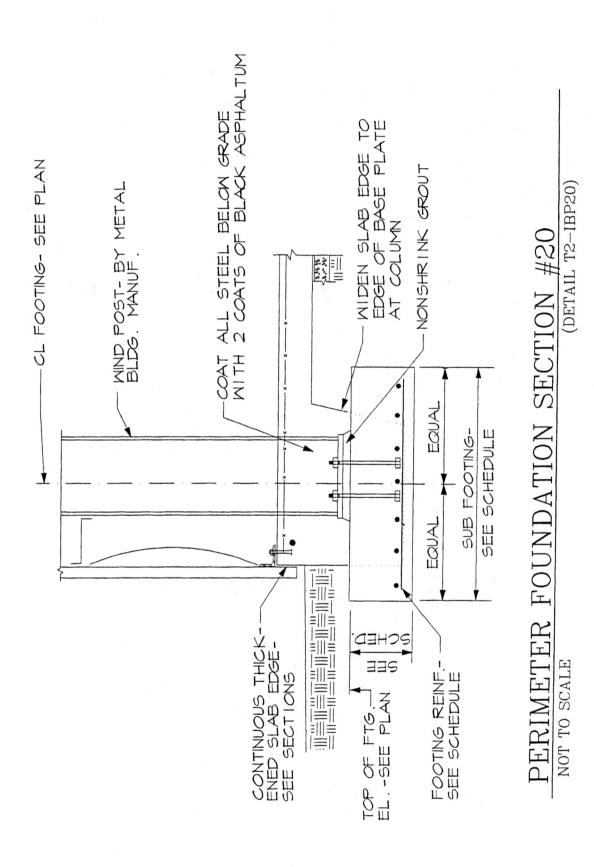

CL FOOTING- SEE PLAN

WIND POST- BY METAL BLDG. MANUF.

COAT ALL STEEL BELOW GRADE WITH 2 COATS OF BLACK ASPHALTUM

WIDEN SLAB EDGE TO EDGE OF BASE PLATE AT COLUMN

NONSHRINK GROUT

CONTINUOUS THICK-ENED SLAB EDGE-SEE SECTIONS

TOP OF FTG. EL.-SEE PLAN

SEE SCHED.

FOOTING REINF.-SEE SCHEDULE

EQUAL

EQUAL

SUB FOOTING-SEE SCHEDULE

PERIMETER FOUNDATION SECTION #20
(DETAIL T2-IBP20)

NOT TO SCALE

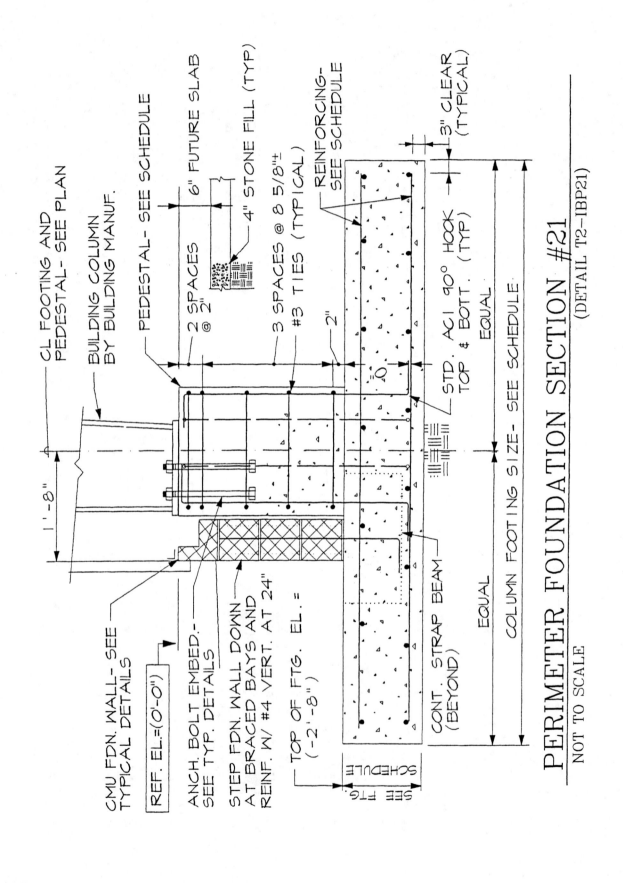

CL FOOTING AND
PEDESTAL- SEE PLAN

BUILDING COLUMN
BY BUILDING MANUF.

PEDESTAL- SEE SCHEDULE

6" FUTURE SLAB

2 SPACES
@ 2"

4" STONE FILL (TYP)

3 SPACES @ 8 5/8"±

#3 TIES (TYPICAL)

REINFORCING-
SEE SCHEDULE

3" CLEAR
(TYPICAL)

2"

REF. EL.=(0'-0")

1'-8"

ANCH. BOLT EMBED.-
SEE TYP. DETAILS

CMU FDN. WALL- SEE
TYPICAL DETAILS

STEP FDN. WALL DOWN
AT BRACED BAYS AND
REINF. W/ #4 VERT. AT 24"

TOP OF FTG. EL. =
(-2'-8")

STD. ACI 90° HOOK
TOP & BOTT. (TYP)

EQUAL

EQUAL

CONT. STRAP BEAM
(BEYOND)

COLUMN FOOTING SIZE- SEE SCHEDULE

SEE FTG.
SCHEDULE

PERIMETER FOUNDATION SECTION #21

NOT TO SCALE

(DETAIL T2-IBP21)

78

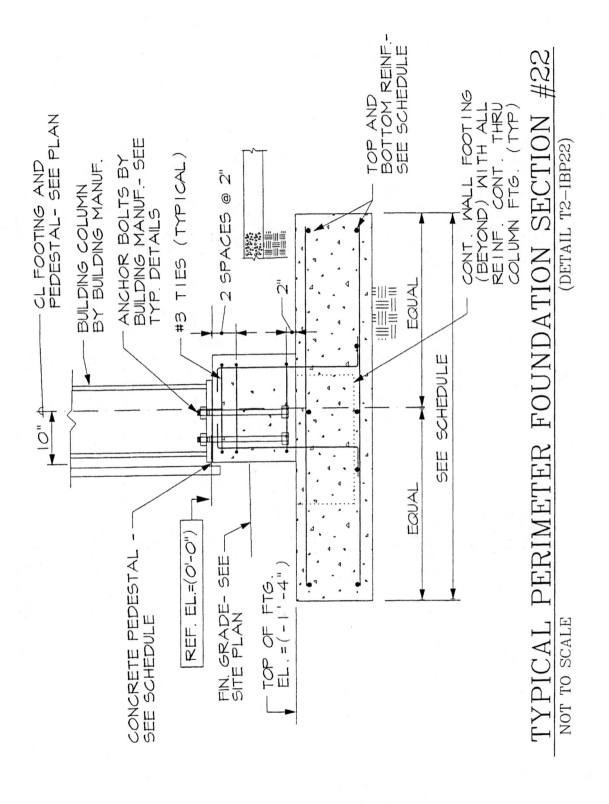

CL FOOTING AND
PEDESTAL- SEE PLAN

BUILDING COLUMN
BY BUILDING MANUF.

ANCHOR BOLTS BY
BUILDING MANUF.- SEE
TYP. DETAILS

#3 TIES (TYPICAL)

2 SPACES @ 2"

2"

10"

CONCRETE PEDESTAL -
SEE SCHEDULE

REF. EL.=(0'-0")

FIN. GRADE- SEE
SITE PLAN

TOP OF FTG.
EL. = (-1'-4")

TOP AND
BOTTOM REINF.-
SEE SCHEDULE

CONT. WALL FOOTING
(BEYOND) WITH ALL
REINF. CONT. THRU
COLUMN FTG. (TYP)

EQUAL

SEE SCHEDULE

EQUAL

TYPICAL PERIMETER FOUNDATION SECTION #22

(DETAIL T2-IBP22)

NOT TO SCALE

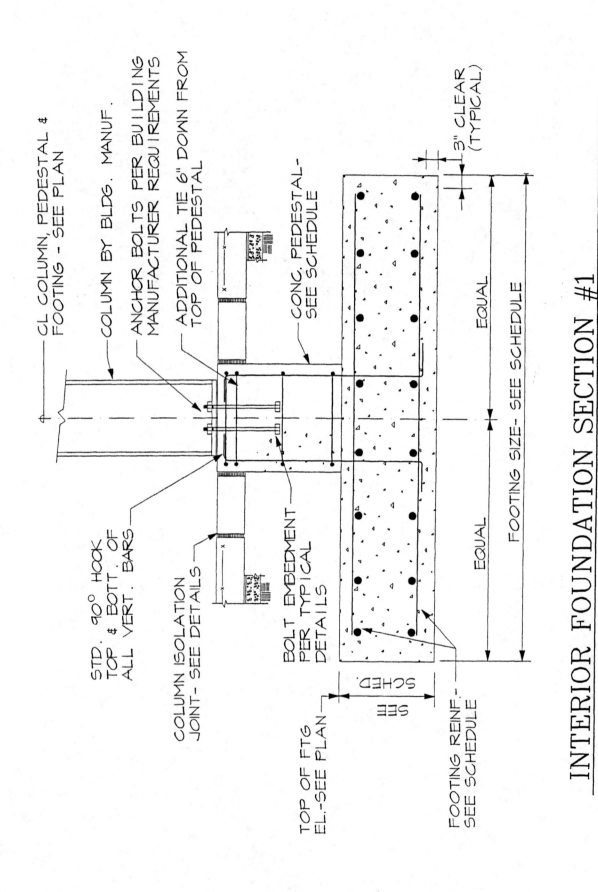

CL COLUMN, PEDESTAL &
FOOTING - SEE PLAN

COLUMN BY BLDG. MANUF.

ANCHOR BOLTS PER BUILDING
MANUFACTURER REQUIREMENTS

ADDITIONAL TIE 6" DOWN FROM
TOP OF PEDESTAL

CONC. PEDESTAL-
SEE SCHEDULE

STD. 90° HOOK
TOP & BOTT. OF
ALL VERT. BARS

COLUMN ISOLATION
JOINT - SEE DETAILS

BOLT EMBEDMENT
PER TYPICAL
DETAILS

TOP OF FTG
EL.-SEE PLAN

FOOTING REINF.-
SEE SCHEDULE

3" CLEAR
(TYPICAL)

EQUAL

FOOTING SIZE- SEE SCHEDULE

EQUAL

SEE
SCHED.

INTERIOR FOUNDATION SECTION #1

NOT TO SCALE

(DETAIL T2-IBIS1)

80

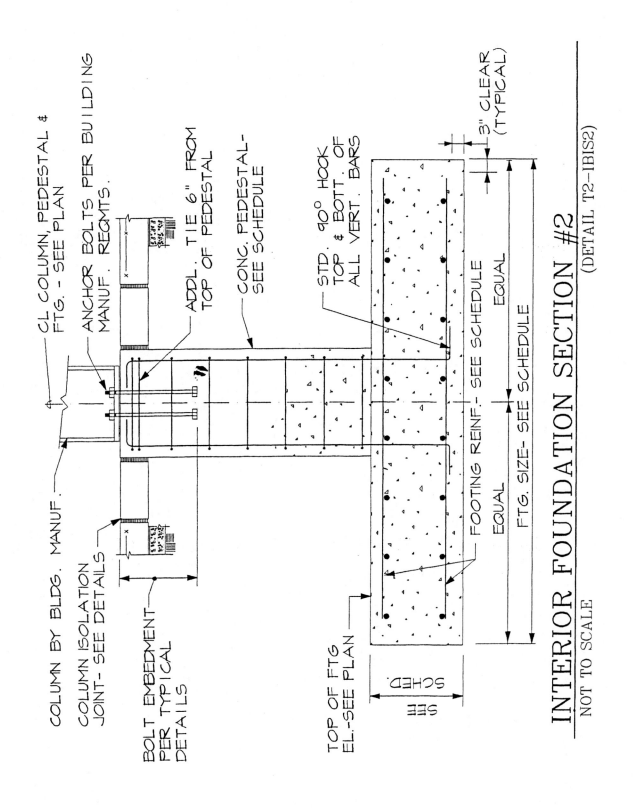

CL COLUMN, PEDESTAL &
FTG. - SEE PLAN

ANCHOR BOLTS PER BUILDING
MANUF. REQMTS.

ADDL. TIE 6" FROM
TOP OF PEDESTAL

CONC. PEDESTAL-
SEE SCHEDULE

STD. 90° HOOK
TOP & BOTT. OF
ALL VERT. BARS

3" CLEAR
(TYPICAL)

COLUMN BY BLDG. MANUF.

COLUMN ISOLATION
JOINT- SEE DETAILS

BOLT EMBEDMENT
PER TYPICAL
DETAILS

FOOTING REINF.- SEE SCHEDULE

EQUAL

EQUAL

FTG. SIZE- SEE SCHEDULE

TOP OF FTG
EL.-SEE PLAN

SEE
SCHED

INTERIOR FOUNDATION SECTION #2

NOT TO SCALE

(DETAIL T2-IBIS2)

81

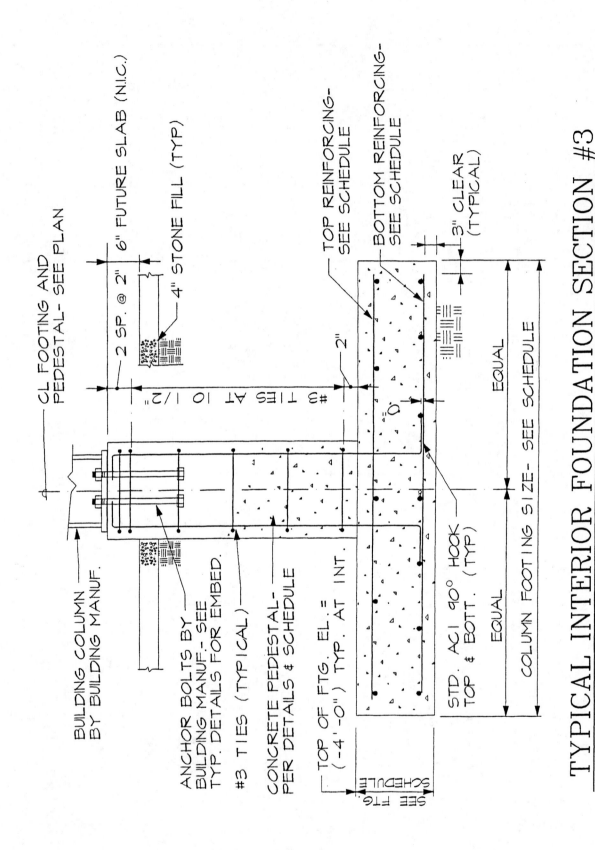

CL FOOTING AND
PEDESTAL- SEE PLAN

6" FUTURE SLAB (N.I.C.)

4" STONE FILL (TYP)

2 SP. @ 2"

TOP REINFORCING-
SEE SCHEDULE

BOTTOM REINFORCING-
SEE SCHEDULE

3" CLEAR
(TYPICAL)

#3 TIES AT 10 1/2"

2"

BUILDING COLUMN
BY BUILDING MANUF.

ANCHOR BOLTS BY
BUILDING MANUF.- SEE
TYP. DETAILS FOR EMBED.

#3 TIES (TYPICAL)

CONCRETE PEDESTAL-
PER DETAILS & SCHEDULE

TOP OF FTG. EL. =
(-4'-0") TYP. AT INT.

STD. ACI 90° HOOK
TOP & BOTT. (TYP)

EQUAL

EQUAL

COLUMN FOOTING SIZE- SEE SCHEDULE

SEE FTG
SCHEDULE

TYPICAL INTERIOR FOUNDATION SECTION #3

NOT TO SCALE

(DETAIL T2-1BIS3)

82

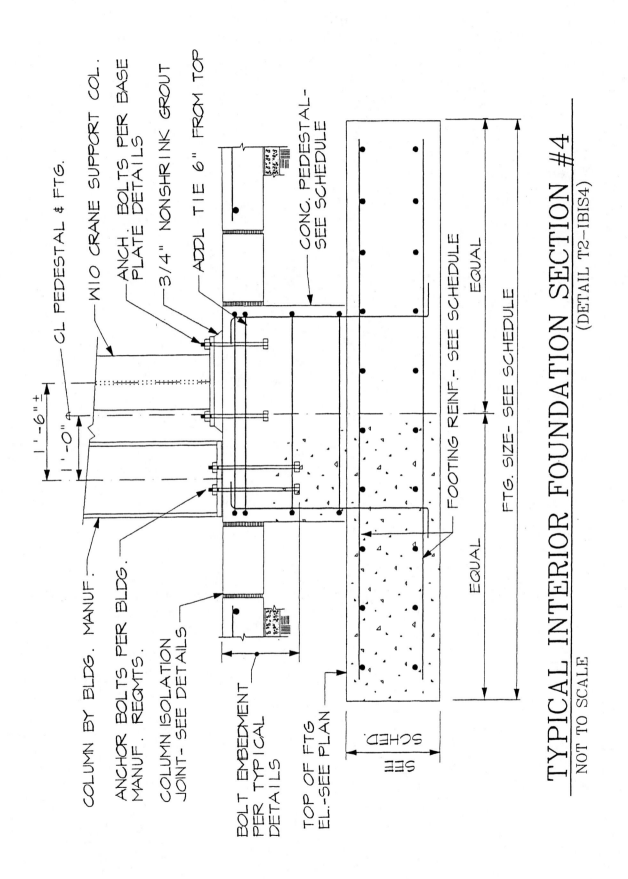

CL PEDESTAL & FTG.

W/O CRANE SUPPORT COL.

ANCH. BOLTS PER BASE PLATE DETAILS

3/4" NONSHRINK GROUT

ADD'L TIE 6" FROM TOP

CONC. PEDESTAL- SEE SCHEDULE

COLUMN BY BLDG. MANUF.

ANCHOR BOLTS PER BLDG. MANUF. REQMTS.

COLUMN ISOLATION JOINT- SEE DETAILS

BOLT EMBEDMENT PER TYPICAL DETAILS

TOP OF FTG EL.-SEE PLAN

1'-6"±

1'-0"

FOOTING REINF.- SEE SCHEDULE

FTG. SIZE- SEE SCHEDULE

EQUAL

EQUAL

SEE SCHED

TYPICAL INTERIOR FOUNDATION SECTION #4

NOT TO SCALE

(DETAIL T2-1BIS4)

83

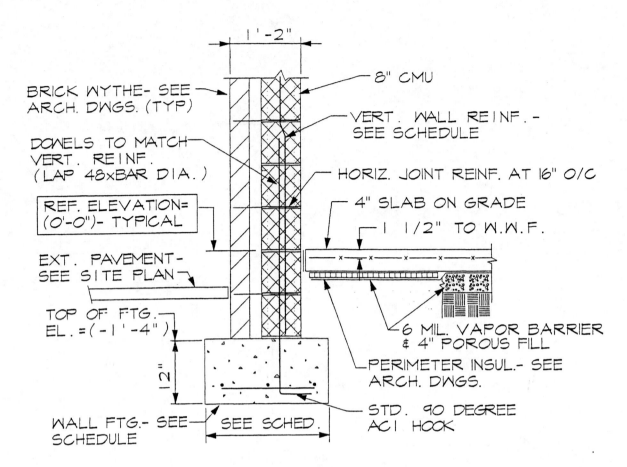

1'-2"

BRICK WYTHE- SEE ARCH. DWGS. (TYP)

8" CMU

VERT. WALL REINF.- SEE SCHEDULE

DOWELS TO MATCH VERT. REINF. (LAP 48xBAR DIA.)

HORIZ. JOINT REINF. AT 16" O/C

REF. ELEVATION= (0'-0")- TYPICAL

4" SLAB ON GRADE

1 1/2" TO W.W.F.

EXT. PAVEMENT- SEE SITE PLAN

TOP OF FTG. EL.=(-1'-4")

6 MIL. VAPOR BARRIER & 4" POROUS FILL

PERIMETER INSUL.- SEE ARCH. DWGS.

12"

STD. 90 DEGREE ACI HOOK

WALL FTG.- SEE SCHEDULE

SEE SCHED.

TYPICAL FOUNDATION SECTION #1

NOT TO SCALE

(DETAIL T2-FS1)

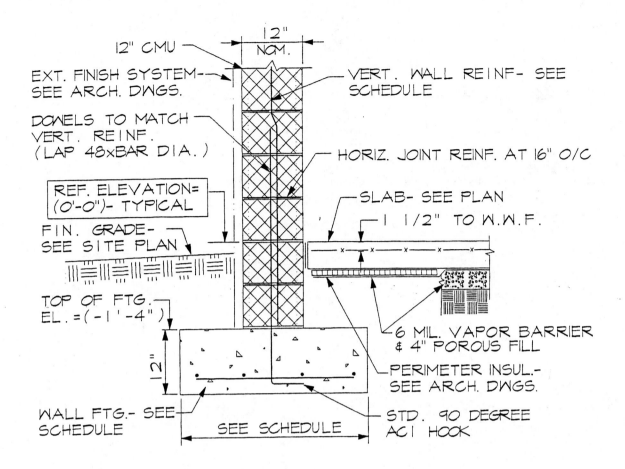

12" CMU

12" NOM.

EXT. FINISH SYSTEM-
SEE ARCH. DWGS.

VERT. WALL REINF- SEE
SCHEDULE

DOWELS TO MATCH
VERT. REINF.
(LAP 48xBAR DIA.)

HORIZ. JOINT REINF. AT 16" O/C

REF. ELEVATION=
(0'-0")- TYPICAL

SLAB- SEE PLAN

1 1/2" TO W.W.F.

FIN. GRADE-
SEE SITE PLAN

TOP OF FTG.
EL.=(-1'-4")

12"

6 MIL. VAPOR BARRIER
& 4" POROUS FILL

PERIMETER INSUL.-
SEE ARCH. DWGS.

WALL FTG.- SEE
SCHEDULE

SEE SCHEDULE

STD. 90 DEGREE
ACI HOOK

TYPICAL FOUNDATION SECTION #2

NOT TO SCALE

(DETAIL T2-FS2)

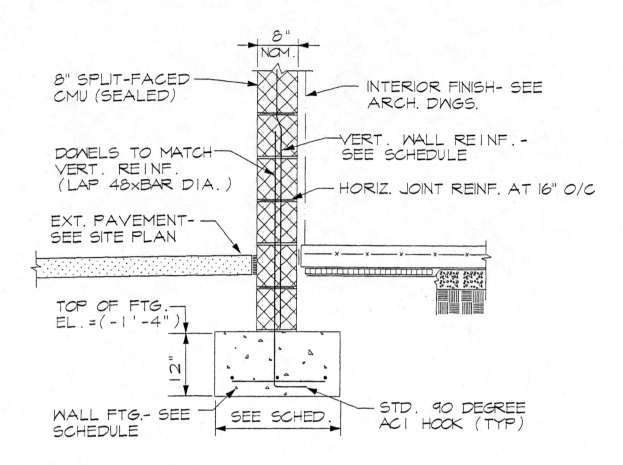

8" SPLIT-FACED CMU (SEALED)

INTERIOR FINISH- SEE ARCH. DWGS.

DONELS TO MATCH VERT. REINF. (LAP 48xBAR DIA.)

VERT. WALL REINF.- SEE SCHEDULE

HORIZ. JOINT REINF. AT 16" O/C

EXT. PAVEMENT- SEE SITE PLAN

TOP OF FTG. EL.=(-1'-4")

12"

WALL FTG.- SEE SCHEDULE

SEE SCHED.

STD. 90 DEGREE ACI HOOK (TYP)

8" NOM.

TYPICAL FOUNDATION SECTION #3

NOT TO SCALE   (DETAIL T2-FS3)

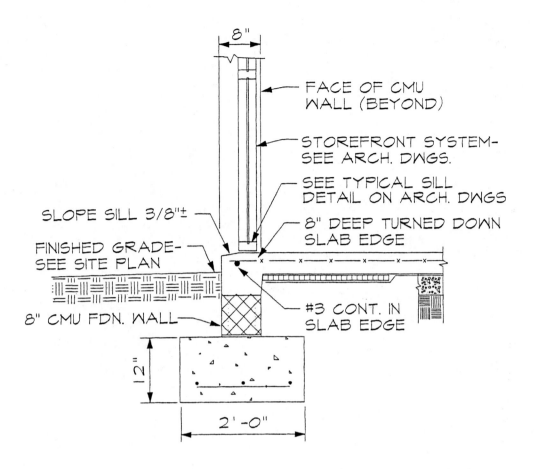

FACE OF CMU
WALL (BEYOND)

STOREFRONT SYSTEM-
SEE ARCH. DWGS.

SEE TYPICAL SILL
DETAIL ON ARCH. DWGS

8" DEEP TURNED DOWN
SLAB EDGE

SLOPE SILL 3/8"±

FINISHED GRADE-
SEE SITE PLAN

8" CMU FDN. WALL

#3 CONT. IN
SLAB EDGE

8"

12"

2'-0"

TYPICAL FOUNDATION SECTION #4
NOT TO SCALE   (DETAIL T2-FS4)

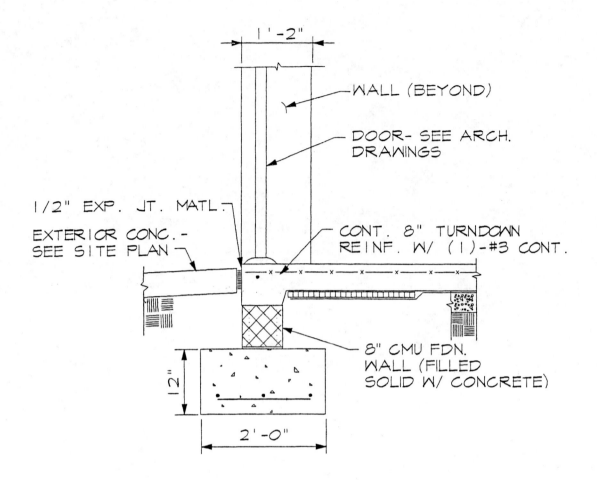

1'-2"

WALL (BEYOND)

DOOR- SEE ARCH. DRAWINGS

1/2" EXP. JT. MATL.

EXTERIOR CONC.- SEE SITE PLAN

CONT. 8" TURNDOWN REINF. W/ (1)-#3 CONT.

8" CMU FDN. WALL (FILLED SOLID W/ CONCRETE)

12"

2'-0"

TYPICAL FOUNDATION SECTION #5

NOT TO SCALE   (DETAIL T2-FS5)

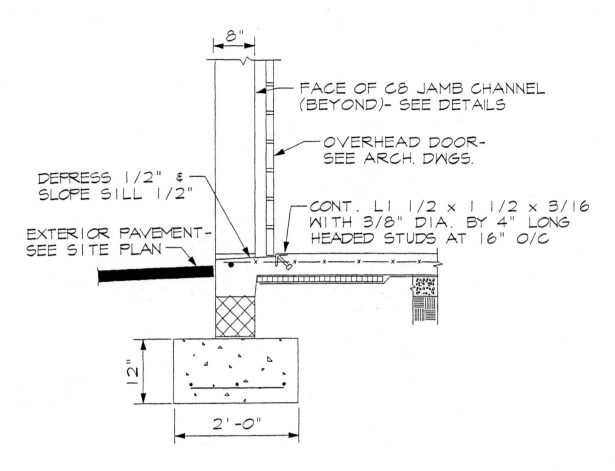

DEPRESS 1/2" &
SLOPE SILL 1/2"

EXTERIOR PAVEMENT-
SEE SITE PLAN

8"

FACE OF C8 JAMB CHANNEL
(BEYOND)- SEE DETAILS

OVERHEAD DOOR-
SEE ARCH. DWGS.

CONT. L1 1/2 x 1 1/2 x 3/16
WITH 3/8" DIA. BY 4" LONG
HEADED STUDS AT 16" O/C

12"

2'-0"

TYPICAL FOUNDATION SECTION #6
NOT TO SCALE   (DETAIL T2-FS6)

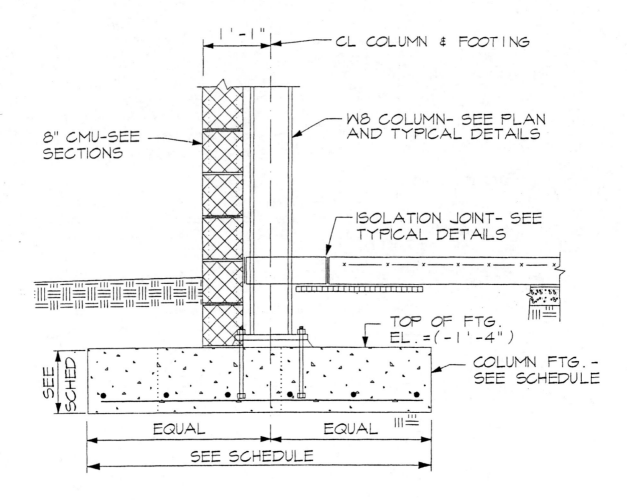

1'-1"

CL COLUMN & FOOTING

8" CMU-SEE SECTIONS

W8 COLUMN- SEE PLAN AND TYPICAL DETAILS

ISOLATION JOINT- SEE TYPICAL DETAILS

TOP OF FTG. EL.=(-1'-4")

COLUMN FTG.- SEE SCHEDULE

SEE SCHED

EQUAL

EQUAL

SEE SCHEDULE

TYPICAL FOUNDATION SECTION #7

NOT TO SCALE   (DETAIL T2-FS7)

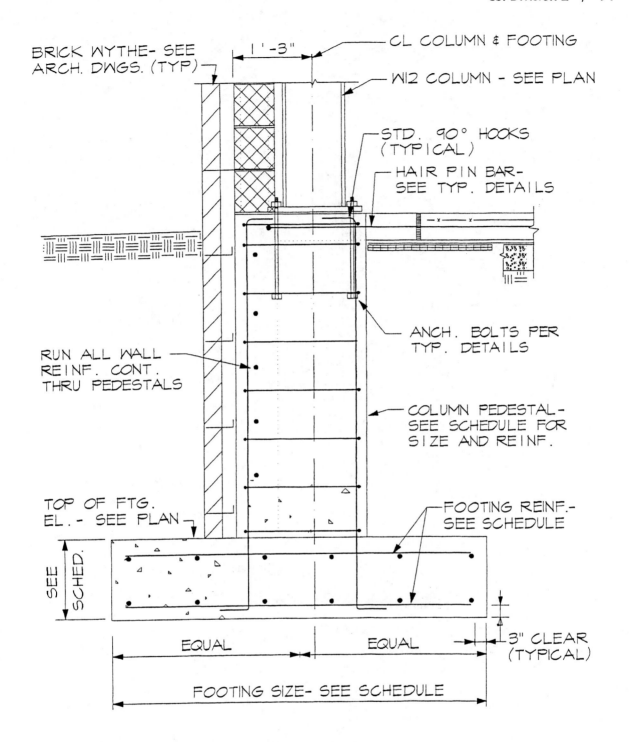

BRICK WYTHE- SEE ARCH. DWGS. (TYP)

1'-3"

CL COLUMN & FOOTING

W12 COLUMN - SEE PLAN

STD. 90° HOOKS (TYPICAL)

HAIR PIN BAR- SEE TYP. DETAILS

RUN ALL WALL REINF. CONT. THRU PEDESTALS

ANCH. BOLTS PER TYP. DETAILS

COLUMN PEDESTAL- SEE SCHEDULE FOR SIZE AND REINF.

TOP OF FTG. EL.- SEE PLAN

FOOTING REINF.- SEE SCHEDULE

SEE SCHED.

3" CLEAR (TYPICAL)

EQUAL

EQUAL

FOOTING SIZE- SEE SCHEDULE

TYPICAL FOUNDATION SECTION #8

NOT TO SCALE

(DETAIL T2-FS8)

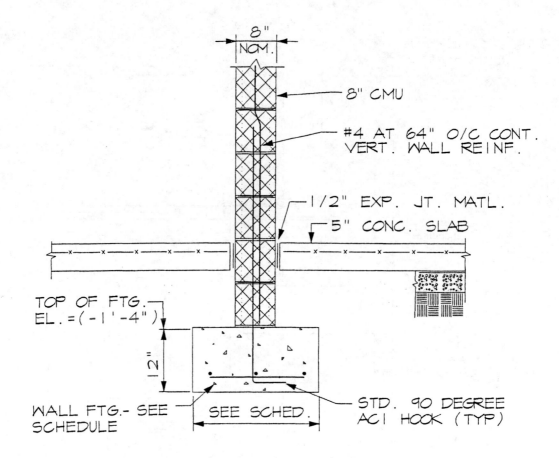

8"
NOM.

8" CMU

#4 AT 64" O/C CONT.
VERT. WALL REINF.

1/2" EXP. JT. MATL.

5" CONC. SLAB

TOP OF FTG.
EL.=(-1'-4")

12"

WALL FTG.- SEE
SCHEDULE

SEE SCHED.

STD. 90 DEGREE
ACI HOOK (TYP)

TYPICAL FOUNDATION SECTION #9

NOT TO SCALE   (DETAIL T2—FS9)

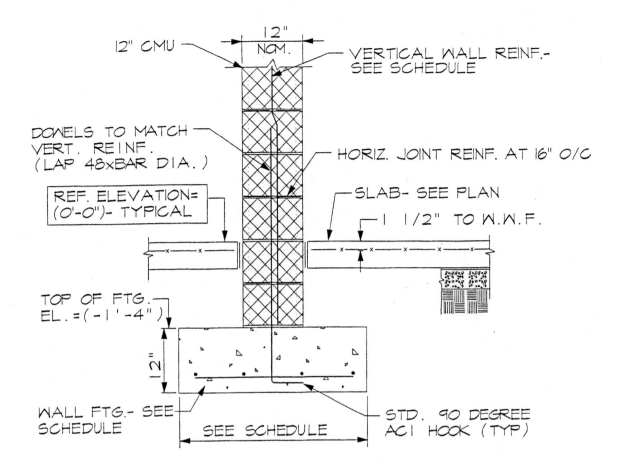

12" CMU

12"
NOM.

VERTICAL WALL REINF.-
SEE SCHEDULE

DOWELS TO MATCH
VERT. REINF.
(LAP 48xBAR DIA.)

HORIZ. JOINT REINF. AT 16" O/C

REF. ELEVATION=
(0'-0")- TYPICAL

SLAB- SEE PLAN

1 1/2" TO W.W.F.

TOP OF FTG.
EL.=(-1'-4")

12"

WALL FTG.- SEE
SCHEDULE

SEE SCHEDULE

STD. 90 DEGREE
ACI HOOK (TYP)

TYPICAL FOUNDATION SECTION #10
NOT TO SCALE   (DETAIL T2-FS10)

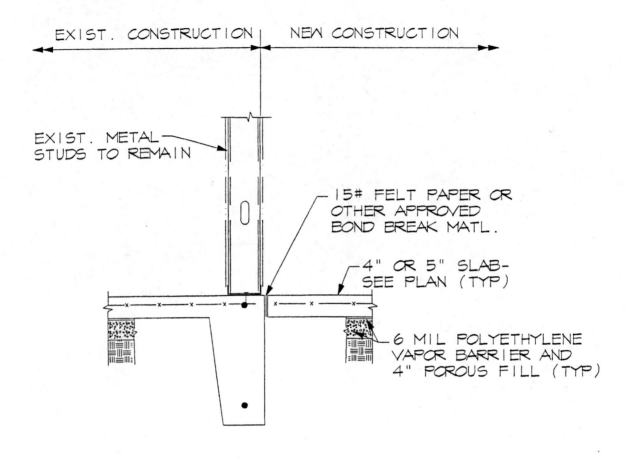

EXIST. CONSTRUCTION | NEW CONSTRUCTION

EXIST. METAL STUDS TO REMAIN

15# FELT PAPER OR OTHER APPROVED BOND BREAK MATL.

4" OR 5" SLAB- SEE PLAN (TYP)

6 MIL POLYETHYLENE VAPOR BARRIER AND 4" POROUS FILL (TYP)

TYPICAL FOUNDATION SECTION #11
NOT TO SCALE   (DETAIL T2–FS11)

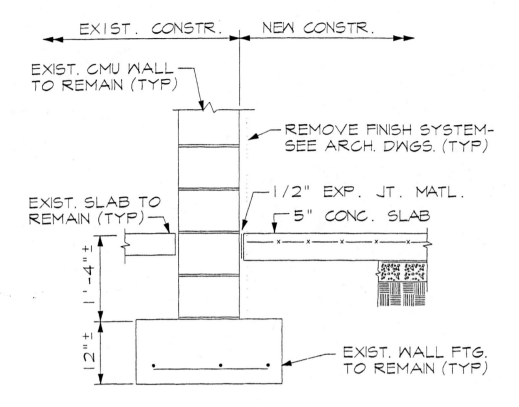

TYPICAL FOUNDATION SECTION #12

NOT TO SCALE   (DETAIL T2–FS12)

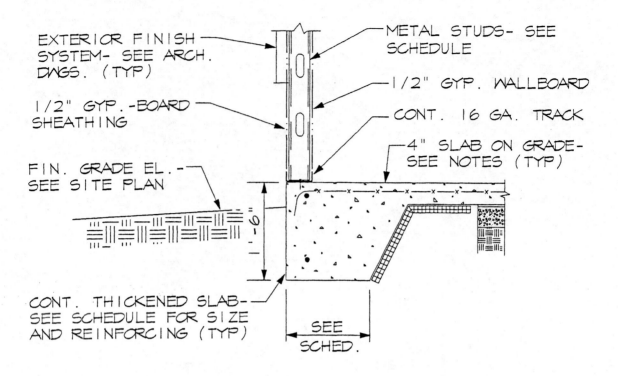

EXTERIOR FINISH SYSTEM- SEE ARCH. DWGS. (TYP)

1/2" GYP.-BOARD SHEATHING

FIN. GRADE EL.- SEE SITE PLAN

CONT. THICKENED SLAB- SEE SCHEDULE FOR SIZE AND REINFORCING (TYP)

METAL STUDS- SEE SCHEDULE

1/2" GYP. WALLBOARD

CONT. 16 GA. TRACK

4" SLAB ON GRADE- SEE NOTES (TYP)

1'-6"

SEE SCHED.

TYPICAL FOUNDATION SECTION #13

NOT TO SCALE   (DETAIL T2-FS13)

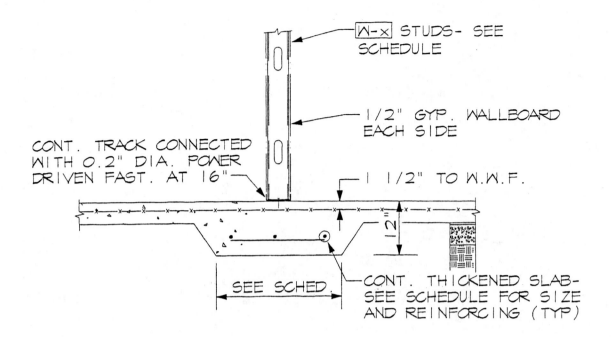

W-x STUDS- SEE SCHEDULE

1/2" GYP. WALLBOARD EACH SIDE

CONT. TRACK CONNECTED WITH 0.2" DIA. POWER DRIVEN FAST. AT 16"

1 1/2" TO W.W.F.

12"

SEE SCHED.

CONT. THICKENED SLAB- SEE SCHEDULE FOR SIZE AND REINFORCING (TYP)

TYPICAL FOUNDATION SECTION #14

NOT TO SCALE   (DETAIL T2-FS14)

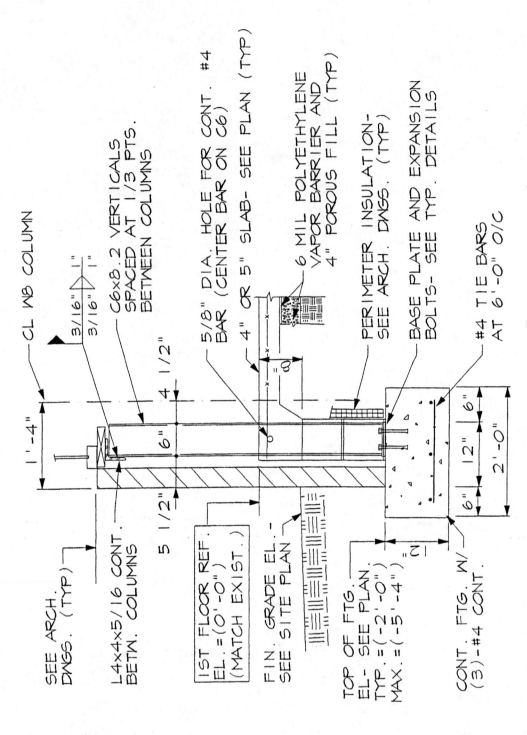

SEE ARCH.
DWGS. (TYP)

L4x4x5/16 CONT.
BETW. COLUMNS

CL W8 COLUMN

3/16" ↓ 1"
3/16" ↓ 1"

C6x8.2 VERTICALS
SPACED AT 1/3 PTS.
BETWEEN COLUMNS

5/8" DIA. HOLE FOR CONT. #4
BAR (CENTER BAR ON C6)

4" OR 5" SLAB- SEE PLAN (TYP)

6 MIL POLYETHYLENE
VAPOR BARRIER AND
4" POROUS FILL (TYP)

PERIMETER INSULATION-
SEE ARCH. DWGS. (TYP)

BASE PLATE AND EXPANSION
BOLTS- SEE TYP. DETAILS

#4 TIE BARS
AT 6'-0" O/C

1'-4"

4 1/2"

6"

5 1/2"

1ST FLOOR REF.
EL.=(0'-0")
(MATCH EXIST.)

FIN. GRADE EL.-
SEE SITE PLAN

TOP OF FTG.
EL- SEE PLAN.
TYP.=(-2'-0")
MAX.=(-5'-4")

CONT. FTG. W/
(3)-#4 CONT.

6" 12" 6"

2'-0"

2"

DETAIL AT CHANNEL STRUTS

TYPICAL FOUNDATION SECTION #15

(DETAIL T2-FS15)

NOT TO SCALE

98

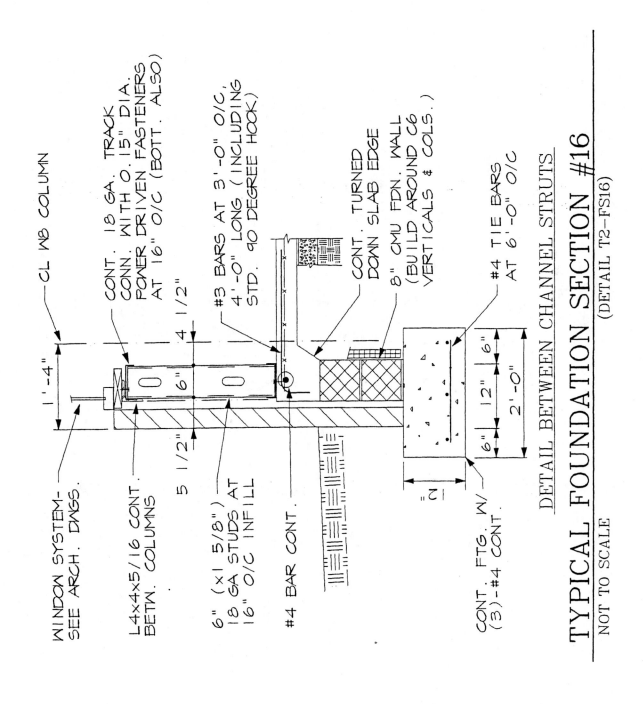

WINDOW SYSTEM-
SEE ARCH. DWGS.

CL W8 COLUMN

CONT. 18 GA. TRACK
CONN. WITH 0.15" DIA.
POWER DRIVEN FASTENERS
AT 16" O/C (BOTT. ALSO)

L4x4x5/16 CONT.
BETN. COLUMNS

#3 BARS AT 3'-0" O/C,
4'-0" LONG (INCLUDING
STD. 90 DEGREE HOOK)

6" (x1 5/8")
18 GA STUDS AT
16" O/C INFILL

CONT. TURNED
DOWN SLAB EDGE

8" CMU FDN. WALL
(BUILD AROUND C6
VERTICALS & COLS.)

#4 BAR CONT.

#4 TIE BARS
AT 6'-0" O/C

CONT. FTG. W/
(3)-#4 CONT.

1'-4"

4 1/2"

6"

5 1/2"

6"   12"   6"

2'-0"

2"

TYPICAL FOUNDATION SECTION #16

DETAIL BETWEEN CHANNEL STRUTS

(DETAIL T2-FS16)

NOT TO SCALE

99

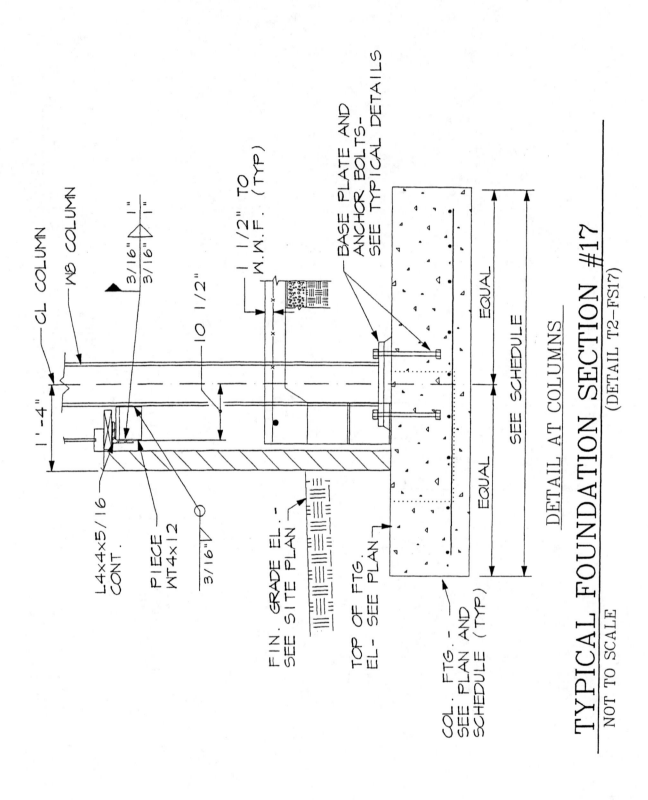

TYPICAL FOUNDATION SECTION #17

DETAIL AT COLUMNS

(DETAIL T2-FS17)

NOT TO SCALE

CL COLUMN

W8 COLUMN

3/16" 1"
3/16" 1"

1 1/2" TO
W.W.F. (TYP)

10 1/2"

1'-4"

L4x4x5/16
CONT.

PIECE
WT4x12

3/16"

FIN. GRADE EL. -
SEE SITE PLAN

TOP OF FTG.
EL- SEE PLAN

BASE PLATE AND
ANCHOR BOLTS-
SEE TYPICAL DETAILS

EQUAL

SEE SCHEDULE

EQUAL

COL. FTG. -
SEE PLAN AND
SCHEDULE (TYP)

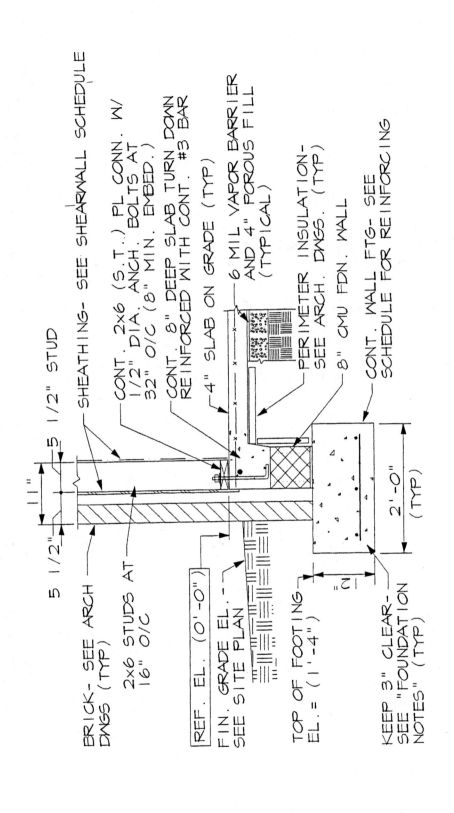

BRICK- SEE ARCH DWGS (TYP)

5 1/2" STUD

SHEATHING- SEE SHEARWALL SCHEDULE

2x6 STUDS AT 16" O/C

CONT. 2x6 (S.T.) PL CONN. W/ 1/2" DIA. ANCH. BOLTS AT 32" O/C (8" MIN. EMBED.)

CONT. 8" DEEP SLAB TURN DOWN REINFORCED WITH CONT. #3 BAR

4" SLAB ON GRADE (TYP)

6 MIL VAPOR BARRIER AND 4" POROUS FILL (TYPICAL)

PERIMETER INSULATION- SEE ARCH. DWGS. (TYP)

8" CMU FDN. WALL

CONT. WALL FTG- SEE SCHEDULE FOR REINFORCING

REF. EL. (0'-0")

FIN. GRADE EL.- SEE SITE PLAN

TOP OF FOOTING EL. = (1'-4")

KEEP 3" CLEAR- SEE "FOUNDATION NOTES" (TYP)

5 1/2"

1"

2'-0" (TYP)

2"

TYPICAL PERIMETER FOUNDATION SECTION #1

NOT TO SCALE

(DETAIL T2-WFS1)

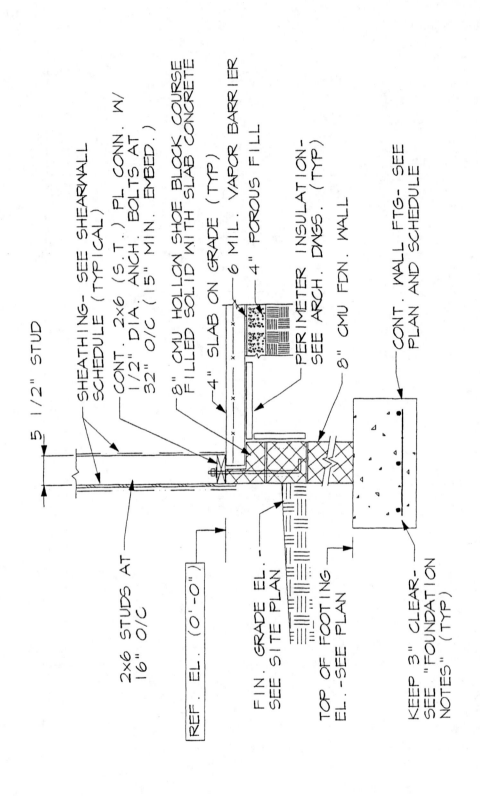

5 1/2" STUD

2x6 STUDS AT 16" O/C

SHEATHING- SEE SHEARWALL SCHEDULE (TYPICAL)

CONT. 2x6 (S.T.) PL CONN. W/ 1/2" DIA. ANCH. BOLTS AT 32" O/C (15" MIN. EMBED.)

8" CMU HOLLOW SHOE BLOCK COURSE FILLED SOLID WITH SLAB CONCRETE

4" SLAB ON GRADE (TYP)

6 MIL. VAPOR BARRIER

4" POROUS FILL

PERIMETER INSULATION- SEE ARCH. DWGS. (TYP)

8" CMU FDN. WALL

CONT. WALL FTG- SEE PLAN AND SCHEDULE

REF. EL. (0'-0")

FIN. GRADE EL.- SEE SITE PLAN

TOP OF FOOTING EL.-SEE PLAN

KEEP 3" CLEAR- SEE "FOUNDATION NOTES" (TYP)

TYPICAL PERIMETER FOUNDATION SECTION #2

NOT TO SCALE

(DETAIL T2-WFS2)

102

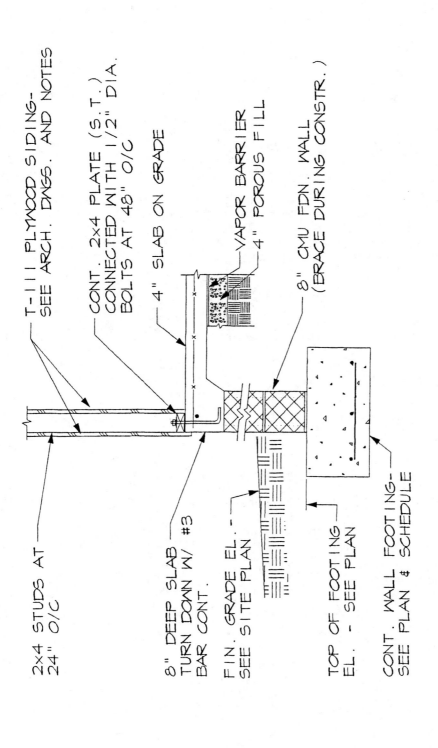

2x4 STUDS AT 24" O/C

T-111 PLYWOOD SIDING- SEE ARCH. DWGS. AND NOTES

CONT. 2x4 PLATE (S.T.) CONNECTED WITH 1/2" DIA. BOLTS AT 48" O/C

4" SLAB ON GRADE

VAPOR BARRIER

4" POROUS FILL

8" CMU FDN. WALL (BRACE DURING CONSTR.)

8" DEEP SLAB TURN DOWN W/ #3 BAR CONT.

FIN. GRADE EL. - SEE SITE PLAN

TOP OF FOOTING EL. - SEE PLAN

CONT. WALL FOOTING- SEE PLAN & SCHEDULE

TYPICAL PERIMETER FOUNDATION SECTION #3
NOT TO SCALE

(DETAIL T2-WFS3)

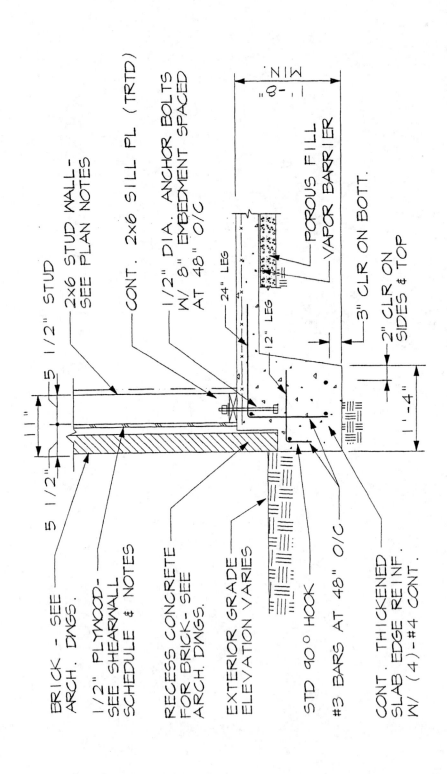

BRICK - SEE ARCH. DWGS.

1/2" PLYWOOD- SEE SHEARWALL SCHEDULE & NOTES

RECESS CONCRETE FOR BRICK- SEE ARCH. DWGS.

EXTERIOR GRADE ELEVATION VARIES

STD 90° HOOK

#3 BARS AT 48" O/C

CONT. THICKENED SLAB EDGE REINF. W/ (4)-#4 CONT.

5 1/2" STUD

2x6 STUD WALL- SEE PLAN NOTES

CONT. 2x6 SILL PL (TRTD)

1/2" DIA. ANCHOR BOLTS W/ 8" EMBEDMENT SPACED AT 48" O/C

24" LEG

12" LEG

1'-8" MIN.

POROUS FILL

VAPOR BARRIER

3" CLR ON BOTT.

2" CLR ON SIDES & TOP

1'-4"

5 1/2"

1"

1"

TYPICAL PERIMETER FOUNDATION SECTION #4

NOT TO SCALE

(DETAIL T2-WFS4)

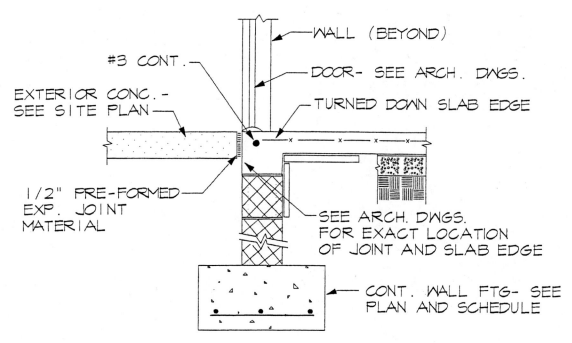

#3 CONT.

WALL (BEYOND)

DOOR- SEE ARCH. DWGS.

EXTERIOR CONC.-
SEE SITE PLAN

TURNED DOWN SLAB EDGE

1/2" PRE-FORMED
EXP. JOINT
MATERIAL

SEE ARCH. DWGS.
FOR EXACT LOCATION
OF JOINT AND SLAB EDGE

CONT. WALL FTG- SEE
PLAN AND SCHEDULE

PERIMETER FOUNDATION DETAIL AT DOORS

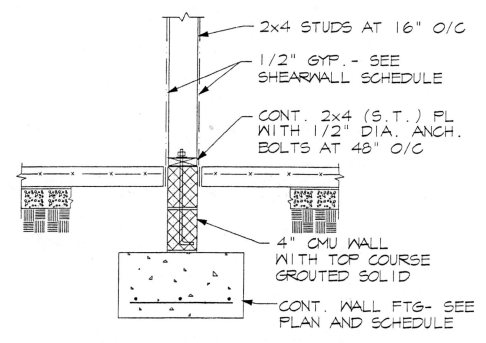

2x4 STUDS AT 16" O/C

1/2" GYP.- SEE
SHEARWALL SCHEDULE

CONT. 2x4 (S.T.) PL
WITH 1/2" DIA. ANCH.
BOLTS AT 48" O/C

4" CMU WALL
WITH TOP COURSE
GROUTED SOLID

CONT. WALL FTG- SEE
PLAN AND SCHEDULE

INTERIOR FOUNDATION DETAIL

MISCELLANEOUS FOUNDATION DETAILS

NOT TO SCALE   (DETAIL T2-WFS5)

NOTE: ALL STEEL AND HARDWARE SHALL BE HOT DIP GALVANIZED.

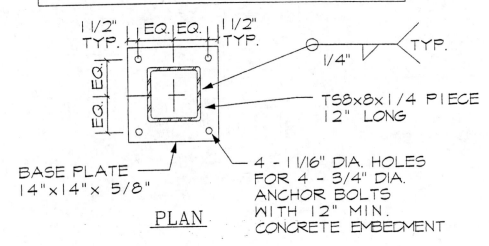

1 1/2" TYP.   EQ. EQ.   1 1/2" TYP.

EQ. EQ.

TS8x8x1/4 PIECE 12" LONG

1/4"   TYP.

BASE PLATE 14"x14"x 5/8"

4 - 1 1/16" DIA. HOLES FOR 4 - 3/4" DIA. ANCHOR BOLTS WITH 12" MIN. CONCRETE EMBEDMENT

<u>PLAN</u>

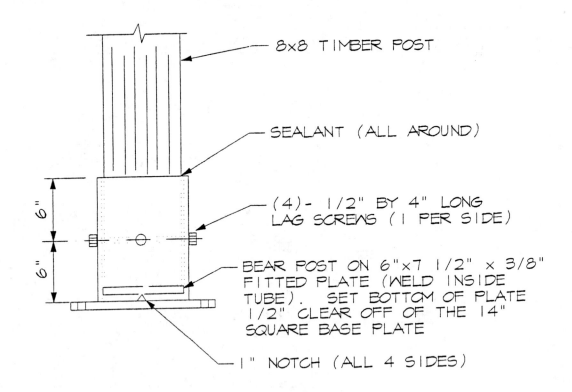

8x8 TIMBER POST

SEALANT (ALL AROUND)

(4)- 1/2" BY 4" LONG LAG SCREWS (1 PER SIDE)

BEAR POST ON 6"x7 1/2" x 3/8" FITTED PLATE (WELD INSIDE TUBE). SET BOTTOM OF PLATE 1/2" CLEAR OFF OF THE 14" SQUARE BASE PLATE

1" NOTCH (ALL 4 SIDES)

6"

6"

<u>ELEVATION</u>

TIMBER POST BASE CONNECTION
NOT TO SCALE   (DETAIL T2-WFS6)

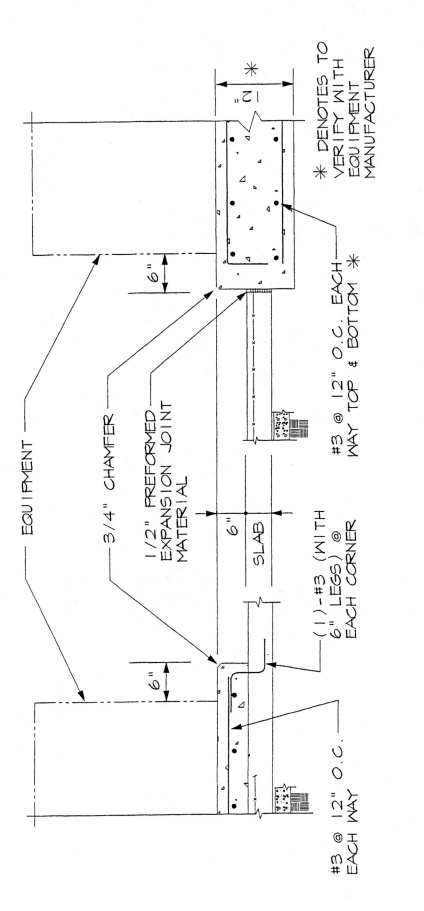

EQUIPMENT

3/4" CHAMFER

1/2" PREFORMED
EXPANSION JOINT
MATERIAL

6"

SLAB

(1)-#3 (WITH
6" LEGS) @
EACH CORNER

#3 @ 12" O.C.
EACH WAY

12"

* DENOTES TO
VERIFY WITH
EQUIPMENT
MANUFACTURER

6"

#3 @ 12" O.C. EACH
WAY TOP & BOTTOM *

VIBRATING EQUIPMENT

NONVIBRATING EQUIPMENT

TYPICAL INTERIOR EQUIPMENT PAD DETAIL
(DETAIL T2—MIPAD)

NOT TO SCALE

NOTE: IF CURB IS NOT PROVIDED WITH
EQUIPMENT, PROVIDE A W8x18 CURB
(GALV) ALL AROUND CONNECTED WITH
3/4" DIA. EXP. BOLTS AT 18" O.C. (5" MIN
EMBEDMENT, STAGGER SPACING)

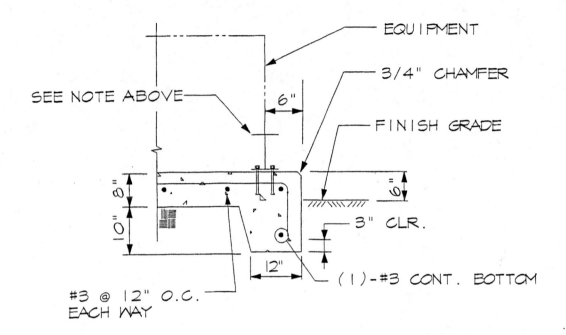

TYPICAL EXTERIOR EQUIPMENT PAD

NOT TO SCALE   (DETAIL T2-MEPAD)

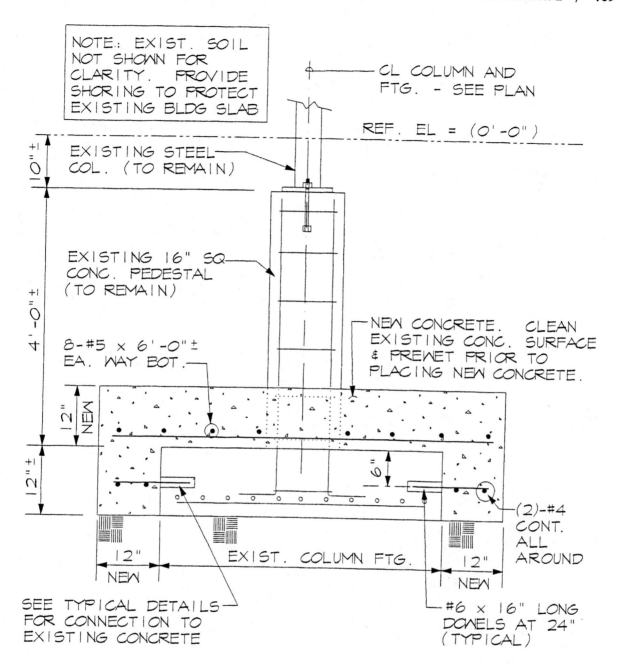

NOTE: EXIST. SOIL NOT SHOWN FOR CLARITY. PROVIDE SHORING TO PROTECT EXISTING BLDG SLAB

CL COLUMN AND FTG. - SEE PLAN

REF. EL = (0'-0")

EXISTING STEEL COL. (TO REMAIN)

EXISTING 16" SQ. CONC. PEDESTAL (TO REMAIN)

NEW CONCRETE. CLEAN EXISTING CONC. SURFACE & PREWET PRIOR TO PLACING NEW CONCRETE.

8-#5 x 6'-0"± EA. WAY BOT.

12" NEW

12"±

(2)-#4 CONT. ALL AROUND

12" NEW

EXIST. COLUMN FTG.

12" NEW

SEE TYPICAL DETAILS FOR CONNECTION TO EXISTING CONCRETE

#6 x 16" LONG DOWELS AT 24" (TYPICAL)

EXIST. FOOTING STRENGTHENING DETAIL

NOT TO SCALE

(DETAIL T2-MREN1)

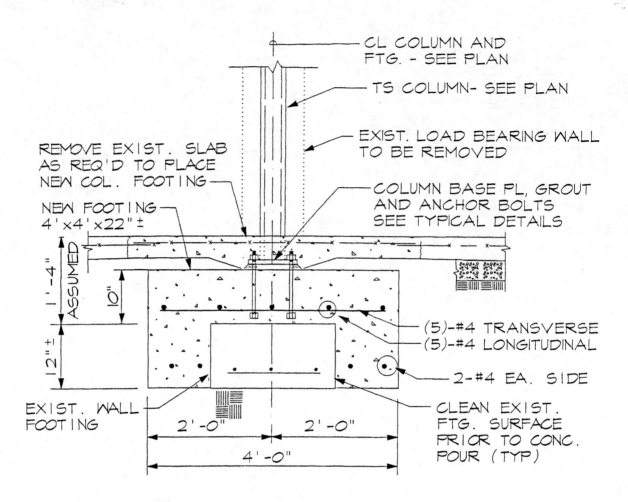

REMOVE EXIST. SLAB AS REQ'D TO PLACE NEW COL. FOOTING

NEW FOOTING 4'x4'x22"±

CL COLUMN AND FTG. - SEE PLAN

TS COLUMN- SEE PLAN

EXIST. LOAD BEARING WALL TO BE REMOVED

COLUMN BASE PL, GROUT AND ANCHOR BOLTS SEE TYPICAL DETAILS

(5)-#4 TRANSVERSE
(5)-#4 LONGITUDINAL

2-#4 EA. SIDE

CLEAN EXIST. FTG. SURFACE PRIOR TO CONC. POUR (TYP)

EXIST. WALL FOOTING

ASSUMED

1'-4"

10"

12"±

2'-0" 2'-0"

4'-0"

EXIST. FOOTING STRENGTHENING DETAIL

NOT TO SCALE   (DETAIL T2-MREN2)

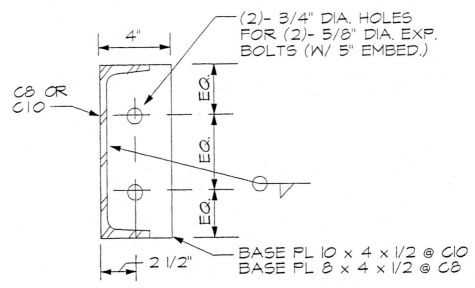

4"

C8 OR C10

(2)- 3/4" DIA. HOLES FOR (2)- 5/8" DIA. EXP. BOLTS (W/ 5" EMBED.)

EQ. EQ. EQ. EQ.

2 1/2"

BASE PL 10 × 4 × 1/2 @ C10
BASE PL 8 × 4 × 1/2 @ C8

TYPICAL JAMB CHANNEL BASE CONNECTION DETAIL

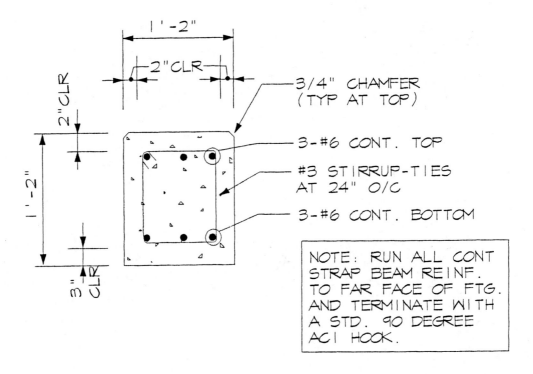

1'-2"

2"CLR

2"CLR

1'-2"

3"CLR

3/4" CHAMFER (TYP AT TOP)

3-#6 CONT. TOP

#3 STIRRUP-TIES AT 24" O/C

3-#6 CONT. BOTTOM

NOTE: RUN ALL CONT STRAP BEAM REINF. TO FAR FACE OF FTG. AND TERMINATE WITH A STD. 90 DEGREE ACI HOOK.

TYPICAL CONCRETE STRAP BEAM DETAIL

MISCELLANEOUS FOUNDATION DETAILS

NOT TO SCALE   (DETAIL T2-MISC1)

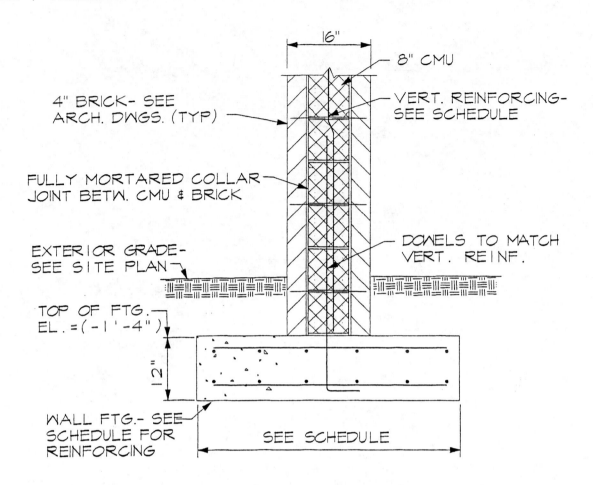

16"

8" CMU

VERT. REINFORCING-
SEE SCHEDULE

4" BRICK- SEE
ARCH. DWGS. (TYP)

FULLY MORTARED COLLAR
JOINT BETW. CMU & BRICK

DOWELS TO MATCH
VERT. REINF.

EXTERIOR GRADE-
SEE SITE PLAN

TOP OF FTG.
EL.=(-1'-4")

12"

WALL FTG.- SEE
SCHEDULE FOR
REINFORCING

SEE SCHEDULE

MASONRY SCREEN WALL DETAIL
NOT TO SCALE   (DETAIL T2-MISC2)

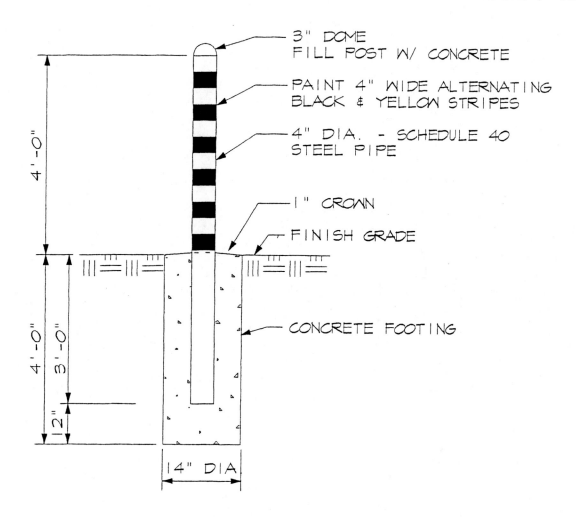

3" DOME
FILL POST W/ CONCRETE

PAINT 4" WIDE ALTERNATING
BLACK & YELLOW STRIPES

4" DIA. - SCHEDULE 40
STEEL PIPE

1" CROWN

FINISH GRADE

CONCRETE FOOTING

4'-0"

4'-0"

3'-0"

12"

14" DIA

TYPICAL PIPE BOLLARD DETAIL
NOT TO SCALE   (DETAIL T2-MBOLL)

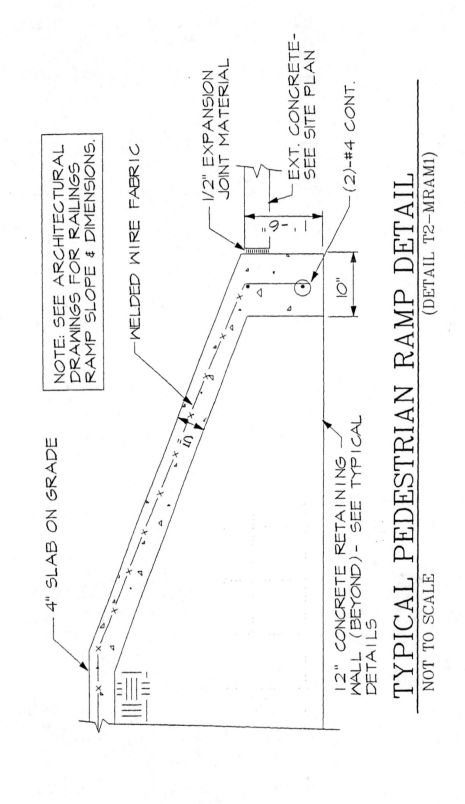

TYPICAL PEDESTRIAN RAMP DETAIL

NOTE: SEE ARCHITECTURAL DRAWINGS FOR RAILINGS RAMP SLOPE & DIMENSIONS.

WELDED WIRE FABRIC

4" SLAB ON GRADE

1/2" EXPANSION JOINT MATERIAL

EXT. CONCRETE— SEE SITE PLAN

(2)-#4 CONT.

1'-6"

10"

12" CONCRETE RETAINING WALL (BEYOND)- SEE TYPICAL DETAILS

(DETAIL T2-MRAM1)

NOT TO SCALE

114

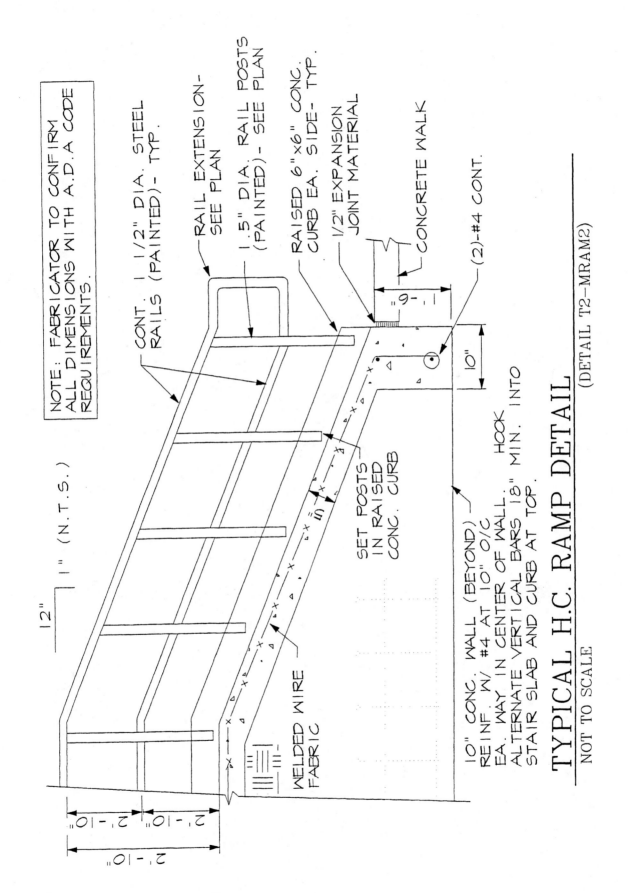

NOTE: FABRICATOR TO CONFIRM ALL DIMENSIONS WITH A.D.A CODE REQUIREMENTS.

RAIL EXTENSION - SEE PLAN

CONT. 1 1/2" DIA. STEEL RAILS (PAINTED) - TYP.

1.5" DIA. RAIL POSTS (PAINTED) - SEE PLAN

RAISED 6"x6" CONC. CURB EA. SIDE - TYP.

1/2" EXPANSION JOINT MATERIAL

CONCRETE WALK

(2)-#4 CONT.

SET POSTS IN RAISED CONC. CURB

WELDED WIRE FABRIC

10" CONC. WALL (BEYOND) REINF. W/ #4 AT 10" O/C EA. WAY IN CENTER OF WALL. HOOK ALTERNATE VERTICAL BARS 18" MIN. INTO STAIR SLAB AND CURB AT TOP.

TYPICAL H.C. RAMP DETAIL

NOT TO SCALE

(DETAIL T2-MRAM2)

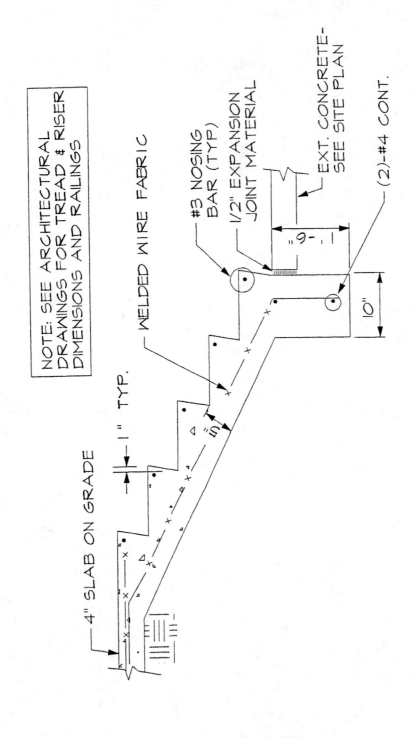

NOTE: SEE ARCHITECTURAL DRAWINGS FOR TREAD & RISER DIMENSIONS AND RAILINGS

WELDED WIRE FABRIC

1" TYP.

#3 NOSING BAR (TYP)

1/2" EXPANSION JOINT MATERIAL

EXT. CONCRETE—SEE SITE PLAN

1'-6"

(2)-#4 CONT.

10"

4" SLAB ON GRADE

5"

TYPICAL CONCRETE STAIR DETAIL

(DETAIL T2—MSTEP)

NOT TO SCALE

** - COORDINATE THE TRENCH DEPTH
AND WIDTH WITH THE ARCH. DWGS
AND THE ACTUAL EQUIPMENT SELECTED

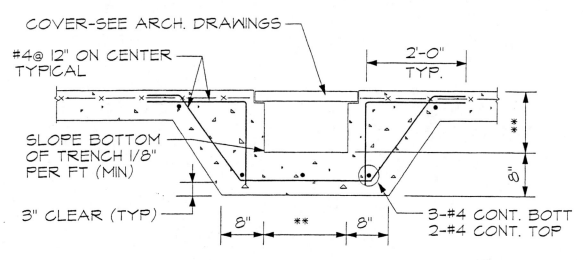

COVER-SEE ARCH. DRAWINGS

#4@ 12" ON CENTER TYPICAL

2'-0" TYP.

SLOPE BOTTOM OF TRENCH 1/8" PER FT (MIN)

3" CLEAR (TYP)

8"   **   8"

8"

3-#4 CONT. BOTT
2-#4 CONT. TOP

CAST-IN-PLACE TRENCH DRAIN

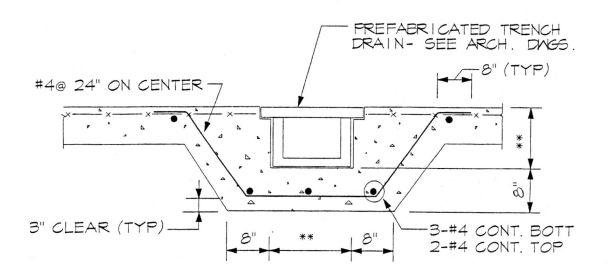

PREFABRICATED TRENCH DRAIN- SEE ARCH. DWGS.

#4@ 24" ON CENTER

8" (TYP)

3" CLEAR (TYP)

8"   **   8"

8"

3-#4 CONT. BOTT
2-#4 CONT. TOP

PREFABRICATED TRENCH DRAIN

TYP. TRENCH DRAIN DETAILS

NOT TO SCALE   (DETAIL T2-TR1)

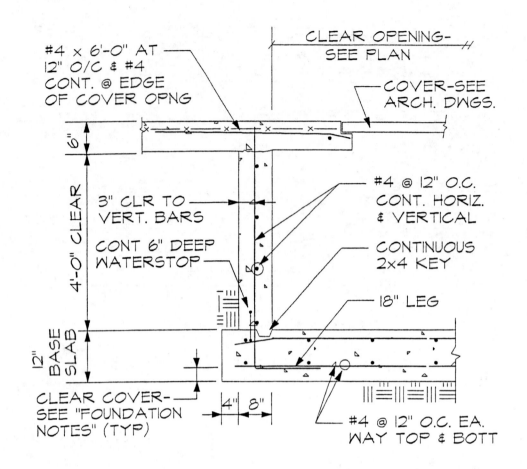

TYPICAL PIT DETAIL

NOT TO SCALE   (DETAIL T2-PIT1)

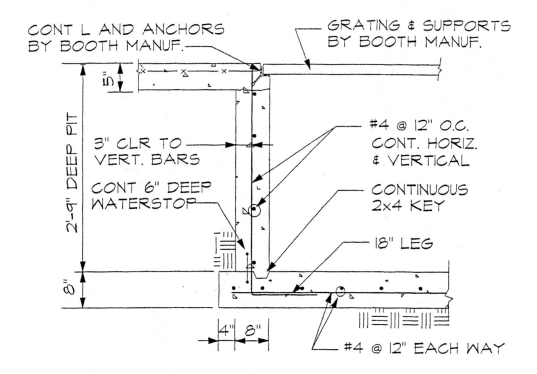

CONT L AND ANCHORS BY BOOTH MANUF.

GRATING & SUPPORTS BY BOOTH MANUF.

2'-9" DEEP PIT

3" CLR TO VERT. BARS

CONT 6" DEEP WATERSTOP

#4 @ 12" O.C. CONT. HORIZ. & VERTICAL

CONTINUOUS 2x4 KEY

18" LEG

#4 @ 12" EACH WAY

5"

8"

4" 8"

TYPICAL SPRAY BOOTH PIT DETAIL

NOT TO SCALE   (DETAIL T2-PIT2)

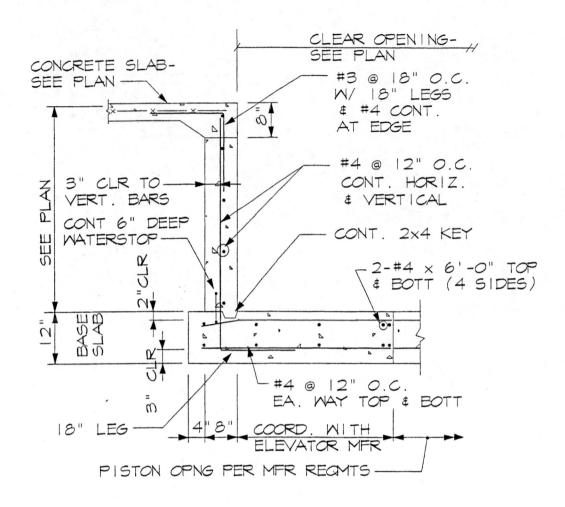

CLEAR OPENING-
SEE PLAN

CONCRETE SLAB-
SEE PLAN

#3 @ 18" O.C.
W/ 18" LEGS
& #4 CONT.
AT EDGE

8"

3" CLR TO
VERT. BARS

#4 @ 12" O.C.
CONT. HORIZ.
& VERTICAL

SEE PLAN

CONT 6" DEEP
WATERSTOP

CONT. 2x4 KEY

2-#4 x 6'-0" TOP
& BOTT (4 SIDES)

2"CLR

12"

BASE
SLAB

3" CLR

#4 @ 12" O.C.
EA. WAY TOP & BOTT

18" LEG

4" 8"

COORD. WITH
ELEVATOR MFR

PISTON OPNG PER MFR REQMTS

SHALLOW FOUNDATION SYSTEM

TYPICAL ELEVATOR PIT DETAIL #1

NOT TO SCALE   (DETAIL T2-PIT3)

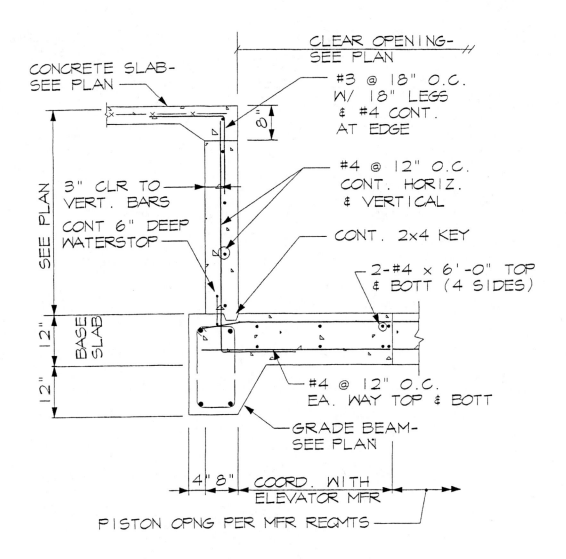

CLEAR OPENING-
SEE PLAN

CONCRETE SLAB-
SEE PLAN

#3 @ 18" O.C.
W/ 18" LEGS
& #4 CONT.
AT EDGE

8"

SEE PLAN

3" CLR TO
VERT. BARS

#4 @ 12" O.C.
CONT. HORIZ.
& VERTICAL

CONT. 6" DEEP
WATERSTOP

CONT. 2x4 KEY

2-#4 x 6'-0" TOP
& BOTT (4 SIDES)

BASE SLAB

12"

12"

#4 @ 12" O.C.
EA. WAY TOP & BOTT

GRADE BEAM-
SEE PLAN

4" 8" COORD. WITH
ELEVATOR MFR

PISTON OPNG PER MFR REQMTS

PILE FOUNDATION SYSTEM

TYPICAL ELEVATOR PIT DETAIL #2

NOT TO SCALE   (DETAIL T2-PIT4)

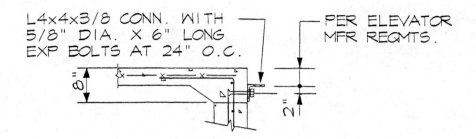

L4x4x3/8 CONN. WITH
5/8" DIA. X 6" LONG
EXP BOLTS AT 24" O.C.

PER ELEVATOR
MFR REQMTS.

8"

2"

DETAIL AT ELEVATOR DOOR SILL

NOTE: POUR PIT MONOLITHIC WITH BASE SLAB

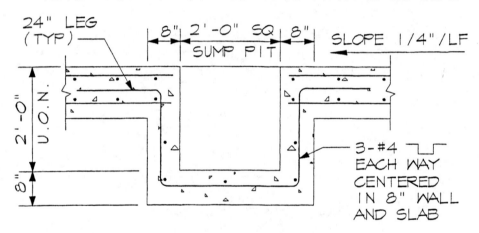

24" LEG
(TYP)

8" | 2'-0" SQ | 8"
SUMP PIT

SLOPE 1/4"/LF

2'-0"
U.O.N.

8"

3-#4
EACH WAY
CENTERED
IN 8" WALL
AND SLAB

DETAIL AT SUMP PIT

MISCELLANEOUS PIT DETAILS

NOT TO SCALE   (DETAIL T2–PIT5)

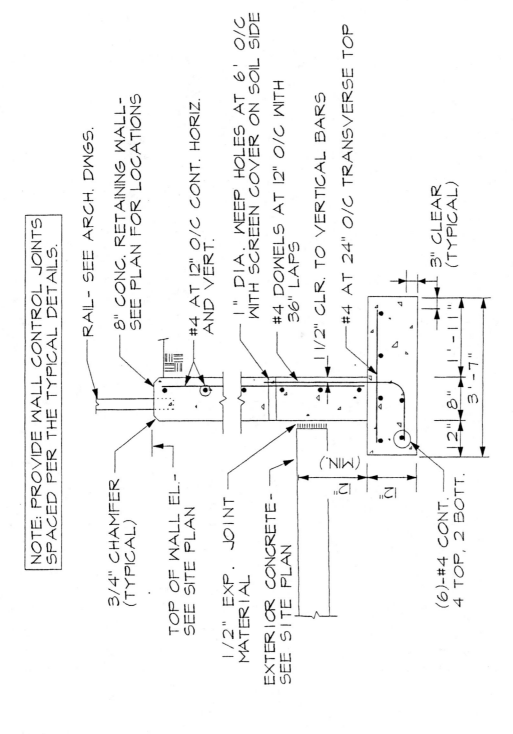

NOTE: PROVIDE WALL CONTROL JOINTS SPACED PER THE TYPICAL DETAILS.

RAIL- SEE ARCH. DWGS.

8" CONC. RETAINING WALL- SEE PLAN FOR LOCATIONS

#4 AT 12" O/C CONT. HORIZ. AND VERT.

1" DIA. WEEP HOLES AT 6' O/C WITH SCREEN COVER ON SOIL SIDE

#4 DOWELS AT 12" O/C WITH 36" LAPS

1 1/2" CLR. TO VERTICAL BARS

#4 AT 24" O/C TRANSVERSE TOP

3" CLEAR (TYPICAL)

3/4" CHAMFER (TYPICAL)

TOP OF WALL EL.- SEE SITE PLAN

1/2" EXP. JOINT MATERIAL

EXTERIOR CONCRETE- SEE SITE PLAN

12" (MIN.)

12"

(6)-#4 CONT. 4 TOP, 2 BOTT.

2" 8" 1'-11"

3'-7"

TYPICAL RETAINING WALL DETAIL #1

(DETAIL T2-RW1)

NOT TO SCALE

123

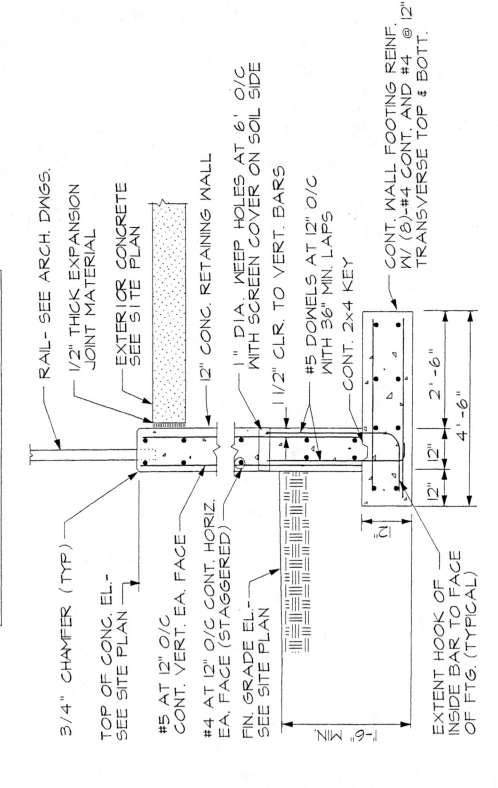

NOTE: PROVIDE WALL CONTROL JOINTS
SPACED PER THE TYPICAL DETAILS.

3/4" CHAMFER (TYP)

RAIL- SEE ARCH. DWGS.

1/2" THICK EXPANSION
JOINT MATERIAL

EXTERIOR CONCRETE
SEE SITE PLAN

TOP OF CONC. EL.-
SEE SITE PLAN

12" CONC. RETAINING WALL

#5 AT 12" O/C
CONT. VERT. EA. FACE

1" DIA. WEEP HOLES AT 6' O/C
WITH SCREEN COVER ON SOIL SIDE

1 1/2" CLR. TO VERT. BARS

#4 AT 12" O/C CONT. HORIZ.
EA. FACE (STAGGERED)

#5 DOWELS AT 12" O/C
WITH 36" MIN. LAPS

CONT. 2x4 KEY

FIN. GRADE EL.-
SEE SITE PLAN

CONT. WALL FOOTING REINF.
W/ (8)-#4 CONT. AND #4 @ 12"
TRANSVERSE TOP & BOTT.

2'-6"

12"

4'-6"

12"

1'-6" MIN.

EXTENT HOOK OF
INSIDE BAR TO FACE
OF FTG. (TYPICAL)

TYPICAL RETAINING WALL DETAIL #2
(DETAIL T2-RW2)

NOT TO SCALE

124

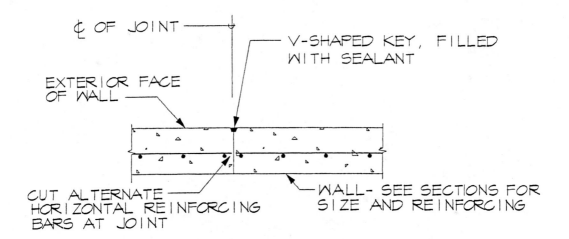

AT SINGLE LAYER OF REINFORCING

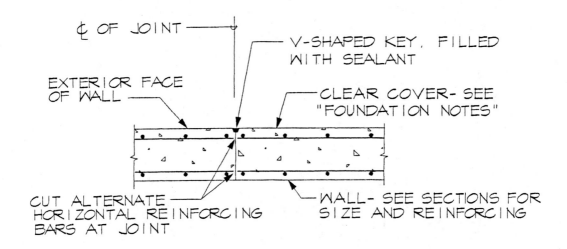

AT DOUBLE LAYER OF REINFORCING

WALL CONTROL JOINT DETAILS
NOT TO SCALE   (DETAIL T2-RWJT1)

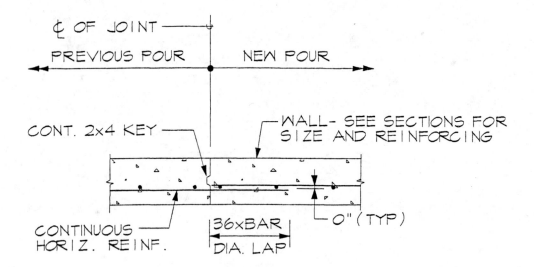

CL OF JOINT

PREVIOUS POUR | NEW POUR

CONT. 2x4 KEY

WALL- SEE SECTIONS FOR SIZE AND REINFORCING

CONTINUOUS HORIZ. REINF.

36xBAR DIA. LAP

0" (TYP)

AT SINGLE LAYER OF REINFORCING

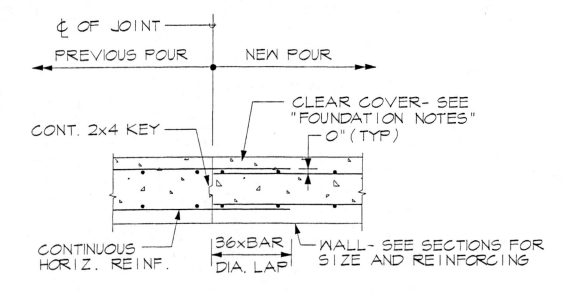

CL OF JOINT

PREVIOUS POUR | NEW POUR

CONT. 2x4 KEY

CLEAR COVER- SEE "FOUNDATION NOTES"

0" (TYP)

CONTINUOUS HORIZ. REINF.

36xBAR DIA. LAP

WALL- SEE SECTIONS FOR SIZE AND REINFORCING

AT DOUBLE LAYER OF REINFORCING

WALL CONSTRUCTION JOINT DETAILS

NOT TO SCALE   (DETAIL T2–RWJT2)

Commentary— Foundation Details

SHALLOW FOUNDATION SYSTEMS—COLUMN AND WALL FOOTINGS

TYPICAL INTERIOR COLUMN / FOOTING DETAIL #1 (DETAIL T2-CFTG1)

The two details shown are essentially the same except for the type of steel column. The footing is centered under the column. The slab is shown as being boxed out at the column. This provides better crack control for exposed slabs and allows the column to be erected after the slab is cast. The column size, top of footing elevation, and footing mark would be shown on the plan. The footing information would be shown in a footing schedule. A minimum of 3″ clear cover is required for all reinforcing steel where concrete is placed directly against earth that is used as a form. The column base plate and anchor bolt details would be shown elsewhere. All steel below grade should be properly coated to prevent corrosion, such as with two coats of heavy black asphaltum. A 1″ thick bed of nonshrink grout is provided below the steel column.

TYPICAL INTERIOR COLUMN / FOOTING DETAIL #2 (DETAIL T2-CFTG2)

The two details shown are essentially the same as the previous one, but the slab is not boxed out at the column. Slab box outs provide better crack control, which is especially important for exposed slabs. For slabs with finishes, such as carpet or certain types of tile, slab box outs may not be necessary.

TYPICAL INTERIOR COLUMN / FOOTING DETAIL #3 (DETAIL T2-CFTG3)

Two details are shown. One is essentially the same as the previous detail, but with a steel pipe column. The other detail shows a concrete column.

TYPICAL WOOD POST / FOOTING DETAIL #1 (DETAIL T2-CFTG4)

This detail shows the footing condition for an exterior wood post. A concrete pedestal is used so that the wood post does not have to extend below grade. The wood post is attached to the pedestal with a prefabricated steel base connector. The wood post is noted as being salt treated (S.T.) because it is exposed to weather.

TYPICAL WOOD POST / FOOTING DETAIL #2
(DETAIL T2-CFTG5)

This detail is similar to the previous one, but it is for an interior glu-laminated wood post. A concrete pedestal is used so that the wood post does not have to extend below grade. The wood post is attached to the pedestal with a fabricated U-shaped steel base connector. A pre-fabricated base connector could also be considered.

TYPICAL WALL FOOTING DETAILS
(DETAIL T2-WFTG1)

Two details are shown: one for a wall footing construction joint and the other for a stepped wall footing. The construction joint detail would be applicable where it is not feasible to pour all of the wall footings in a single monolithic pour. The footing reinforcing is run continuous through the construction joint. The stepped wall footing detail is applicable where the footing elevation must change, such as due to changes in the exterior finished grade elevation.

PIPE SLEEVE THRU WALL FOOTING DETAILS
(DETAIL T2-WFTG2)

Two pipe sleeve details are shown. One is for where the pipe would not affect the reinforcing, but the pipe is still too close to the footing base. The other detail shows the condition where the pipe affects the footing bottom reinforcing, so the reinforcing is bent to clear the top of the pipe. The pipes are sleeved and wrapped with loose insulation. This allows the footing to settle slightly without harming the pipe. These details would not apply where the pipes are a sufficient distance below the footing base.

TYPICAL DETAIL SHOWING CONTINUOUS
REINFORCING AT CORNERS AND INTERSECTIONS
(DETAIL T2-FTGI)

Steel reinforcing is typically noted as being continuous. Bars are lapped at splices. This detail shows the condition for lapping continuous reinforcing at corners and intersections.

TYPICAL DETAIL SHOWING CONTINUOUS
REINFORCING AT CORNERS (DETAIL T2-FTGI2)

This detail is the same as the previous one, but only a corner condition is shown. This detail would suffice for many small jobs that do not have the intersection condition.

DEEP FOUNDATION SYSTEMS—PILE CAPS AND GRADE BEAMS

TYPICAL TIMBER PILE EMBEDMENT DETAIL (DETAIL T2-PEMB1)

This detail is applicable for a timber pile under compression loading. The pile is cut off and embedded 4″ into the concrete cap. The reinforcing is located 3″ clear above the pile cutoff point.

TYPICAL TIMBER TENSION PILE EMBEDMENT DETAILS (DETAIL T2-PEMB2)

Two timber tension pile details are shown. One is for a straight pile and the other is for a battered pile. The battered pile would provide a higher lateral load resistance. Tension is transferred from the pile caps to the piles through the steel bars, which are field bolted to the piles. These bars are galvanized to prevent corrosion. Top reinforcing is provided in the cap to match the bottom reinforcing.

TYPICAL COLUMN / PILE CAP DETAIL #1 (DETAIL T2-PCAP1)

This detail is similar to detail T2-CFTG1, but a pile cap is shown in lieu of the column footing. The pile cap is centered under the column. The slab is shown as being boxed out at the column. This provides better crack control for exposed slabs and allows the column to be erected after the slab is cast. The column size, top of cap elevation, and cap mark would be shown on the plan. The pile cap information would be shown via typical details. A minimum of 3″ clear cover is required for all reinforcing steel where concrete is placed directly against earth used as a form. The column base plate and anchor bolt details would be shown elsewhere. All steel below grade should be properly coated to prevent corrosion, such as with two heavy coats of black asphaltum. A 1″ thick bed of nonshrink grout is provided below the steel column.

TYPICAL COLUMN / PILE CAP DETAIL #2 (DETAIL T2-PCAP2)

This detail is the same as the previous one, but with a steel tube column.

TYPICAL COLUMN / PILE CAP DETAIL #3 (DETAIL T2-PCAP3)

This detail is the same as the previous one, but with a concrete column.

TYPICAL P-1 & P-2 PILE CAP DETAILS
(DETAIL T2-P1 & P2)

The pile cap reinforcing and dimensions are shown in plan view. This detail shows the dimensions and reinforcing for a single-pile (P-1) cap and two-pile (P-2) cap. The reinforcing and thicknesses shown are for a low-capacity timber pile and must be investigated for other cases.

TYPICAL P-3 & P-4 PILE CAP DETAILS
(DETAIL T2-P3 & P4)

This detail is similar to the previous one, but it shows the dimensions and reinforcing for a three-pile (P-3) and four-pile (P-4) cap.

TYPICAL P-5 & P-6 PILE CAP DETAILS
(DETAIL T2-P5 & P6)

This detail is similar to detail T2-P1 & P2, but it shows the dimensions and reinforcing for a five-pile (P-5) and six-pile (P-6) cap.

PERIMETER FOUNDATION SECTION #1
(DETAIL T2-PERM1)

Continuous load-bearing masonry walls can be pile supported in several ways. One way is to use pile groups, such as P-2 caps, spaced at a certain distance and to provide nominal width grade beams to span in between. Another way, shown here, is to provide piles at a certain spacing, staggered as shown. A continuous grade beam serves as the pile cap. This simplifies the grade beam formwork, which is a significant cost factor. The pile staggering provides stability; otherwise, transverse strap beams would be required. The grade beam size and reinforcing are shown in a schedule.

PERIMETER FOUNDATION SECTION #2
(DETAIL T2-PERM2)

Two details are shown. One is for a load-bearing metal stud wall and the other is at a perimeter door. Both details show the slab as being turned down on top of the masonry foundation wall. Masonry shoe blocks could be considered in lieu of turning the slab edge down, thus eliminating concrete formwork. However, the slab turndown provides a more significant structural tie between the slab and foundation wall.

PERIMETER FOUNDATION SECTION #3
(DETAIL T2-PERM3)

This detail is a typical perimeter condition where the first-floor slab is structurally supported on piles as opposed to a *floating* (not structurally supported) slab on grade. The building is pre-engineered. The floor slab is a two-way spanning system supported by a continuous grade beam at the perimeter. The slab and grade beam are not poured monolithically. The grade beams are supported by pile caps lowered below the frostline. It is not recommended that the pile caps and grade beams be poured monolithic with the slab, due to complications in the field.

PERIMETER FOUNDATION SECTION #4 (DETAIL T2-PERM4)

This detail is similar to the previous one, but it shows the condition at a column. The vertical dowels between the grade beam and pile cap transfer uplift loads to the cap. In this case, tension piles are not used because there is sufficient dead load in the slab, grade beam, pile cap, and soil weight to resist the wind uplift.

PERIMETER FOUNDATION SECTION #5 (DETAIL T2-PERM5)

This detail shows the condition for a pile-supported floor slab as it abuts an existing building. The P-1 cap shown is located several feet away from the existing building face due to clearances required for the pile driving rig. The two-way slab cantilevers to the face of the existing building.

TYPICAL GRADE BEAM ELEVATION DETAIL (DETAIL T2-PGBME)

Concrete grade beams generally have shear and flexural reinforcing, which is described in a schedule for each particular grade beam. This detail shows an elevation view of how that reinforcing is arranged. Grade beams that have special reinforcing requirements should be detailed elsewhere.

TYPICAL GRADE BEAM SECTION AND CONSTRUCTION JOINT DETAIL (DETAIL T2-PGBMS)

Two details are shown. One is a typical grade beam section and the other is a grade beam construction joint. The grade beam construction joint detail would be applicable where it is not feasible to pour all of the grade beams in a single monolithic pour.

TYPICAL STEPPED GRADE BEAM DETAIL (DETAIL T2-PGBSF)

The stepped grade beam detail is applicable where the beam elevation must change, such as due to changes in the exterior finished grade elevation. The reinforcing is run continuous.

PIPE SLEEVE THRU GRADE BEAM DETAILS (DETAIL T2-PGBSL)

Two details are shown. One is for where the pipe would not affect the reinforcing, but the pipe is still too close to the grade beam bottom. The other detail shows the condition where the pipe does affect the bottom reinforcing, so the reinforcing is bent to clear the top of the pipe and additional reinforcing is provided. The pipes are sleeved and wrapped with loose insulation. This allows the foundation system to settle slightly without harming the pipe. These details would not apply where the pipes are a sufficient distance below the grade beam bottom.

INTERIOR FOUNDATION SECTION #1 (DETAIL T2-PINT1)

This detail is essentially the same as detail T2-PERM1, but it shows an interior wall condition. The slab on grade is "floating" (not structurally supported).

INTERIOR FOUNDATION SECTION #2 (DETAIL T2-PINT2)

This detail shows the condition for a pile-supported floor slab at a typical interior P-1 cap. A two-way slab supported by interior P-1 caps is generally more economical than a one-way slab and grade beam system because grade beam formwork is so costly. Vertical dowels extend from the P-1 cap providing a physical tie between the slab and cap. These dowels also provide stability to the top of the pile cap. The slab is not poured monolithic with the cap due to complications in the field.

INTERIOR FOUNDATION SECTION #3 (DETAIL T2-PINT3)

This detail is similar to the previous one, but it shows the condition at an interior pre-engineered metal building column. The vertical dowels over the piles are provided to brace the weak axis of the P-2 pile cap. Tension piles are used to resist the uplift force from the metal building column.

COLUMN BASE PLATES AND ANCHOR BOLTS

COLUMN BASE PLATE AND ANCHOR BOLT LAYOUT DETAILS (DETAIL T2-BP1)

This detail shows a plan and elevation of a steel tube column, base plate, and anchor bolts. The base plate size, base plate thickness, and anchor bolt requirements would be shown in a schedule. This detail and the schedule allow a single plan view to suffice for a variety of column sizes. A continuous fillet weld is provided, even though an intermittent weld would suffice for most loading conditions.

COLUMN BASE PLATE DETAIL #2 (DETAIL T2-BP2)

Two specific base plate details are shown: one for a 5″ tube column and the other for a 6″ standard pipe column. The base plate size, base plate thickness, and anchor bolt requirements are called out on the details. A continuous weld is provided.

COLUMN BASE PLATE DETAIL #3 (DETAIL T2-BP3)

Two specific base plate details are shown: one for a 6″ wide-flange (WF) column and the other for an 8″ WF column. The base plate size, base plate thickness, and anchor bolt requirements are called out on the details. It is more economical to weld the base plate to wide-flange columns from one direction as shown for the W6 column so that the steel fabricator does not have to rotate the column while welding. The weld at the W8 column is shown as being continuous because this column has more anchor bolts and the weld is more critical.

COLUMN BASE PLATE DETAIL #4 (DETAIL T2-BP4)

The two specific base plate details shown are both for 10″ WF columns. One detail is similar to the previous two. The other detail shows a condition for a column in a freezer building, where the column is thermally isolated from the footing by means of an oak block piece.

COLUMN BASE PLATE DETAIL #5 (DETAIL T2-BP5)

The two specific base plate details shown are both at braced bay columns. One detail shows a WF column that has a braced bay diagonal and gusset plate on two faces. The other detail shows a smaller TS column with a braced bay diagonal and gusset plate on one face.

TYPICAL ANCHOR BOLT DETAIL #1
(DETAIL T2-AB1)

Two typical anchor bolt details are shown: one for a pre-engineered metal building and the other for a conventional steel framed building. As described in the Discussion, headed anchor bolts are recommended. J- and L-type bolts do not perform as well under uplift loading conditions. Since pre-engineered metal building systems are so light, their columns generally have significant wind uplift, and a 12 to 15″ minimum bolt embedment is recommended. The other detail shows two different anchor bolt types. Type 1 bolts would be for columns with minimal loading requirements, such as mezzanine columns.

Type 2 bolts would be for columns with more significant loading requirements, such as at braced bays. It is recommended that the amount of bolt projection above the foundation be left up to the steel fabricator.

TYPICAL ANCHOR BOLT DETAIL #2
(DETAIL T2-AB2)

Two typical anchor bolt details are shown. The J-bolt detail would be for columns with minimal loading requirements, such as mezzanine columns. The other anchor bolt allows threaded rod material to be used in lieu of headed anchor bolts.

TYPICAL ANCHOR BOLT DETAIL #3
(DETAIL T2-AB3)

The two details shown are both similar to detail T2-AB1, but one detail shows a concrete pedestal foundation and the other detail shows a condition at a freezer building column.

TYPICAL ANCHOR BOLT DETAIL #4
(DETAIL T2-AB4)

The two details shown are similar to the previous one, but one detail shows a pre-engineered metal building column and the other detail shows a TS canopy column. Both details show a concrete pedestal foundation.

SLABS ON GRADE

SLAB CONTRACTION JOINT DETAIL
(DETAIL T2-SCJ)

This detail is used to create an intentionally weakened plane in the concrete slab so that shrinkage cracks occur at these control joint locations instead of at random. Simply saw-cutting the slab is not sufficient to create an effective weakened plane. The wire reinforcing also needs to be intentionally weakened prior to pouring the concrete. This can be accomplished by precutting the alternate wires of the welded wire fabric.

KEYED CONSTRUCTION JOINT DETAIL #1
(DETAIL T2-SKCJ)

This detail is used where the slab area is too large for a single pour. The slab should be thickened as shown so that concrete spalling does not occur above the key. This detail is appropriate for many conditions; however, it is not recommended where hard wheeled forklift traffic will be encountered, for very high-loading requirements (such as for rack storage loading), nor for high-performance requirements (such as for "superflat" slabs). The key material can be made of wood or metal. It is recommended that all key material be removed prior to pouring the adjacent slab. Galvanized metal keys are available; however, even galvanized material will eventually corrode.

KEYED CONSTRUCTION JOINT DETAIL #2
(DETAIL T2-SKCJ2)

This detail is similar to the previous one, but the slab is not thickened at the key. This detail should be used with caution. It would possibly be appropriate for slabs that are covered by finishes and where minor spalling above the key would not be a concern. Also, some key manufacturers have developed products where the top of the key is lowered a sufficient amount so that the likelihood of concrete spalling is minimized.

KEYED CONSTRUCTION JOINT DETAIL #3
(DETAIL T2-SKCJ3)

This detail is similar to detail T2-SKCJ, but the slab joint is located at an interior cmu partition wall. The slab is thickened and reinforced similar to a typical thickened slab detail. The turndown width must be adequate to support the wall weight.

DOWELED CONSTRUCTION JOINT DETAIL
(DETAIL T2-SDCJ)

This detail is also used where the slab area is too large for a single pour. It provides much better shear transfer between slab segments. It can be used where hard wheeled traffic will be encountered, for high-loading requirements, and high-performance requirements.

TYPICAL HAIR PIN BAR DETAILS
(DETAIL T2-SHPIN)

Hair pin bars are primarily used with pre-engineered metal buildings. A horizontal thrust occurs at the bases of the main frame perimeter columns. The interior frame columns of pre-engineered metal building columns do not have thrusts. The hair pin bars distribute these forces back into the slab, which can be transferred to the other opposing end column. Two details are shown: one for a concrete pedestal and the other for a monolithic slab/footing foundation.

TYPICAL COLUMN ISOLATION JOINT DETAIL #1
(DETAIL T2-SIS01)

This detail shows how to isolate the slab at pre-engineered metal building columns, which are resting on concrete column pedestals. The isolation joint prevents bonding between the pedestal and slab, allowing the slab to move or settle without damaging the slab. It also provides better crack control and allows the slab pour to occur after the steel has been erected.

TYPICAL COLUMN ISOLATION JOINT DETAIL #2 (DETAIL T2-SIS02)

This detail is similar to the previous one, but it shows the condition for non-pre-engineered columns.

THICKENED SLAB DETAIL #1 (DETAIL T2-STS1)

Two thickened slab details are shown: one at a stair post and another at a construction joint. The 2'-6" square footprint below the stair post is a nominal area and would need to be investigated for the actual post reaction and allowable soil bearing capacity.

THICKENED SLAB DETAIL #2 (DETAIL T2-STS2)

Two thickened slab details are shown: one at a stair stringer and another at a non-load-bearing masonry partition wall. Again, the width of thickening below the masonry wall is nominal and would need to be investigated for the actual wall weight and allowable soil bearing capacity.

THICKENED SLAB DETAIL #3 (DETAIL T2-STS3)

This detail is similar to the previous one, but the partition wall is cantilevered from the slab and a slab joint occurs on one side of the wall. The masonry wall is cantilevered from the slab because the masonry wall is not supported at the top (where the metal studwall begins).

MISCELLANEOUS SLAB DETAIL #1 (DETAIL T2-SMIS1)

Two slab details are shown. One detail is used where the top surface of the slab needs to be depressed, such as due to thick-set tile floor finishes. The wire reinforcing should be continuous as shown to help inhibit the slab cracking that will tend to develop at the slab transition. The other detail shows the condition at a concrete bank vault.

INDUSTRIAL BUILDING FOUNDATION SECTIONS

PERIMETER FOUNDATION SECTION #1 (DETAIL T2-IBPS1)

The foundation system cost for industrial buildings is a major consideration. This detail shows a typical perimeter foundation condition, where finished grade is 4′ below finished floor. The 4′ is a fairly standard height used for truck loading/unloading. The wall system above the slab is an insulated metal panel, which could be subject to damage from forklift use. To avoid this problem, an 8 to 10′ high masonry skirt wall could be considered. A 6″ slab on grade is shown, fairly standard for normal warehouse projects. Thicker slabs would be required for special conditions, such as rack loading. The slab is thickened to 8″ at the perimeter to agree with masonry coursing. The 8″ masonry foundation wall is cantilevered from the footing during construction and reinforced vertically. In some cases, it may be more economical to brace the wall during construction in lieu of cantilevering. After the slab "sets," the wall is braced at the top and no longer cantilevers. The top of footing elevation is based on minimal frost penetration requirements.

PERIMETER FOUNDATION SECTION #2 (DETAIL T2-IBPS2)

This detail is similar to the previous one, but the condition at an interior truck dock well is shown. A more durable concrete foundation wall is provided in lieu of masonry. A continuous angle is provided at the top edge to prevent concrete spalling. The architectural drawings are referred to for the removable guardrail.

PERIMETER FOUNDATION SECTION #3 (DETAIL T2-IBPS3)

This detail is similar to the previous one, but the condition at a perimeter truck dock well is shown. A concrete foundation wall is provided in lieu of masonry. A continuous angle is provided at the door sill to provide impact resistance. The architectural drawings are referred to for the overhead door and sill detail. It is recommended that the door sill be depressed and sloped to prevent wind driven rain from entering the building.

PERIMETER FOUNDATION SECTION #4 (DETAIL T2-IBPS4) .

This detail is similar to the previous one, but it shows the condition at a perimeter dock leveler. A concrete foundation wall is provided in lieu of masonry. Several of the critical dimensions are noted as being per the dock leveler equipment selected. Continuous angles are required around the top per the dock leveler manufacturer's recommendations.

PERIMETER FOUNDATION SECTION #5 (DETAIL T2-IBPS5)

This detail is similar to the previous one, but it shows the condition at a perimeter metal building column. A concrete pedestal supports the column.

The column footing is centered on the pedestal. Since the metal building column size is not typically known by the foundation design professional during the foundation design stage, the pedestal size will need to be coordinated with the metal building column after the metal building is designed. This will ensure that the column fits on the pedestal. Since the building column and footing are not exactly centered, some slight eccentric footing load should be accounted for. However, the hair pin bar will tend to resist this eccentric loading condition. The hair pin bar is also used to resist the horizontal thrust at the base of the column. Hair pin bars should be sized for the maximum loading condition; however, for many situations, the maximum sustained load case (dead load plus roof/snow load) is the most practical to consider. The maximum wind load case is very temporary and passive soil pressures may resist this temporary load. All wall footing reinforcing is run continuous through the column footing. This provides better wind uplift resistance.

PERIMETER FOUNDATION SECTION #6 (DETAIL T2-IBPS6)

This detail is similar to the previous one, but a masonry skirt wall is shown.

PERIMETER FOUNDATION SECTION #7 (DETAIL T2-IBPS7)

This detail is similar to the previous two, but a concrete foundation wall is indicated beyond. The footing is dropped due to a truck dock well beyond.

PERIMETER FOUNDATION SECTION #8 (DETAIL T2-IBPS8)

Many industrial buildings have small areas of office space that will require a finished wall system, such as metal studs. The wall system and metal stud requirements are noted. A shoe block course is provided above the 12″ masonry first course. The shoe block eliminates a turned down slab edge and the related formwork. The stud wall is non-load-bearing and wind loads are the main concern. Therefore, the stud wall is connected to the slab as opposed to the shoe block, providing a more significant wind resistant connection. Again, the top of footing elevation is based on minimal frost penetration requirements.

PERIMETER FOUNDATION SECTION #9 (DETAIL T2-IBPS9)

This detail is similar to the previous one, but a storefront (glass wall) condition is shown. Although this design is structurally satisfactory, a more weather resistant detail would be to pull the storefront in and use a turned down slab edge. The outer edge of the turned down slab edge could then be depressed or sloped to inhibit water from entering below the storefront sill.

PERIMETER FOUNDATION SECTION #10 (DETAIL T2-IBP10)

This detail is similar to the previous one, but a metal siding wall condition is shown. The top of footing is lowered due to the exterior grade elevation. Depending on the actual height of the masonry wall, vertical wall reinforcing may be required.

PERIMETER FOUNDATION SECTION #11 (DETAIL T2-IBP11)

This detail is similar to detail T2-IBPS3, but the exterior grade is higher and a masonry foundation wall is shown. Depending on the actual height of the masonry wall, vertical wall reinforcing may be required.

PERIMETER FOUNDATION SECTION #12 & #12A (DETAIL T2-IBP12 & T2-IB12A)

The first detail is similar to detail T2-IBPS7, but the exterior grade is higher. The second detail shows a dropped column base; however, metal building manufacturers prefer that the column base be at finished floor elevation.

PERIMETER FOUNDATION SECTION #13 (DETAIL T2-IBP13)

This detail is similar to the previous two, but it occurs at an office area where there is a brick façade. The column pedestal is recessed so that the brick can extend below finished grade.

PERIMETER FOUNDATION SECTION #14 (DETAIL T2-IBP14)

A monolithic slab/footing system is ideal for many industrial-related or pre-engineered metal buildings, particularly where there are minimal frost penetration requirements. The slab edge is thickened and poured monolithic with the slab. Perimeter formwork is required. The depth of slab edge and reinforcing will depend on the thickened slab edge design as required to span between main building columns and resist the net wind uplift. A 24″ deep turndown is shown; however, an 18″ deep turndown is a recommended minimum. Perimeter insulation is difficult to deal with in the field, especially on the sloping surface of the slab edge. If horizontal insulation alone will suffice, it will be much easier in the field.

PERIMETER FOUNDATION SECTION #15
(DETAIL T2-IBP15)

The two details shown are similar to the previous detail. One shows a store-front (glass wall) condition. The other shows increased slab edge reinforcing at a braced bay location.

PERIMETER FOUNDATION SECTION #16
(DETAIL T2-IBP16)

A perimeter slab edge condition can also be used in combination with a low masonry perimeter wall. In this case, the masonry wall was desired for architectural reasons. The wall is cantilevered from the slab because the top of the masonry is not braced by the metal building system. An 18″ deep turndown and horizontal perimeter insulation were used.

PERIMETER FOUNDATION SECTION #17
(DETAIL T2-IBP17)

This detail is similar to the previous one, but a metal stud wall is used in lieu of the metal siding.

PERIMETER FOUNDATION SECTION #18
(DETAIL T2-IBP18)

This detail is similar to the previous one, but it shows the thickened slab condition at a metal building column. The slab edge is widened at the column, sized to support the column reaction. The hair pin bar is used to resist the horizontal force from the metal building column. The depth of turndown at the monolithic footing should match the typical slab edge.

PERIMETER FOUNDATION SECTION #19
(DETAIL T2-IBP19)

This detail is similar to the previous one, but the column reaction is too large for a monolithic footing. Therefore, a subfooting is provided, tied to the slab edge to resist wind uplift.

PERIMETER FOUNDATION SECTION #20
(DETAIL T2-IBP20)

Perimeter metal building columns can be lowered to bear directly on the footings, although this technique is not preferred by most building manufacturers. It does hide the column base plate and make the column anchorage less prone to forklift damage.

PERIMETER FOUNDATION SECTION #21
(DETAIL T2-IBP21)

Many industrial buildings are constructed as *shell buildings*. Shell buildings consist of the structural and perimeter architectural systems. Minimal interior, electrical, plumbing, and mechanical systems are initially provided. These systems are designed and installed after the tenant is selected. Due to the difficulty of running underground utilities, the slab on grade can be initially omitted. This complicates the foundation design, especially for pre-engineered metal buildings.

PERIMETER FOUNDATION SECTION #22
(DETAIL T2-IBP22)

This detail is similar to the previous one, but the top of footing elevation is higher and the perimeter foundation wall butts into the pedestal in lieu of bypassing the pedestal.

INTERIOR FOUNDATION SECTION #1
(DETAIL T2-IBIS1)

This detail is similar to the previous details, but a typical interior column condition is shown. Interior metal building columns can have significant wind uplift. This uplift is resisted by lowering the footing and relying on the footing, soil, and slab weights. The pedestal vertical bars are hooked at the top and bottom. These hooks help develop the full capacity of the vertical bars, which transfer the wind uplift from the anchor bolts, through the vertical bars, and into the footing.

INTERIOR FOUNDATION SECTION #2
(DETAIL T2-IBIS2)

This detail is similar to the previous one, but the wind uplift is much more significant. The footing is lowered to resist the wind uplift.

INTERIOR FOUNDATION SECTION #3
(DETAIL T2-IBIS3)

This detail is similar to the previous two, but it is for a shell building, and the slab is not being provided until tenant build-out.

INTERIOR FOUNDATION SECTION #4
(DETAIL T2-IBIS4)

Overhead cranes are frequently required for industrial buildings, including pre-engineered metal buildings. In this detail, the metal building columns are not used to provide vertical support for the crane system. This support is provided by an independently framed steel system. The crane support columns are aligned with the metal building columns and sit on the same pedestal and footing. The footing is centered under the center of load.

MISCELLANEOUS FOUNDATION SECTIONS AND DETAILS

TYPICAL FOUNDATION SECTION #1 (DETAIL T2-FS1)

This detail shows a typical perimeter load-bearing masonry wall condition. The brick is a nonstructural element. The vertical wall reinforcing and footing information would be shown in a separate schedule. The top of footing elevation is based on minimal frost penetration requirements.

TYPICAL FOUNDATION SECTION #2 (DETAIL T2-FS2)

This detail is similar to the previous one, but the perimeter wall is 12″ thick masonry. The architectural drawings are referred to for the exterior finish system.

TYPICAL FOUNDATION SECTION #3 (DETAIL T2-FS3)

This detail is similar to the previous two, but the perimeter wall is 8″ split-faced masonry. The exterior of this wall must be properly sealed to prevent moisture penetration.

TYPICAL FOUNDATION SECTION #4 (DETAIL T2-FS4)

This detail is similar to the previous one, but it shows a condition at a storefront system.

TYPICAL FOUNDATION SECTION #5 (DETAIL T2-FS5)

This detail is similar to the previous one, but it shows a condition at an exterior door.

TYPICAL FOUNDATION SECTION #6 (DETAIL T2-FS6)

This detail is similar to the previous one, but it shows a condition at an overhead door.

TYPICAL FOUNDATION SECTION #7
(DETAIL T2-FS7)

This detail shows the condition at a perimeter steel column and masonry wall.

TYPICAL FOUNDATION SECTION #8
(DETAIL T2-FS8)

This detail is similar to detail T2-IBPS6, but it shows a condition at an exterior brick finish. The column is not pre-engineered.

TYPICAL FOUNDATION SECTION #9
(DETAIL T2-FS9)

This detail shows the condition at an interior load-bearing 8″ thick masonry wall. A bond break material should be placed between the slab and wall. In this detail, ½″ expansion joint material provides the bond break. The vertical reinforcing is called out on the detail in lieu of showing it in a schedule.

TYPICAL FOUNDATION SECTION #10
(DETAIL T2-FS10)

This detail is similar to the previous one, but it shows a 12″ thick masonry wall.

TYPICAL FOUNDATION SECTION #11
(DETAIL T2-FS11)

A bond break material should be provided between new slabs and existing construction. In this case, 15# felt paper provides the bond break. The delineation between new and existing construction is clearly shown.

TYPICAL FOUNDATION SECTION #12
(DETAIL T2-FS12)

This detail is similar to the previous one, but it shows a 12″ thick existing masonry wall. The existing finish is noted to be removed.

TYPICAL FOUNDATION SECTION #13
(DETAIL T2-FS13)

Load-bearing metal stud framing presents several advantages, including non-combustibility and high strength. This detail shows a turned down slab edge foundation system to support the perimeter metal studwall. The stud and slab edge information would be shown in schedules.

TYPICAL FOUNDATION SECTION #14
(DETAIL T2-FS14)

This detail is similar to the previous one, but it shows an interior load-bearing metal studwall.

TYPICAL FOUNDATION SECTIONS #15, #16, & #17
(DETAILS T2-FS15, T2-FS16, & T2-FS17)

These three details pertain to supporting a lower level wall system below a continuous window system. The window sill elevation is too high to cantilever the metal studs from the slab; therefore, a continuous angle is provided at the sill for support. The continuous angle is supported by vertical channels spaced at one-third points between the columns. These channels are tied to the slab and cantilever to the continuous angle. 6″ steel framing is shown; however, 4″ metal studs and channels could be considered.

TYPICAL PERIMETER FOUNDATION SECTION #1
(DETAIL T2-WFS1)

Many small commercial buildings can be economically framed with wood. This detail shows a typical foundation condition with a perimeter 2 × 6 wood studwall and a brick façade. A shoe block could be considered in lieu of using the turned down slab edge. Again, the top of footing elevation is based on minimal frost penetration requirements.

TYPICAL PERIMETER FOUNDATION SECTION #2
(DETAIL T2-WFS2)

This detail is similar to the previous one, but a hollow shoe block is used in lieu of the slab turndown. The top of footing elevation has also been lowered. Filling the hollow shoe block with the slab concrete tends to lock it in with the slab.

TYPICAL PERIMETER FOUNDATION SECTION #3
(DETAIL T2-WFS3)

This detail is similar to the previous two, but a 2 × 4 studwall is used.

TYPICAL PERIMETER FOUNDATION SECTION #4
(DETAIL T2-WFS4)

Monolithic foundation systems can also be used with wood framing systems, but only for very lightly loaded walls. The recess for the brick is usually obtained by providing a box out tied to the perimeter formwork.

MISCELLANEOUS FOUNDATION DETAILS
(DETAIL T2-WFS5)

Two details are shown. One is a typical perimeter condition at a door and the other is an interior load-bearing wood studwall.

TIMBER POST BASE CONNECTION
(DETAIL T2-WFS6)

Wood column bases can be secured by several means. In this case, a steel fabrication is used. Another option would be a standard prefabricated post base connector.

TYPICAL INTERIOR EQUIPMENT PAD DETAIL
(DETAIL T2-MIPAD)

This detail shows two different options for providing interior equipment pads. For nonvibrating equipment, the pad can be poured over the slab on grade. For vibrating equipment, it is recommended that the pad be isolated from the slab on grade.

TYPICAL EXTERIOR EQUIPMENT PAD
(DETAIL T2-MEPAD)

Exterior equipment typically requires concrete pads for support. This detail shows a turned down slab edge foundation.

EXISTING FOOTING STRENGTHENING DETAIL
(DETAIL T2-MREN1)

Existing foundations can be strengthened in several ways. In this case, the existing column footing is being reinforced by adding 12″ of new concrete on all four sides of the footing. Twelve inches of concrete are also added on top of the existing footing as a redundancy so that the #6 bar doweling is not solely relied on. The existing column and pedestal did not need strengthening.

EXISTING FOOTING STRENGTHENING DETAIL
(DETAIL T2-MREN2)

This detail is similar to the previous one, but a new column is being added over an existing wall footing. The existing wall footing is left in place and a new column footing is provided.

MISCELLANEOUS FOUNDATION DETAILS
(DETAIL T2-MISC1)

Two details are shown: one for a steel jamb channel base connection and the other for a concrete strap beam. Strap beams are typically used at braced bays.

MASONRY SCREEN WALL DETAIL
(DETAIL T2-MISC2)

Masonry screen walls are typically cantilevered from the footing. In this case, several different wall heights and footing sizes were required. Top reinforcing is provided in the footing because the soil weight provides overturning resistance.

TYPICAL PIPE BOLLARD DETAIL
(DETAIL T2-MBOLL)

Pipe bollards are used to protect doorjambs from potential forklift or truck damage. They are set in a concrete footing and filled with concrete.

TYPICAL PEDESTRIAN RAMP DETAIL
(DETAIL T2-MRAM1)

This detail shows a concrete ramp. The architectural drawings are referred to for the railings, ramp slope, and dimensions.

TYPICAL HANDICAPPED RAMP DETAIL
(DETAIL T2-MRAM2)

This detail is similar to the previous one, but it pertains to a handicapped accessible ramp.

TYPICAL CONCRETE STAIR DETAIL
(DETAIL T2-MSTEP)

This detail shows a concrete stair. The architectural drawings are referred to for the railings and dimensions.

TYPICAL TRENCH DRAIN DETAILS
(DETAIL T2-TR1)

Two details are shown. One shows a prefabricated trench. The other detail shows a trench essentially made entirely out of concrete. Only its cover is prefabricated.

TYPICAL PIT DETAIL (DETAIL T2-PIT1)

This detail shows a cast-in-place concrete pit. The 12″ base slab thickness could possibly be reduced depending on the size of the pit and the soil bearing pressure. The architectural drawings are referred to for the removable cover.

TYPICAL SPRAY BOOTH PIT DETAIL (DETAIL T2-PIT2)

This detail is similar to the previous one, but an 8″ thick base slab is used.

TYPICAL ELEVATOR PIT DETAIL #1 (DETAIL T2-PIT3)

Elevator pits are similar to the previous details, but the top of the pit is open. In this detail, an opening is provided in the base slab for elevator piston.

TYPICAL ELEVATOR PIT DETAIL #2 (DETAIL T2-PIT4)

This detail is similar to the previous one, but the pit base is pile supported.

MISCELLANEOUS PIT DETAILS (DETAIL T2-PIT5)

Two details are shown. One is a detail at a sump pit and the other is a detail at an elevator door sill.

TYPICAL RETAINING WALL DETAILS #1 & #2 (DETAILS T2-RW1 & T2-RW2)

Cantilevered retaining walls need to be designed for the specific loading conditions. Detail T2-RW1 has soil on its high side and concrete paving on its low side. Detail T2-RW2 has concrete paving on its high side and soil on its low side. The architectural drawings are referred to for the railings.

WALL CONTROL JOINT DETAILS (DETAIL T2-RWJT1)

Control joints are intentionally weakened planes in the concrete wall so that shrinkage cracks occur at these control joint locations instead of at random.

WALL CONSTRUCTION JOINT DETAILS (DETAIL T2-RWJT2)

These details are used where the wall area is too large for a single pour. All reinforcing is run continuous through the construction joints.

Concrete

(CSI Division 3)

Discussion—Concrete

TYPES OF DETAILS INCLUDED

CONCRETE DETAILS ARE PROVIDED IN THIS
CHAPTER FOR THE FOLLOWING:

- ▶ *Supported Slabs*
- ▶ *Beams and Columns*
- ▶ *Miscellaneous Details*

Details for concrete footings, pile caps, grade beams, slabs on grade, and wall
joints are included in the previous chapter.

SUPPORTED SLABS

There are several different types of supported concrete slabs. One-way slabs are designed to span in a single direction. They are very simple to design and construct and are ideally suited for buildings where continuous wall supports occur at regular intervals, such as with hotel and apartment buildings. Two-way slabs are designed to span in two directions. They are more difficult to design and build; however, they are frequently more economical where larger open spaces are required, such as with office buildings.

The American Concrete Institute (ACI) gives recommendations for minimum slab thicknesses for both types of slabs. A minimum 28-day concrete slab compressive strength (f'_c) of 3000 psi is recommended. For two-way slabs, higher strengths may be required.

Concrete is strong in compression and weak in tension. Therefore, flexural reinforcing is provided in regions where the tensile stresses exceed the allowable stresses of unreinforced concrete. Shear reinforcing is normally avoided in slabs by using a sufficient thickness. A variety of design aids and computer programs are available to assist in the structural design of concrete slabs. ACI provides design criteria.

The preferred method of describing the slab thickness and reinforcing is to show them in a schedule on the plan. The slab mark (relationship to the schedule) should be shown on the plan.

BEAMS AND COLUMNS

Concrete beams and columns are designed as one-way systems. Again, a variety of design aids and computer programs are available to assist in the structural design of concrete beams and columns.

The preferred way to describe the beam size and reinforcing is to show them in a beam schedule on the plan. The beam mark should be shown on the plan. Columns can be shown in the typical details or in a column schedule.

Cast-in-place concrete beam and column construction can be expensive, primarily due to on-site formwork costs. For this reason, precast concrete is a viable alternative. All of the major concrete elements are plant fabricated, transported to the site, and field erected. The connections between major elements are made in the field. However, these connections can be complicated and frequently result in less ductility as compared to cast-in-place concrete. Therefore, precast concrete may not be suited in areas with high seismic activity.

Precast concrete is also used for specific types of structural elements that can be combined with other types of systems. A variety of floor systems have been successfully precast, including solid slabs, hollow core slabs, and double tees.

MISCELLANEOUS DETAILS

Miscellaneous details are also included for a slab on grade replacement, doweling to existing concrete, precast double tee sections, precast double tee flange connections, wall control joints, and wall construction joints.

GENERAL COMMENT

Practices vary from region to region and the details that follow have generally been taken from specific applications. Any member sizes or connections shown were intended for that particular application. Accordingly, the applicability of all details should be investigated and all member sizes and connections should be determined prior to use.

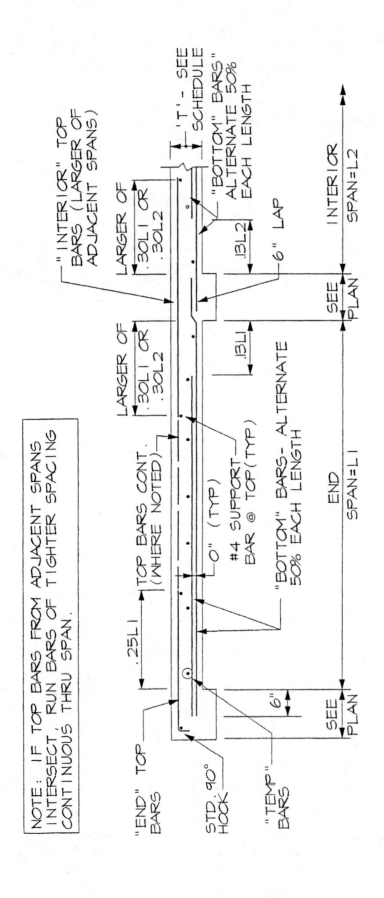

NOTE: IF TOP BARS FROM ADJACENT SPANS INTERSECT, RUN BARS OF TIGHTER SPACING CONTINUOUS THRU SPAN.

"INTERIOR" TOP BARS (LARGER OF ADJACENT SPANS)

LARGER OF .30L1 OR .30L2

'T' - SEE SCHEDULE

"BOTTOM" BARS ALTERNATE 50% EACH LENGTH

.13L2

6" LAP

INTERIOR SPAN=L2

SEE PLAN

.13L1

LARGER OF .30L1 OR .30L2

.25L1

TOP BARS CONT. (WHERE NOTED)

0" (TYP)

#4 SUPPORT BAR @ TOP (TYP)

"BOTTOM" BARS- ALTERNATE 50% EACH LENGTH

END SPAN=L1

"END" TOP BARS

STD. 90° HOOK

"TEMP" BARS

6"

SEE PLAN

TYP. ONE-WAY CONCRETE SLAB REINFORCING DETAIL
(DETAIL T3-SLAB1)

NOT TO SCALE

152

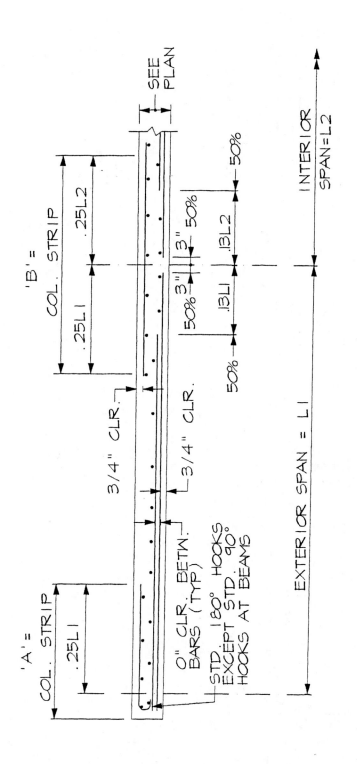

SECTION THRU MIDDLE STRIP

TYP. TWO-WAY CONCRETE SLAB REINFORCING DETAIL

NOT TO SCALE

(DETAIL T3-2WAYM)

153

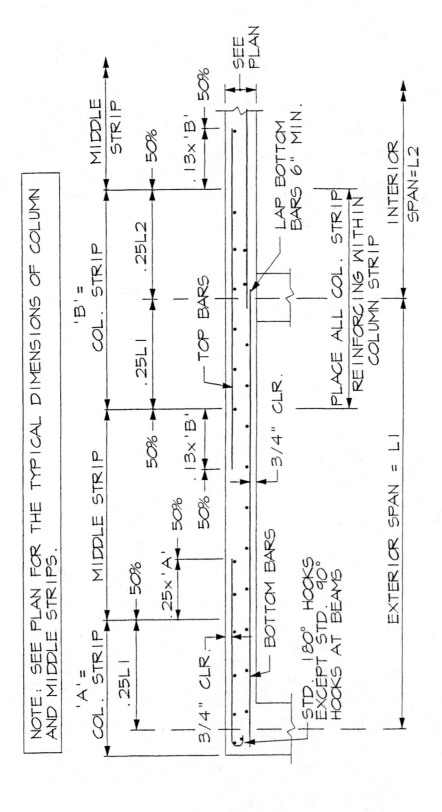

SECTION THRU COLUMN STRIP

TYP. TWO-WAY CONCRETE SLAB REINFORCING DETAIL

(DETAIL T3-2WAYC)

NOT TO SCALE

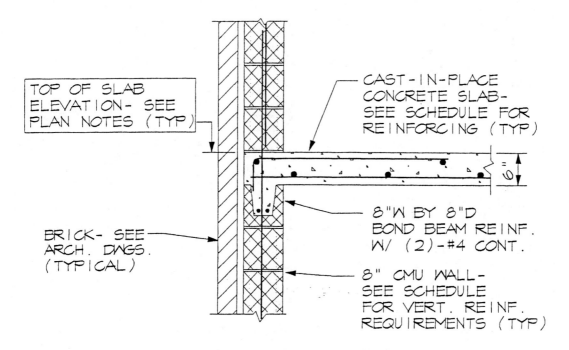

TOP OF SLAB ELEVATION- SEE PLAN NOTES (TYP)

CAST-IN-PLACE CONCRETE SLAB- SEE SCHEDULE FOR REINFORCING (TYP)

8"W BY 8"D BOND BEAM REINF. W/ (2)-#4 CONT.

BRICK- SEE ARCH. DWGS. (TYPICAL)

8" CMU WALL- SEE SCHEDULE FOR VERT. REINF. REQUIREMENTS (TYP)

6"

BEARING DETAIL AT ONE-WAY SLAB

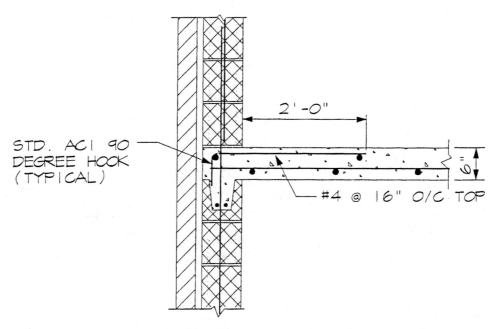

2'-0"

STD. ACI 90 DEGREE HOOK (TYPICAL)

#4 @ 16" O/C TOP

6"

NONBEARING DETAIL AT ONE-WAY SLAB

SLAB BEARING ON MASONRY DETAIL 1

NOT TO SCALE   (DETAIL T3-CMU1)

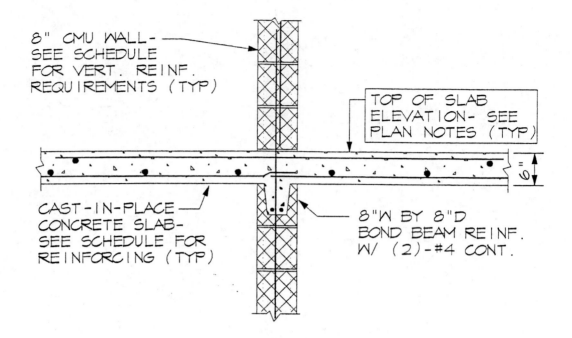

8" CMU WALL-
SEE SCHEDULE
FOR VERT. REINF.
REQUIREMENTS (TYP)

TOP OF SLAB
ELEVATION- SEE
PLAN NOTES (TYP)

6"

CAST-IN-PLACE
CONCRETE SLAB-
SEE SCHEDULE FOR
REINFORCING (TYP)

8"W BY 8"D
BOND BEAM REINF.
W/ (2)-#4 CONT.

INTERIOR BEARING DETAIL

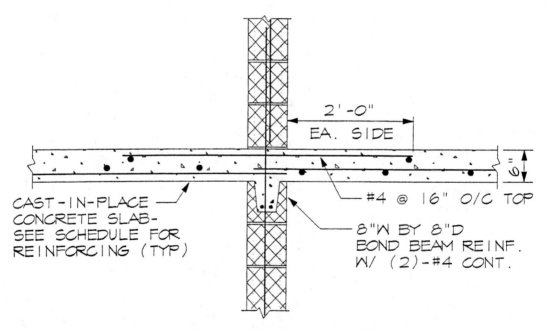

2'-0"

EA. SIDE

6"

CAST-IN-PLACE
CONCRETE SLAB-
SEE SCHEDULE FOR
REINFORCING (TYP)

#4 @ 16" O/C TOP

8"W BY 8"D
BOND BEAM REINF.
W/ (2)-#4 CONT.

INTERIOR SLAB DIRECTION CHANGE

SLAB BEARING ON MASONRY DETAIL 2
NOT TO SCALE   (DETAIL T3-CMU2)

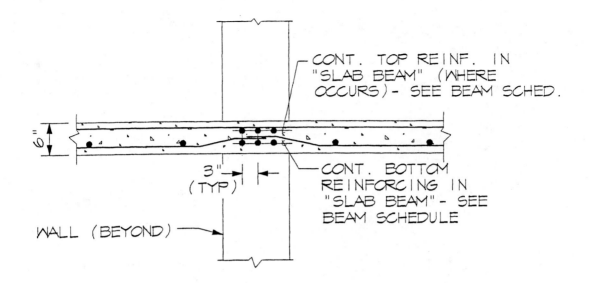

CONT. TOP REINF. IN "SLAB BEAM" (WHERE OCCURS)- SEE BEAM SCHED.

CONT. BOTTOM REINFORCING IN "SLAB BEAM"- SEE BEAM SCHEDULE

6"

3" (TYP)

WALL (BEYOND)

DETAIL AT 'SLAB BEAMS'

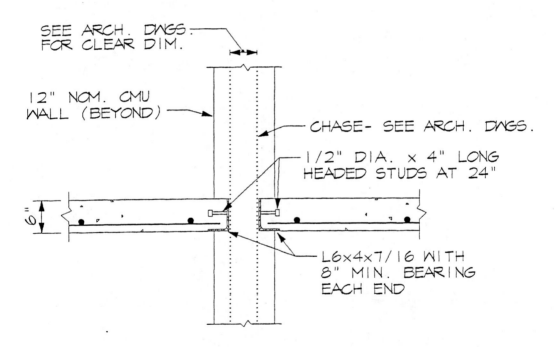

SEE ARCH. DWGS. FOR CLEAR DIM.

12" NOM. CMU WALL (BEYOND)

CHASE- SEE ARCH. DWGS.

1/2" DIA. x 4" LONG HEADED STUDS AT 24"

6"

L6x4x7/16 WITH 8" MIN. BEARING EACH END

DETAIL AT INTERIOR HVAC CHASE

SLAB BEARING ON MASONRY DETAIL 3

NOT TO SCALE   (DETAIL T3-CMU3)

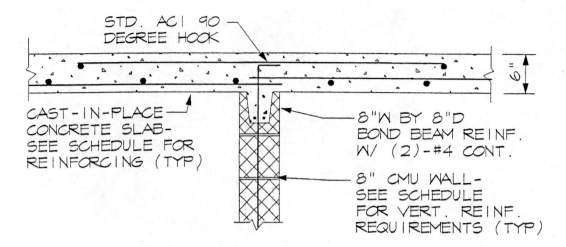

STD. ACI 90 DEGREE HOOK

CAST-IN-PLACE CONCRETE SLAB- SEE SCHEDULE FOR REINFORCING (TYP)

8"W BY 8"D BOND BEAM REINF. W/ (2)-#4 CONT.

8" CMU WALL- SEE SCHEDULE FOR VERT. REINF. REQUIREMENTS (TYP)

6"

SLAB DIRECTION CHANGE AT ROOF LEVEL

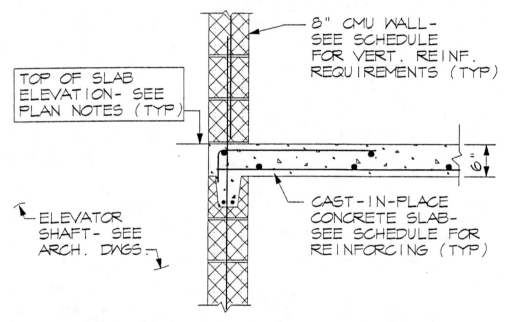

8" CMU WALL- SEE SCHEDULE FOR VERT. REINF. REQUIREMENTS (TYP)

TOP OF SLAB ELEVATION- SEE PLAN NOTES (TYP)

CAST-IN-PLACE CONCRETE SLAB- SEE SCHEDULE FOR REINFORCING (TYP)

6"

ELEVATOR SHAFT- SEE ARCH. DWGS.

BEARING DETAIL AT ELEVATOR SHAFT WALL

SLAB BEARING ON MASONRY DETAIL 4
NOT TO SCALE   (DETAIL T3-CMU4)

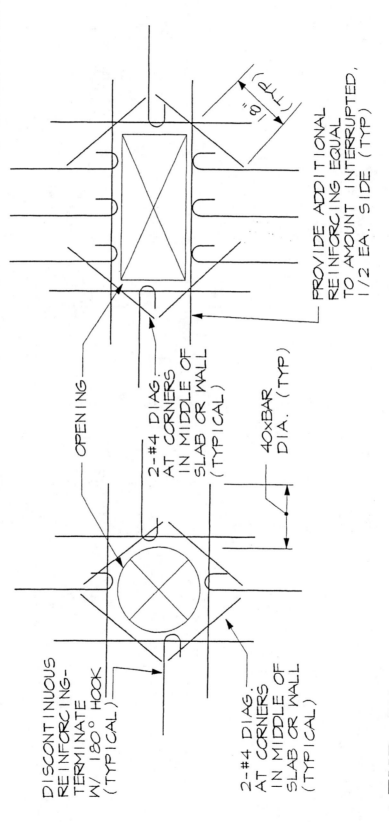

NOTE: WHERE OPENINGS ARE SUCH THAT THE REINFORCING STEEL IS NOT INTERRUPTED, SUCH AS AT CONDUITS AND SMALL PIPES, NO ADDITIONAL REINFORCING IS REQUIRED. WHERE OPENINGS ARE LARGER THAN THE SPACING BETWEEN THE BARS AND ARE SUCH THAT THE REINFORCING CAN NOT BE SHIFTED TO CLEAR, THEN THIS DETAIL SHALL APPLY.

DISCONTINUOUS REINFORCING- TERMINATE W/ 180° HOOK (TYPICAL)

2-#4 DIAG. AT CORNERS IN MIDDLE OF SLAB OR WALL (TYPICAL)

OPENING

2-#4 DIAG. AT CORNERS IN MIDDLE OF SLAB OR WALL (TYPICAL)

40xBAR DIA. (TYP)

"a" (TYP)

PROVIDE ADDITIONAL REINFORCING EQUAL TO AMOUNT INTERRUPTED, 1/2 EA. SIDE (TYP)

TYP. CONCRETE WALL OR SLAB OPENING DETAIL

NOT TO SCALE

(DETAIL T3-OPNG)

159

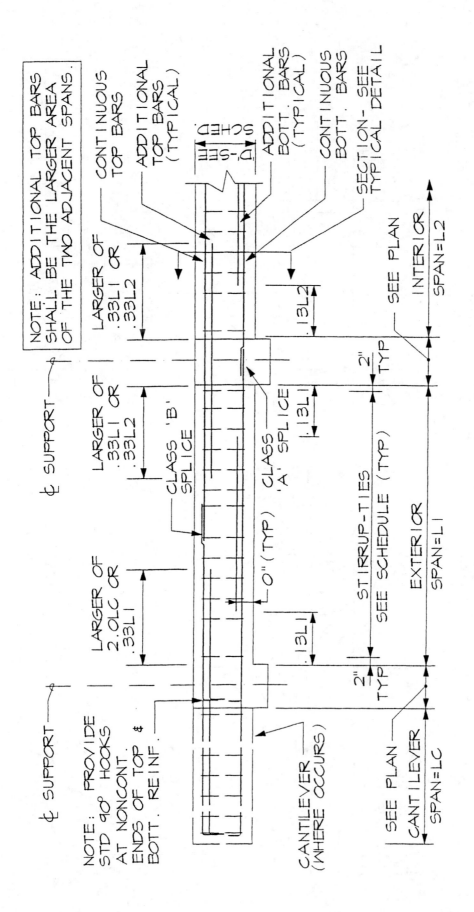

TYP. CONCRETE BEAM ELEVATION DETAIL

NOT TO SCALE

(DETAIL T3-BMEL)

160

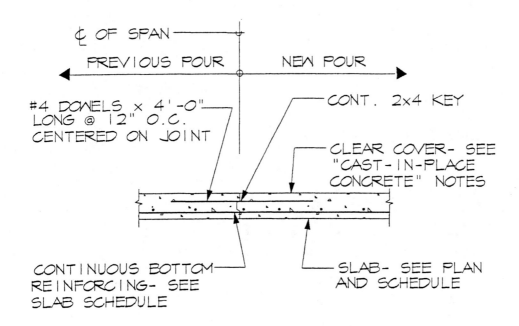

ELEVATED SLAB CONSTRUCTION JOINT
NOT TO SCALE

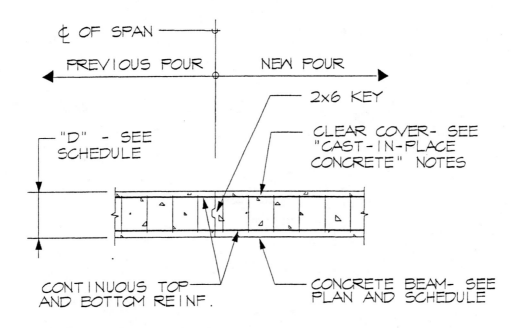

BEAM CONSTRUCTION JOINT DETAIL
NOT TO SCALE   (DETAIL T3–BMJT)

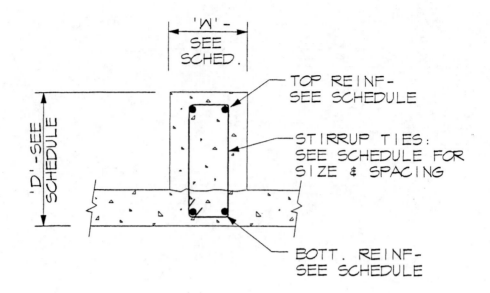

'W' –
SEE
SCHED.

TOP REINF-
SEE SCHEDULE

STIRRUP TIES:
SEE SCHEDULE FOR
SIZE & SPACING

'D'-SEE
SCHEDULE

BOTT. REINF-
SEE SCHEDULE

UPTURNED BEAM

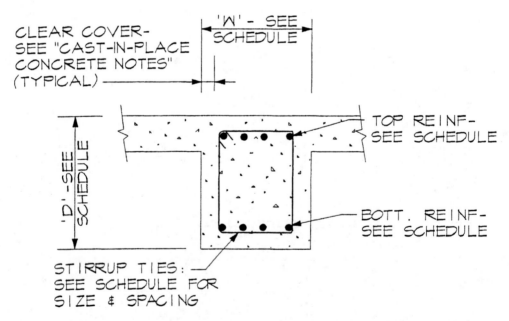

CLEAR COVER-
SEE "CAST-IN-PLACE
CONCRETE NOTES"
(TYPICAL)

'W' – SEE
SCHEDULE

TOP REINF-
SEE SCHEDULE

'D'-SEE
SCHEDULE

BOTT. REINF-
SEE SCHEDULE

STIRRUP TIES:
SEE SCHEDULE FOR
SIZE & SPACING

DOWNTURNED BEAM

TYPICAL CONCRETE BEAM SECTIONS

NOT TO SCALE   (DETAIL T3-BMSEC)

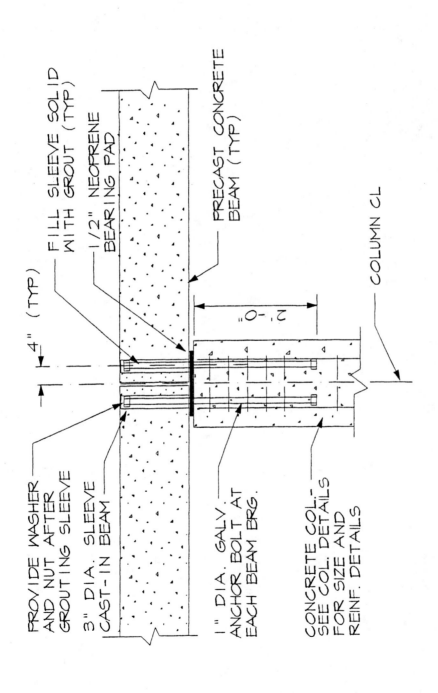

PROVIDE WASHER
AND NUT AFTER
GROUTING SLEEVE

FILL SLEEVE SOLID
WITH GROUT (TYP)

1/2" NEOPRENE
BEARING PAD

PRECAST CONCRETE
BEAM (TYP)

4" (TYP)

3" DIA. SLEEVE
CAST-IN BEAM

1" DIA. GALV.
ANCHOR BOLT AT
EACH BEAM BRG.

2'-0"

COLUMN CL

CONCRETE COL.-
SEE COL. DETAILS
FOR SIZE AND
REINF. DETAILS

TYPICAL PRECAST CONCRETE BEAM BEARING DETAIL
(DETAIL T3-PCBM)

NOT TO SCALE

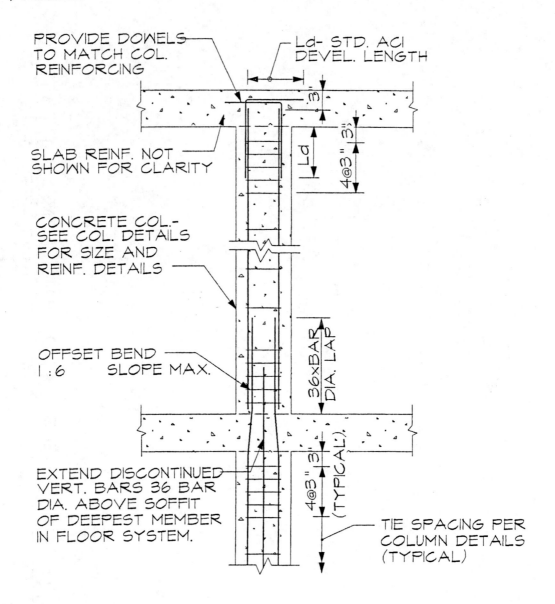

PROVIDE DOWELS TO MATCH COL. REINFORCING

Ld- STD. ACI DEVEL. LENGTH

SLAB REINF. NOT SHOWN FOR CLARITY

CONCRETE COL.- SEE COL. DETAILS FOR SIZE AND REINF. DETAILS

OFFSET BEND 1:6 SLOPE MAX.

EXTEND DISCONTINUED VERT. BARS 36 BAR DIA. ABOVE SOFFIT OF DEEPEST MEMBER IN FLOOR SYSTEM.

3"

Ld

4@3" 3"

36×BAR DIA. LAP

4@3" 3" (TYPICAL)

TIE SPACING PER COLUMN DETAILS (TYPICAL)

CONCRETE COLUMN DETAIL #1

NOT TO SCALE   (DETAIL T3-COL)

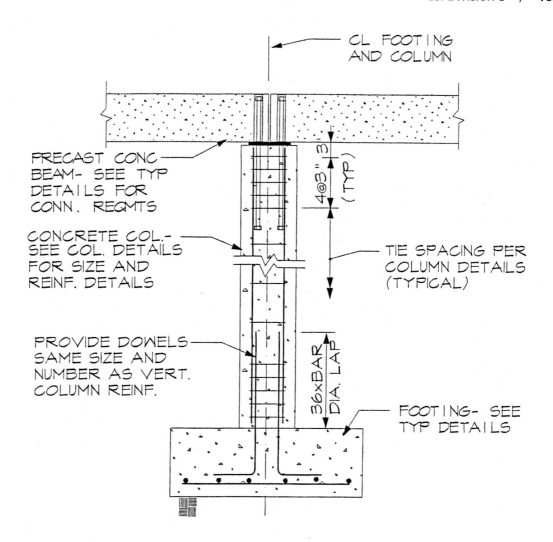

CL FOOTING
AND COLUMN

PRECAST CONC
BEAM- SEE TYP
DETAILS FOR
CONN. REQMTS

CONCRETE COL.-
SEE COL. DETAILS
FOR SIZE AND
REINF. DETAILS

PROVIDE DOWELS
SAME SIZE AND
NUMBER AS VERT.
COLUMN REINF.

4@3" 3 (TYP)

TIE SPACING PER
COLUMN DETAILS
(TYPICAL)

36xBAR DIA. LAP

FOOTING- SEE
TYP DETAILS

CONCRETE COLUMN DETAIL #2

NOT TO SCALE   (DETAIL T3-COL2)

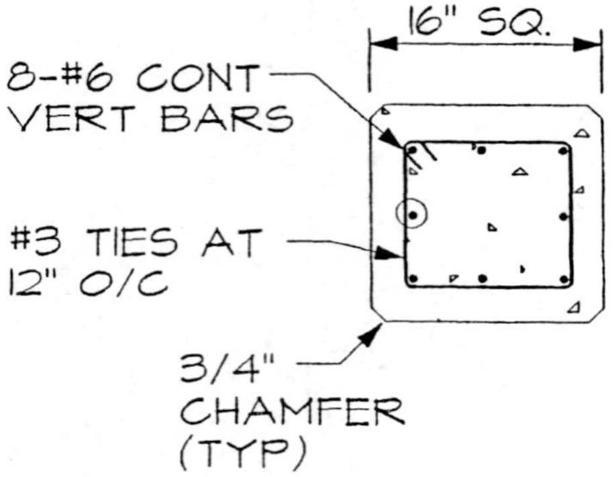

12-#6 CONT
VERT BARS

1 1/2" CLEAR
TO TIE (TYP)

#3 TIES
AT 12" O/C

32"

6"

3/4"
CHAMFER
(TYP)

COLUMN C-2

16" SQ.

8-#6 CONT
VERT BARS

#3 TIES AT
12" O/C

3/4"
CHAMFER
(TYP)

COLUMN C-1

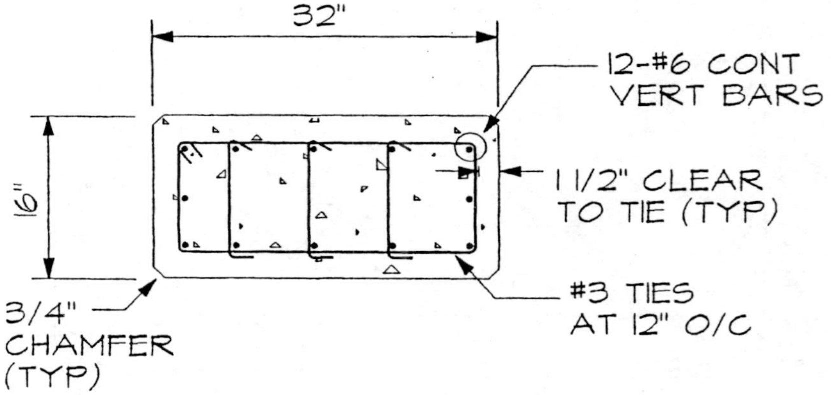

CONCRETE COLUMN REINFORCING DETAILS

(DETAIL T3-CPLAN)

NOT TO SCALE

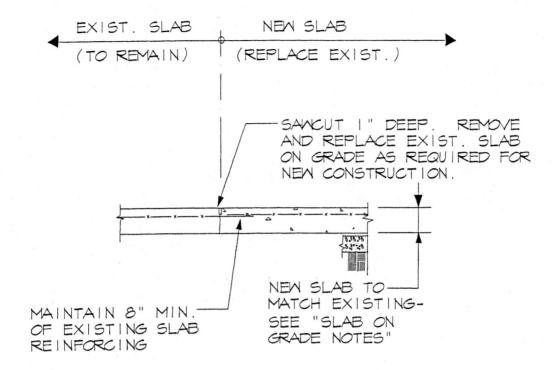

EXIST. SLAB
(TO REMAIN)

NEW SLAB
(REPLACE EXIST.)

SAWCUT 1" DEEP. REMOVE AND REPLACE EXIST. SLAB ON GRADE AS REQUIRED FOR NEW CONSTRUCTION.

MAINTAIN 8" MIN. OF EXISTING SLAB REINFORCING

NEW SLAB TO MATCH EXISTING- SEE "SLAB ON GRADE NOTES"

SLAB ON GRADE REPLACEMENT DETAIL

NOT TO SCALE

(DETAIL T3-SEXST)

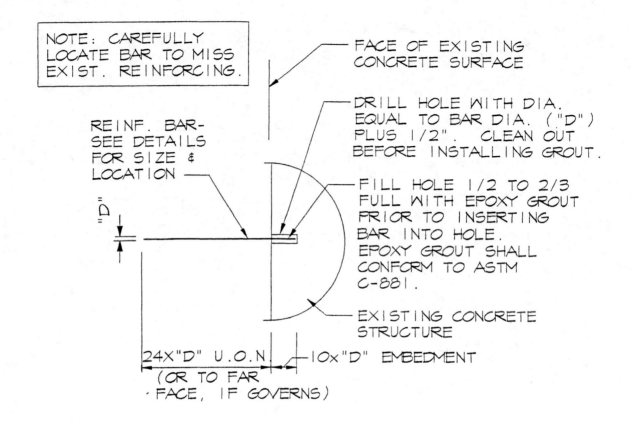

NOTE: CAREFULLY LOCATE BAR TO MISS EXIST. REINFORCING.

REINF. BAR- SEE DETAILS FOR SIZE & LOCATION

"D"

FACE OF EXISTING CONCRETE SURFACE

DRILL HOLE WITH DIA. EQUAL TO BAR DIA. ("D") PLUS 1/2". CLEAN OUT BEFORE INSTALLING GROUT.

FILL HOLE 1/2 TO 2/3 FULL WITH EPOXY GROUT PRIOR TO INSERTING BAR INTO HOLE. EPOXY GROUT SHALL CONFORM TO ASTM C-881.

EXISTING CONCRETE STRUCTURE

24X"D" U.O.N. (OR TO FAR FACE, IF GOVERNS)

10x"D" EMBEDMENT

REINF. BAR CONNECTION TO EXISTING
NOT TO SCALE   (DETAIL T3—CEXST)

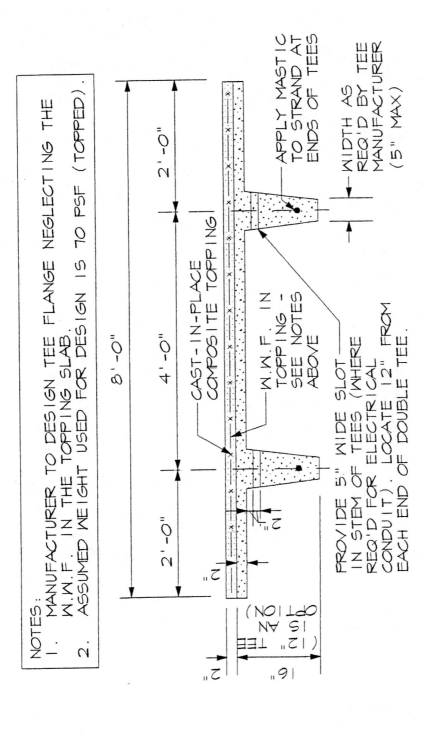

NOTES:
1. MANUFACTURER TO DESIGN TEE FLANGE NEGLECTING THE W.W.F. IN THE TOPPING SLAB.
2. ASSUMED WEIGHT USED FOR DESIGN IS 70 PSF (TOPPED).

8'-0"

2'-0"   4'-0"   2'-0"

2"

CAST-IN-PLACE COMPOSITE TOPPING

W.W.F. IN TOPPING - SEE NOTES ABOVE

APPLY MASTIC TO STRAND AT ENDS OF TEES

WIDTH AS REQ'D BY TEE MANUFACTURER (5" MAX)

PROVIDE 5" WIDE SLOT IN STEM OF TEES (WHERE REQ'D FOR ELECTRICAL CONDUIT). LOCATE 12" FROM EACH END OF DOUBLE TEE.

2"

(12" AT TEE OPTION)

16"

TYP. PRECAST DOUBLE TEE SECTION

NOT TO SCALE

(DETAIL T3-TSECT)

NOTE: SPACE CONNECTIONS 4'-0" FROM EACH END
AND AT A MAXIMUM SPACING OF 8'-0" IN BETWEEN.

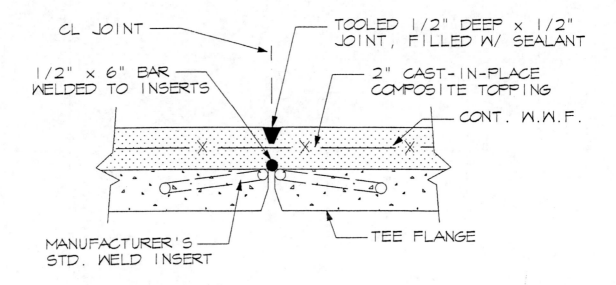

CL JOINT

TOOLED 1/2" DEEP x 1/2"
JOINT, FILLED W/ SEALANT

1/2" x 6" BAR
WELDED TO INSERTS

2" CAST-IN-PLACE
COMPOSITE TOPPING

CONT. W.W.F.

MANUFACTURER'S
STD. WELD INSERT

TEE FLANGE

DOUBLE TEE FLANGE CONNECTION
NOT TO SCALE   (DETAIL T3—TFLG)

Commentary—Concrete Details

SUPPORTED SLABS

TYPICAL ONE-WAY CONCRETE SLAB REINFORCING DETAIL (DETAIL T3-SLAB1)

One-way slabs have flexural reinforcing, which is described in a schedule for each particular slab. This detail shows an elevation view of how that reinforcing is arranged. Reinforcement perpendicular to the slab span direction is required for temperature crack control. Slabs that have special reinforcing requirements should be detailed elsewhere.

TYPICAL TWO-WAY CONCRETE SLAB REINFORCING DETAIL (DETAILS T3-2WAYM & T3-2WAYC)

Two-way slabs have flexural reinforcing in two different directions. Reinforcing is provided in a middle strip and column strip. These two details show an elevation view of how that reinforcing is arranged.

SLAB BEARING ON MASONRY DETAIL 1 (DETAIL T3-CMU1)

The two details shown are both for one-way slabs. One detail is at a perimeter masonry bearing wall condition and the other is a perimeter detail at a nonbearing wall.

SLAB BEARING ON MASONRY DETAIL 2 (DETAIL T3-CMU2)

The two details shown are both for one-way slabs. One detail is at an interior masonry bearing wall condition and the other is an interior bearing wall condition where the slab span direction changes.

SLAB BEARING ON MASONRY DETAIL 3 (DETAIL T3-CMU3)

The two details shown are both for one-way slabs. One detail is at an interior *slab beam condition* that occurs at corridors, where a beam is required to span a short distance between walls. A slab beam is used, and the beam is designed within the slab depth to avoid complications resulting from a beam turndown. The other detail shows a condition at a mechanical chase. Slab support is provided by means of steel angles.

SLAB BEARING ON MASONRY DETAIL 4 (DETAIL T3-CMU4)

The two details shown are both for one-way slabs. One detail is an interior bearing wall condition at an elevator shaft wall. The other is a roof level interior bearing wall condition where the slab span direction changes.

TYPICAL CONCRETE WALL OR SLAB OPENING DETAIL (DETAIL T3-OPNG)

This detail shows the special reinforcing required for openings in walls or slabs. The effect of openings needs to be investigated on a case-by-case basis. For two-way slabs, openings near the column supports should be avoided.

BEAMS AND COLUMNS

TYPICAL CONCRETE BEAM ELEVATION DETAIL (DETAIL T3-BMEL)

Concrete beams generally have shear and flexural reinforcing, which is described in a schedule for each particular beam. This detail shows an elevation view of how that reinforcing is arranged. Beams that have special reinforcing requirements should be detailed elsewhere.

ELEVATED SLAB CONSTRUCTION JOINT AND BEAM CONSTRUCTION JOINT DETAIL (DETAIL T3-BMJT)

Two details are shown. One is an elevated slab construction joint and the other is a beam construction joint. These types of joints are applicable where it is not practical to pour the concrete in a single monolithic pour.

TYPICAL CONCRETE BEAM SECTIONS (DETAIL T3-BMSEC)

Two details are shown. One is a typical upturned beam section and the other is a typical downturned beam section. Upturned beams are more costly to make; however, downturned beams may not be practical for other reasons, such as headroom clearance, aesthetics, and so forth.

TYPICAL PRECAST CONCRETE BEAM BEARING DETAIL (DETAIL T3-PCBM)

This detail shows an elevation view of a precast concrete beam connection to a cast-in-place concrete column. Anchors are set in the column. Sleeves are set in the precast beam. The beam is placed into position and the sleeves are filled with grout. This detail provides a minimal physical tie between the column and beam; however, it does not provide much continuity or ductility. Hence, it would not be recommended in areas with high seismic activity.

CONCRETE COLUMN DETAIL #1 (DETAIL T3-COL)

This detail shows an elevation view of a cast-in-place concrete column. The reinforcing can be shown in plan details or in a schedule.

CONCRETE COLUMN DETAIL #2 (DETAIL T3-COL2)

This detail is a combination of the previous two. It also shows the footing below the column.

CONCRETE COLUMN REINFORCING DETAILS (DETAIL T3-CPLAN)

This detail shows a simple way to describe the size and reinforcing for cast-in-place concrete columns. Two different plan details are shown: one for a 16 in^2 column and the other for a $16 \times 32''$ column.

MISCELLANEOUS DETAILS

SLAB ON GRADE REPLACEMENT DETAIL (DETAIL T3-SEXST)

Many building renovation projects involve replacing existing slabs on grade. This detail shows the procedure for slab replacement.

REINFORCING BAR CONNECTION TO EXISTING (DETAIL T3-CEXST)

This detail shows the procedure for doweling reinforcing bars to existing concrete structures. In general, an embedment of 10 times the bar diameter is sufficient to develop the bar's shear capacity.

TYPICAL PRECAST DOUBLE TEE SECTION (DETAIL T3-TSECT)

Precast double tees are designed by the precast manufacturer; however, a typical detail showing the geometry and topping slab is recommended.

DOUBLE TEE FLANGE CONNECTION
(DETAIL T3-TFLG)

This detail shows a method for connecting the flanges of precast double tees. The topping is poured after the tee flanges are connected. A tooled joint is recommended in the topping as shown.

Masonry

(CSI Division 4)

Discussion—Masonry

TYPES OF DETAILS INCLUDED

MASONRY DETAILS ARE PROVIDED IN THIS
CHAPTER FOR THE FOLLOWING:

- ► *Masonry Walls*
- ► *Lintels*
- ► *Steel Bar Joist Framing*
- ► *Miscellaneous Details*

MASONRY WALLS

Load-bearing masonry is typically associated with concrete masonry units (cmu). Accordingly, masonry design is very similar to reinforced concrete design. One difference is that cmu is primarily used for vertical elements, such as columns and walls. One major advantage is noncombustibility. Brick can also be used in load-bearing applications; however, it is primarily used nonstructurally as a facia element.

Masonry units come in 4, 6, 8, 10, and 12″ nominal widths. The standard (nominal) length of masonry units is 16″. The standard (nominal) height is 8″. All actual dimensions are ⅜″ smaller than the nominal dimensions. Mortar joints are typically ⅜″ thick.

Masonry walls can be constructed with *low-lift* or *high-lift* procedures. The low-lift procedure is simpler; however, the high-lift procedure is more expedient. Due to the specialized nature of high-lift masonry construction, it should only be approved for qualified contractors. Masonry is normally laid in running bond. Stacked bond is permissible; however, running bond is preferred structurally.

Masonry strength is related to the strength of the masonry unit in combination with the mortar. Being a concrete product, cmu is strong in compression and weak in tension. Therefore, flexural reinforcing is provided in regions where the tensile stresses exceed the allowable stresses of unreinforced masonry. A variety of design aids and computer programs are available to assist in the structural design of masonry walls. ACI provides design criteria.

Most single wythe cmu walls are designed with the reinforcing centered in the wall. One exception would be for a masonry retaining wall, where the reinforcing can be located more closely to the tension face. In general, it is not practical to provide more than a single layer of reinforcing in walls up to 12″ in width. The vertical wall reinforcing can be shown in the sections, details, or in a schedule.

One consideration frequently overlooked with masonry construction is crack control. Concrete-based products tend to shrink; hence, masonry crack control joints are required. Brick, being a clay-based material, tends to expand; hence, expansion joints are required.

LINTELS

Lintels are essentially beams. They are used to support masonry at openings. A variety of different types of lintels can be used, including reinforced masonry, precast concrete, and steel beams. The depth of masonry lintels is typically chosen such that shear reinforcing is not required. Lintels for brick façades typically consist of steel angles. Walls laid in running bond tend to support themselves over openings as long as there is a sufficient amount of masonry above the lintel.

STEEL BAR JOIST FRAMING

A combination of masonry and steel bar joist systems is a practical and economical arrangement. Masonry walls function well as vertically spanning elements. Bar joists function well as horizontally spanning elements. Masonry walls can also serve as shearwalls.

Joists must be properly anchored to their masonry wall supports. The Steel Joist Institute gives prescriptive requirements for anchorage. Roof joist uplift must be carefully addressed.

MISCELLANEOUS DETAILS

Miscellaneous details are also included for wall to girt connections, masonry columns, knockout panels, masonry wall removal, and masonry partition wall bracing.

GENERAL COMMENT

Practices vary from region to region and the details that follow have generally been taken from specific applications. Any member sizes or connections shown were intended for that particular application. Accordingly, the applicability of all details should be investigated and all member sizes and connections should be determined prior to use.

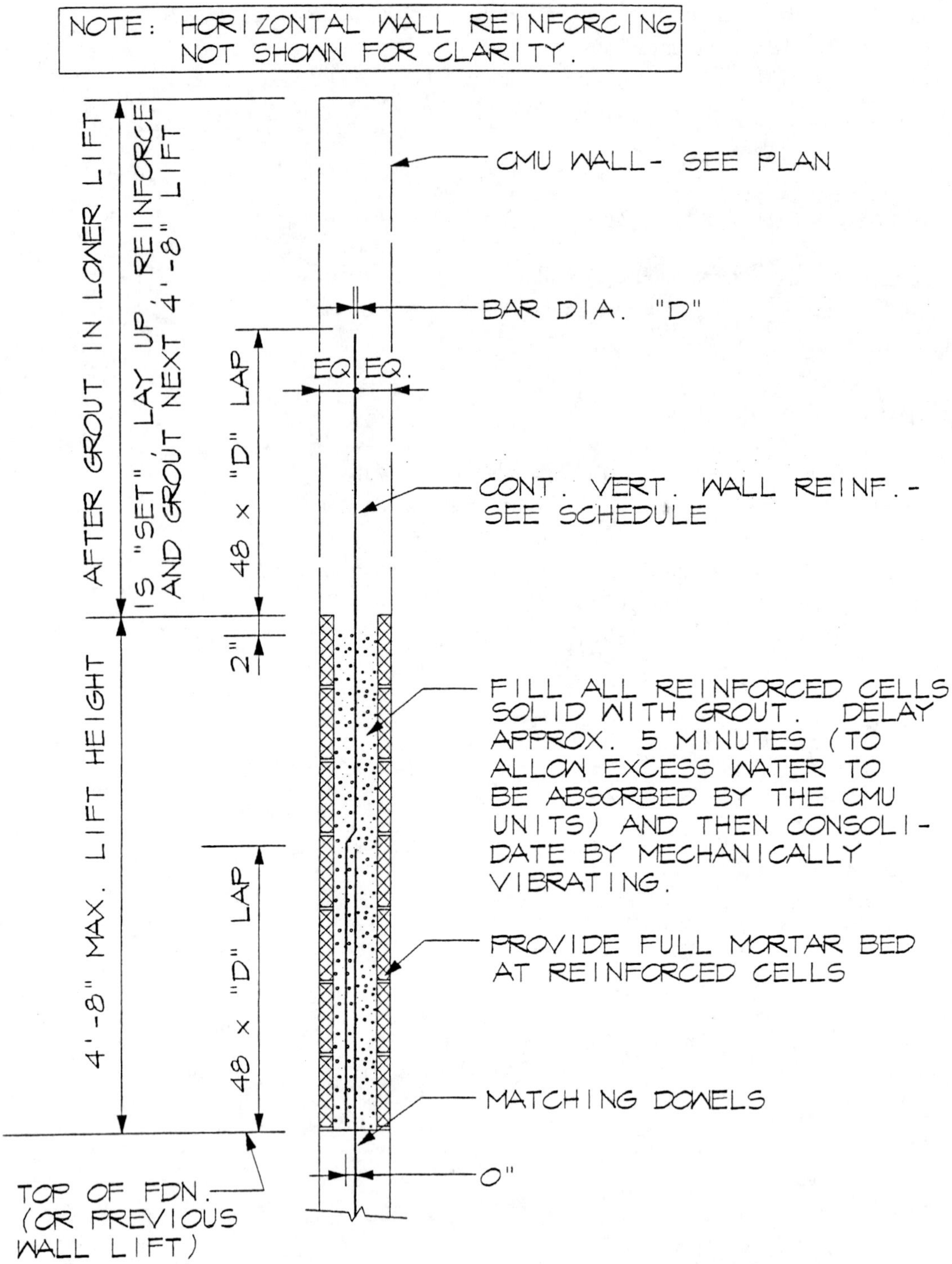

NOTE: HORIZONTAL WALL REINFORCING NOT SHOWN FOR CLARITY.

CMU WALL- SEE PLAN

BAR DIA. "D"

EQ. EQ.

CONT. VERT. WALL REINF.- SEE SCHEDULE

AFTER GROUT IN LOWER LIFT IS "SET", LAY UP, REINFORCE AND GROUT NEXT 4'-8" LIFT

48 x "D" LAP

2"

FILL ALL REINFORCED CELLS SOLID WITH GROUT. DELAY APPROX. 5 MINUTES (TO ALLOW EXCESS WATER TO BE ABSORBED BY THE CMU UNITS) AND THEN CONSOLI-DATE BY MECHANICALLY VIBRATING.

4'-8" MAX. LIFT HEIGHT

48 x "D" LAP

PROVIDE FULL MORTAR BED AT REINFORCED CELLS

MATCHING DOWELS

0"

TOP OF FDN. (OR PREVIOUS WALL LIFT)

TYP. LOW-LIFT WALL CONSTRUCTION

NOT TO SCALE

(DETAIL T4-WLOW)

NOTE: "HIGH-LIFT" MASONRY WALL CONSTRUCTION IS LIMITED TO SPECIALLY QUALIFIED CONTRACTORS. HORIZONTAL REINFORCING IS NOT SHOWN FOR CLARITY.

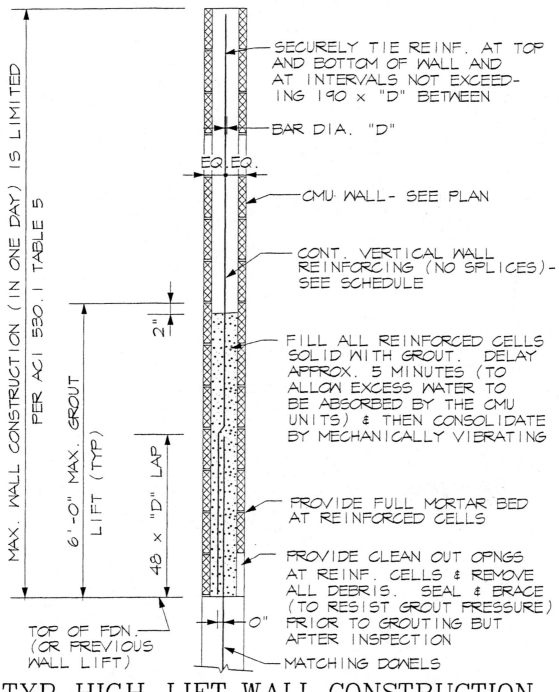

MAX. WALL CONSTRUCTION (IN ONE DAY) IS LIMITED PER ACI 530.1 TABLE 5

6'-0" MAX. GROUT LIFT (TYP)

48 x "D" LAP

2"

SECURELY TIE REINF. AT TOP AND BOTTOM OF WALL AND AT INTERVALS NOT EXCEEDING 190 x "D" BETWEEN

BAR DIA. "D"

EQ. EQ.

CMU WALL- SEE PLAN

CONT. VERTICAL WALL REINFORCING (NO SPLICES)- SEE SCHEDULE

FILL ALL REINFORCED CELLS SOLID WITH GROUT. DELAY APPROX. 5 MINUTES (TO ALLOW EXCESS WATER TO BE ABSORBED BY THE CMU UNITS) & THEN CONSOLIDATE BY MECHANICALLY VIBRATING

PROVIDE FULL MORTAR BED AT REINFORCED CELLS

PROVIDE CLEAN OUT OPNGS AT REINF. CELLS & REMOVE ALL DEBRIS. SEAL & BRACE (TO RESIST GROUT PRESSURE) PRIOR TO GROUTING BUT AFTER INSPECTION

0"

TOP OF FDN. (OR PREVIOUS WALL LIFT)

MATCHING DOWELS

TYP. HIGH-LIFT WALL CONSTRUCTION
NOT TO SCALE   (DETAIL T4-WHIGH)

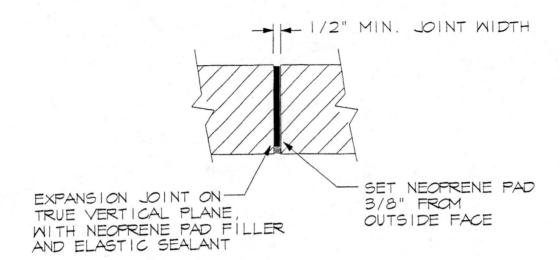

1/2" MIN. JOINT WIDTH

EXPANSION JOINT ON TRUE VERTICAL PLANE, WITH NEOPRENE PAD FILLER AND ELASTIC SEALANT

SET NEOPRENE PAD 3/8" FROM OUTSIDE FACE

TYP. BRICK EXPANSION JOINT DETAIL
NOT TO SCALE

NOTE: SEE ARCH. DRAWINGS FOR WALL CONTROL JOINT LOCATIONS.

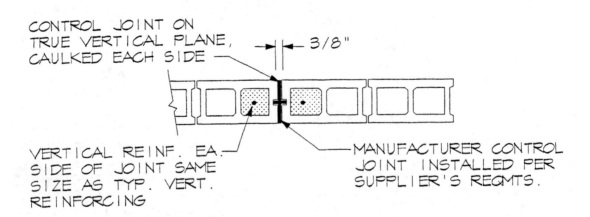

CONTROL JOINT ON TRUE VERTICAL PLANE, CAULKED EACH SIDE

3/8"

VERTICAL REINF. EA. SIDE OF JOINT SAME SIZE AS TYP. VERT. REINFORCING

MANUFACTURER CONTROL JOINT INSTALLED PER SUPPLIER'S REQMTS.

TYP. CMU CONTROL JOINT DETAIL
NOT TO SCALE

(DETAIL T4-WJT)

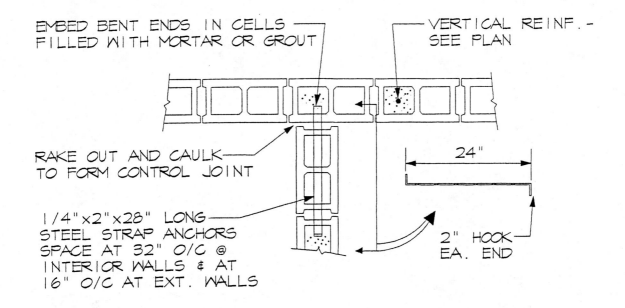

EMBED BENT ENDS IN CELLS FILLED WITH MORTAR OR GROUT

VERTICAL REINF.- SEE PLAN

RAKE OUT AND CAULK TO FORM CONTROL JOINT

1/4"x2"x28" LONG STEEL STRAP ANCHORS SPACE AT 32" O/C @ INTERIOR WALLS & AT 16" O/C AT EXT. WALLS

24"

2" HOOK EA. END

DETAIL AT WALL INTERSECTION
NOT TO SCALE   (DETAIL T4—WTIE)

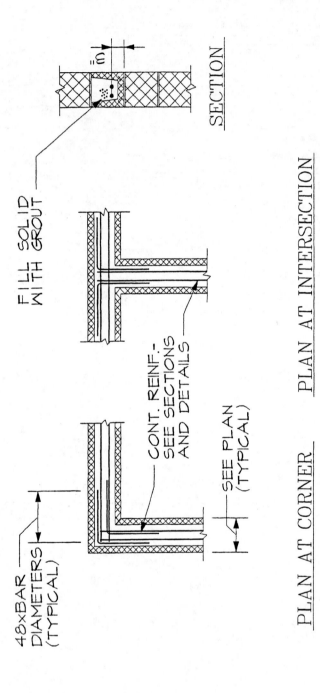

FILL SOLID WITH GROUT

48×BAR DIAMETERS (TYPICAL)

CONT. REINF.- SEE SECTIONS AND DETAILS

SEE PLAN (TYPICAL)

SECTION

PLAN AT INTERSECTION

PLAN AT CORNER

TYP. BOND BEAM REINFORCING DETAILS
(DETAIL T4−BBM1)

NOT TO SCALE

BOND BEAM LINTEL SCHEDULE		
CLR. SPAN	'D'	REINFORCING
0" TO 3'-4"	8"	2-#3 BOTT.
3'-5" TO 5'-4"	8"	2-#4 BOTT.
5'-5" TO 6'-8"	8"	2-#5 BOTT.

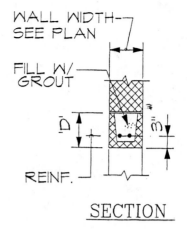

WALL WIDTH—
SEE PLAN

FILL W/
GROUT

REINF.

<u>SECTION</u>

NOTES:

1. SEE THE ARCHITECTURAL DRAWINGS FOR SIZE AND LOCATION OF ALL OPENINGS.

2. PROVIDE 8" MINIMUM BEARING EACH END.

3. SCHEDULE APPLIES ONLY TO LINTELS NOT OTHERWISE SHOWN.

4. CONTRACTOR TO PROVIDE TEMPORARY SHORING UNTIL MASONRY HAS PROPERLY SET (3 DAY MIN.)

BOND BEAM LINTEL DETAILS

NOT TO SCALE   (DETAIL T4—LNTL1)

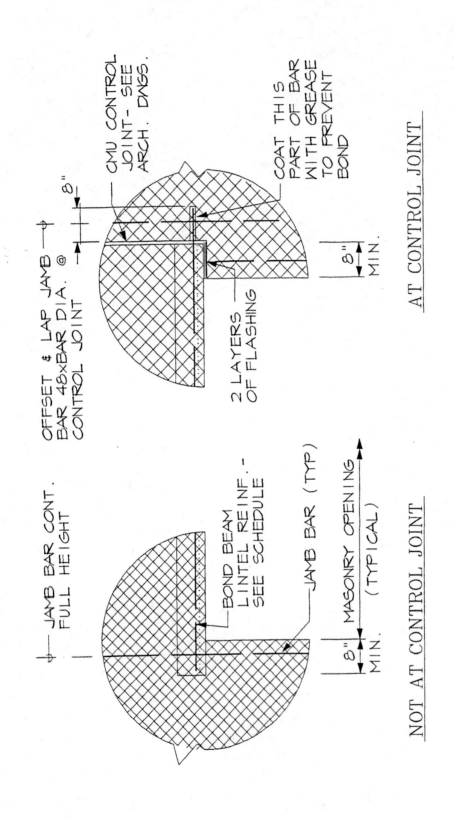

JAMB BAR CONT. FULL HEIGHT

OFFSET & LAP JAMB BAR 48×BAR DIA. @ CONTROL JOINT

CMU CONTROL JOINT- SEE ARCH. DWGS.

8"

COAT THIS PART OF BAR WITH GREASE TO PREVENT BOND

2 LAYERS OF FLASHING

8" MIN.

AT CONTROL JOINT

BOND BEAM LINTEL REINF. - SEE SCHEDULE

JAMB BAR (TYP)

MASONRY OPENING (TYPICAL)

8" MIN. (TYPICAL)

NOT AT CONTROL JOINT

TYP. BOND BEAM LINTEL BEARING DETAILS

(DETAIL T4—LNTL2)

NOT TO SCALE

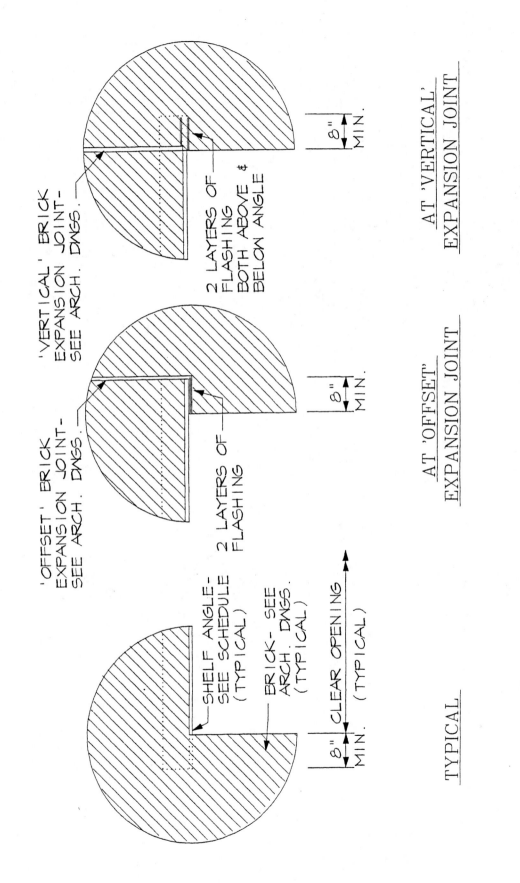

'VERTICAL' BRICK
EXPANSION JOINT-
SEE ARCH. DWGS.

2 LAYERS OF
FLASHING
BOTH ABOVE &
BELOW ANGLE

8" MIN.

'OFFSET' BRICK
EXPANSION JOINT-
SEE ARCH. DWGS.

2 LAYERS OF
FLASHING

8" MIN.

SHELF ANGLE-
SEE SCHEDULE
(TYPICAL)

BRICK- SEE
ARCH. DWGS.
(TYPICAL)

CLEAR OPENING
(TYPICAL)

8" MIN. (TYPICAL)

AT 'VERTICAL'
EXPANSION JOINT

AT 'OFFSET'
EXPANSION JOINT

TYPICAL

TYP. BRICK SHELF ANGLE BEARING DETAILS
(DETAIL T4-SHELF)

NOT TO SCALE

185

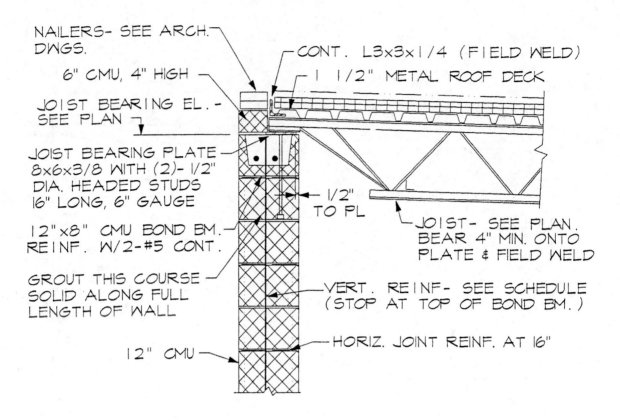

NAILERS- SEE ARCH. DWGS.

6" CMU, 4" HIGH

JOIST BEARING EL.- SEE PLAN

JOIST BEARING PLATE 8x6x3/8 WITH (2)- 1/2" DIA. HEADED STUDS 16" LONG, 6" GAUGE

12"x8" CMU BOND BM. REINF. W/2-#5 CONT.

GROUT THIS COURSE SOLID ALONG FULL LENGTH OF WALL

12" CMU

CONT. L3x3x1/4 (FIELD WELD)

1 1/2" METAL ROOF DECK

1/2" TO PL

JOIST- SEE PLAN. BEAR 4" MIN. ONTO PLATE & FIELD WELD

VERT. REINF- SEE SCHEDULE (STOP AT TOP OF BOND BM.)

HORIZ. JOINT REINF. AT 16"

STEEL BAR JOIST FRAMING DETAIL #1

NOT TO SCALE   (DETAIL T4—JOIS1)

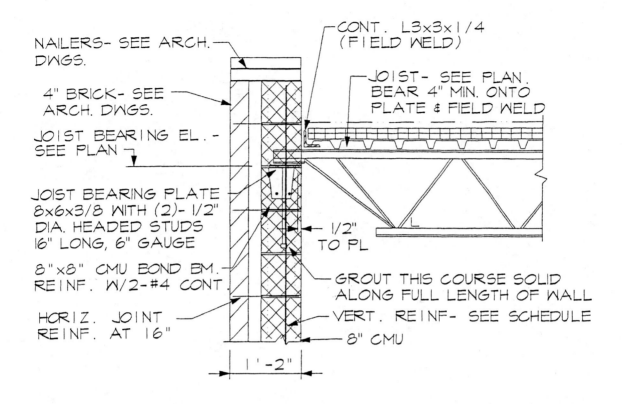

NAILERS- SEE ARCH. DWGS.

4" BRICK- SEE ARCH. DWGS.

JOIST BEARING EL.- SEE PLAN

JOIST BEARING PLATE 8x6x3/8 WITH (2)- 1/2" DIA. HEADED STUDS 16" LONG, 6" GAUGE

8"x8" CMU BOND BM. REINF. W/2-#4 CONT.

HORIZ. JOINT REINF. AT 16"

CONT. L3x3x1/4 (FIELD WELD)

JOIST- SEE PLAN. BEAR 4" MIN. ONTO PLATE & FIELD WELD

1/2" TO PL

GROUT THIS COURSE SOLID ALONG FULL LENGTH OF WALL

VERT. REINF- SEE SCHEDULE

8" CMU

1'-2"

STEEL BAR JOIST FRAMING DETAIL #2
NOT TO SCALE   (DETAIL T4-JOIS2)

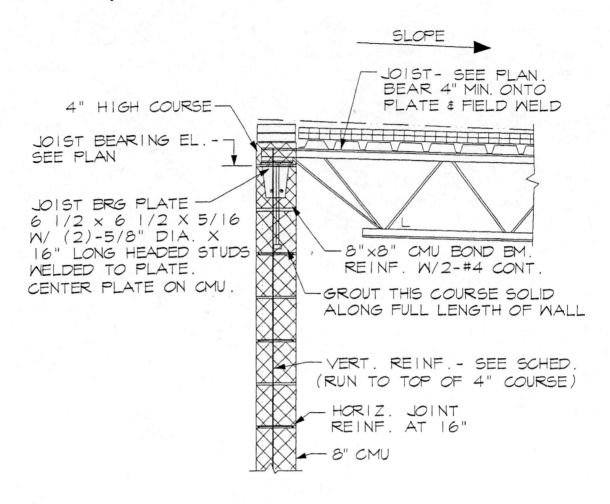

SLOPE

JOIST- SEE PLAN.
BEAR 4" MIN. ONTO
PLATE & FIELD WELD

4" HIGH COURSE

JOIST BEARING EL. -
SEE PLAN

JOIST BRG PLATE
6 1/2 X 6 1/2 X 5/16
W/ (2)-5/8" DIA. X
16" LONG HEADED STUDS
WELDED TO PLATE.
CENTER PLATE ON CMU.

8"x8" CMU BOND BM.
REINF. W/2-#4 CONT.

GROUT THIS COURSE SOLID
ALONG FULL LENGTH OF WALL

VERT. REINF. - SEE SCHED.
(RUN TO TOP OF 4" COURSE)

HORIZ. JOINT
REINF. AT 16"

8" CMU

STEEL BAR JOIST FRAMING DETAIL #3

NOT TO SCALE   (DETAIL T4-JOIS3)

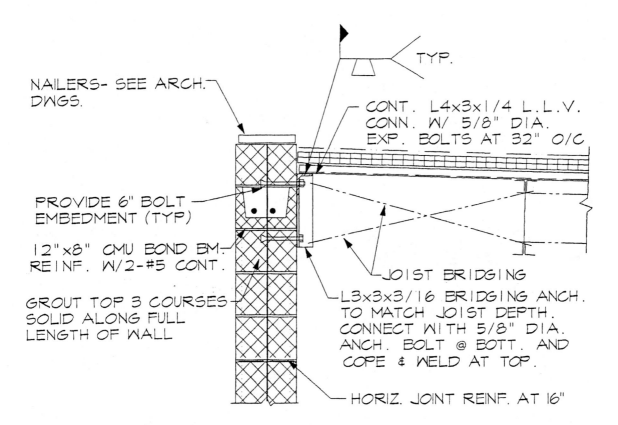

NAILERS- SEE ARCH. DWGS.

TYP.

CONT. L4x3x1/4 L.L.V. CONN. W/ 5/8" DIA. EXP. BOLTS AT 32" O/C

PROVIDE 6" BOLT EMBEDMENT (TYP)

12"x8" CMU BOND BM. REINF. W/2-#5 CONT.

GROUT TOP 3 COURSES SOLID ALONG FULL LENGTH OF WALL

JOIST BRIDGING

L3x3x3/16 BRIDGING ANCH. TO MATCH JOIST DEPTH. CONNECT WITH 5/8" DIA. ANCH. BOLT @ BOTT. AND COPE & WELD AT TOP.

HORIZ. JOINT REINF. AT 16"

NOTE: EXP. BOLTS SHALL BE 'DYNABOLTS' PER RAMSET/RED HEAD OR APPROVED EQUAL.

STEEL BAR JOIST FRAMING DETAIL #4
NOT TO SCALE   (DETAIL T4-JOIS4)

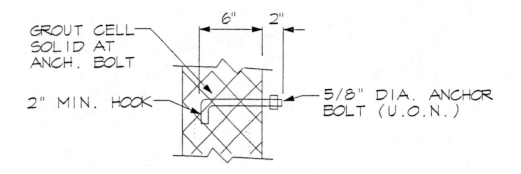

TYPICAL CMU ANCHOR BOLT DETAIL
NOT TO SCALE

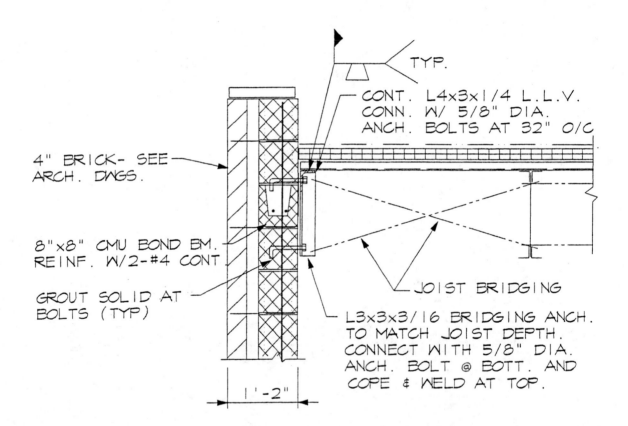

STEEL BAR JOIST FRAMING DETAIL #5
NOT TO SCALE
(DETAIL T4–JOIS5)

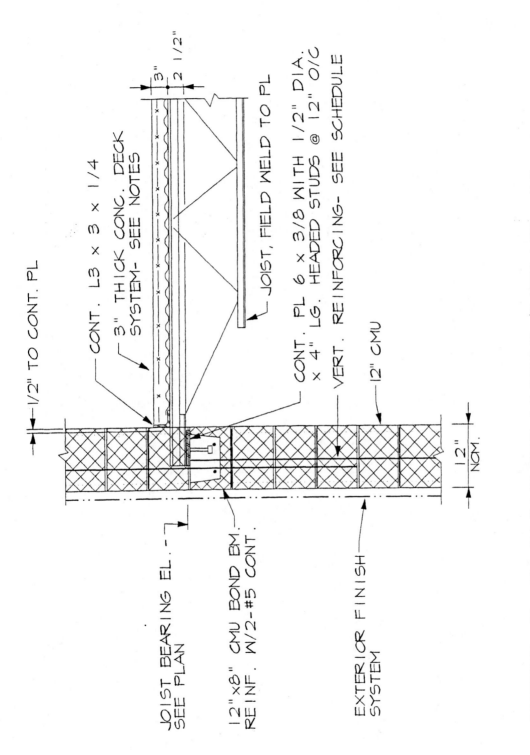

1/2" TO CONT. PL

CONT. L3 × 3 × 1/4

3" THICK CONC. DECK SYSTEM- SEE NOTES

3"

2 1/2"

JOIST, FIELD WELD TO PL

CONT. PL 6 × 3/8 WITH 1/2" DIA. × 4" LG. HEADED STUDS @ 12" O/C

VERT. REINFORCING- SEE SCHEDULE

12" CMU

JOIST BEARING EL. - SEE PLAN

12"×8" CMU BOND BM. REINF. W/2-#5 CONT.

EXTERIOR FINISH SYSTEM

12" NOM.

STEEL BAR JOIST FRAMING DETAIL #6
(DETAIL T4-JOIS6)

NOT TO SCALE

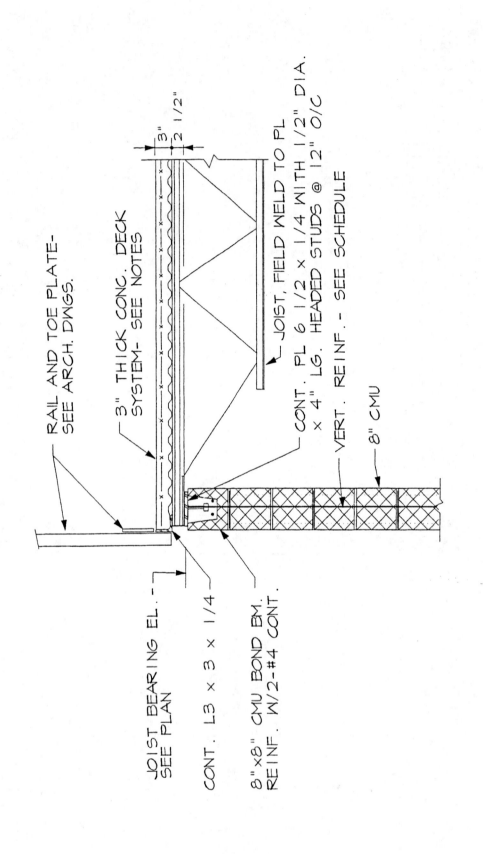

RAIL AND TOE PLATE-
SEE ARCH. DWGS.

3" THICK CONC. DECK
SYSTEM- SEE NOTES

3"

2 1/2"

JOIST, FIELD WELD TO PL

CONT. PL 6 1/2 × 1/4 WITH 1/2" DIA.
× 4" LG. HEADED STUDS @ 12" O/C

VERT. REINF.- SEE SCHEDULE

8" CMU

JOIST BEARING EL.-
SEE PLAN

CONT. L3 × 3 × 1/4

8"×8" CMU BOND BM.
REINF. W/2-#4 CONT.

STEEL BAR JOIST FRAMING DETAIL #7
(DETAIL T4-JOIS7)

NOT TO SCALE

192

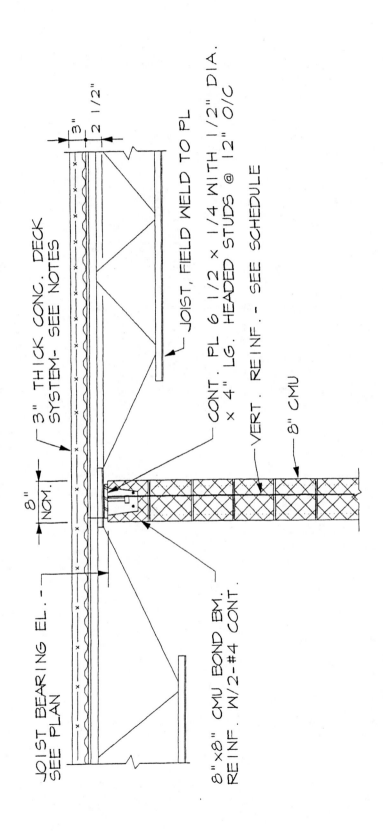

JOIST BEARING EL. -
SEE PLAN

3" THICK CONC. DECK
SYSTEM- SEE NOTES

3"

2 1/2"

8"
NOM.

JOIST, FIELD WELD TO PL

CONT. PL 6 1/2 × 1/4 WITH 1/2" DIA.
× 4" LG. HEADED STUDS @ 12" O/C

VERT. REINF. - SEE SCHEDULE

8" CMU

8"×8" CMU BOND BM.
REINF. W/ 2-#4 CONT.

STEEL BAR JOIST FRAMING DETAIL #8
(DETAIL T4-JOIS8)

NOT TO SCALE

193

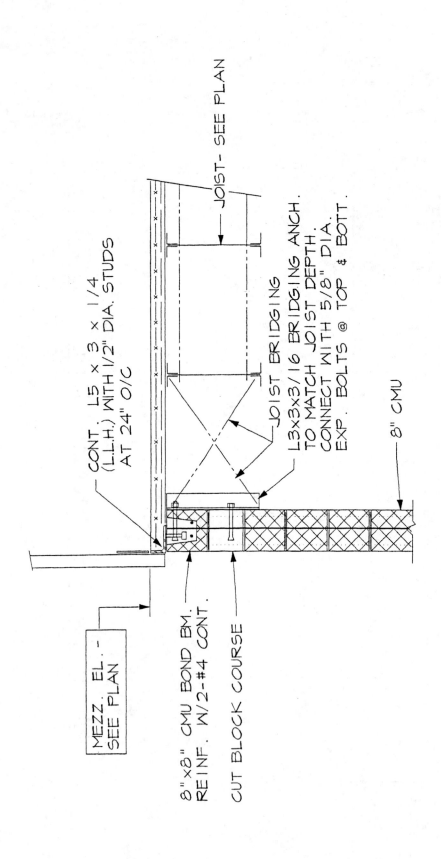

MEZZ. EL. -
SEE PLAN

CONT. L5 × 3 × 1/4
(L.L.H.) WITH 1/2" DIA. STUDS
AT 24" O/C

8"×8" CMU BOND BM.
REINF. W/2-#4 CONT.

CUT BLOCK COURSE

JOIST- SEE PLAN

JOIST BRIDGING

L3×3×3/16 BRIDGING ANCH.
TO MATCH JOIST DEPTH.
CONNECT WITH 5/8" DIA.
EXP. BOLTS @ TOP & BOTT.

8" CMU

STEEL BAR JOIST FRAMING DETAIL #9

(DETAIL T4—JOIS9)

NOT TO SCALE

194

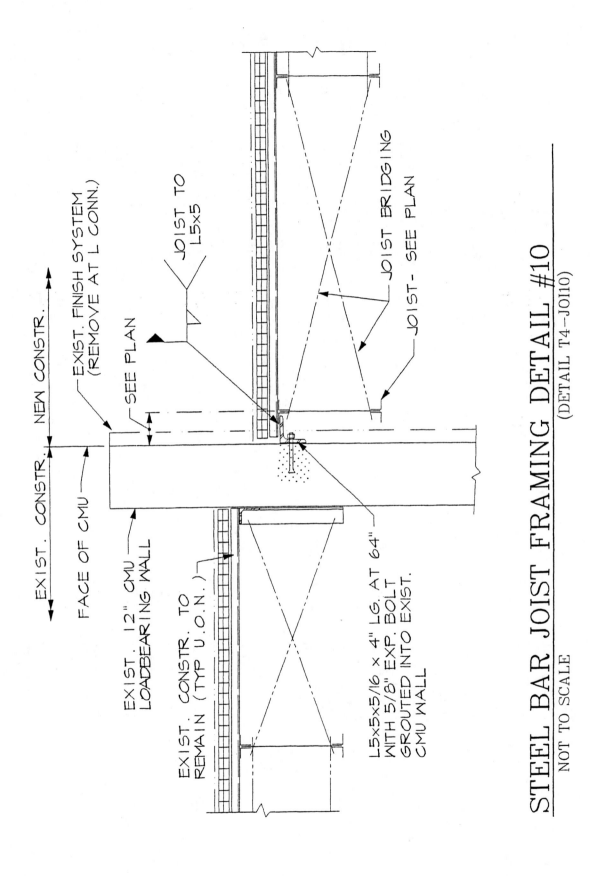

EXIST. CONSTR. | NEW CONSTR.

EXIST. FINISH SYSTEM
(REMOVE AT L CONN.)

FACE OF CMU

SEE PLAN

JOIST TO
L5x5

JOIST BRIDGING

JOIST- SEE PLAN

EXIST. 12" CMU
LOADBEARING WALL

EXIST. CONSTR. TO
REMAIN (TYP U.O.N.)

L5x5x5/16 x 4" LG. AT 64"
WITH 5/8" EXP. BOLT
GROUTED INTO EXIST.
CMU WALL

STEEL BAR JOIST FRAMING DETAIL #10
(DETAIL T4−JOI10)

NOT TO SCALE

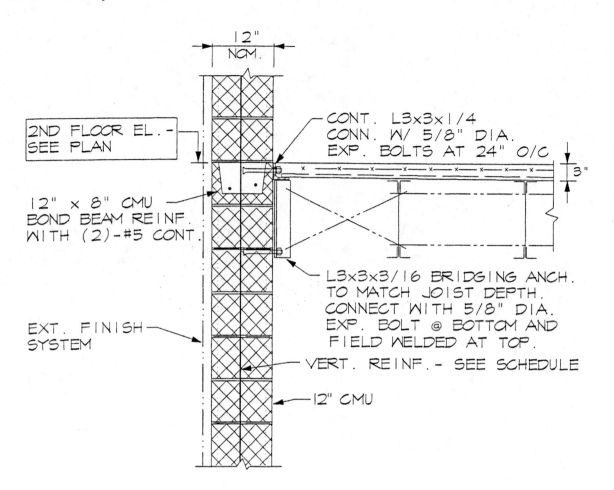

12"
NOM.

2ND FLOOR EL.-
SEE PLAN

CONT. L3x3x1/4
CONN. W/ 5/8" DIA.
EXP. BOLTS AT 24" O/C

3"

12" x 8" CMU
BOND BEAM REINF.
WITH (2)-#5 CONT.

L3x3x3/16 BRIDGING ANCH.
TO MATCH JOIST DEPTH.
CONNECT WITH 5/8" DIA.
EXP. BOLT @ BOTTOM AND
FIELD WELDED AT TOP.

EXT. FINISH
SYSTEM

VERT. REINF. - SEE SCHEDULE

12" CMU

STEEL BAR JOIST FRAMING DETAIL #11
NOT TO SCALE   (DETAIL T4-JOI11)

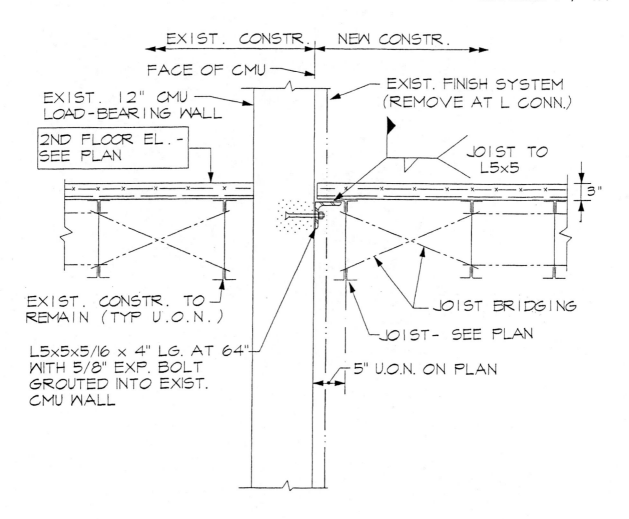

EXIST. CONSTR. | NEW CONSTR.

FACE OF CMU

EXIST. 12" CMU LOAD-BEARING WALL

EXIST. FINISH SYSTEM (REMOVE AT L CONN.)

2ND FLOOR EL.- SEE PLAN

JOIST TO L5x5

3"

EXIST. CONSTR. TO REMAIN (TYP U.O.N.)

JOIST BRIDGING

JOIST- SEE PLAN

L5x5x5/16 x 4" LG. AT 64" WITH 5/8" EXP. BOLT GROUTED INTO EXIST. CMU WALL

5" U.O.N. ON PLAN

STEEL BAR JOIST FRAMING DETAIL #12

NOT TO SCALE   (DETAIL T4-JOI12)

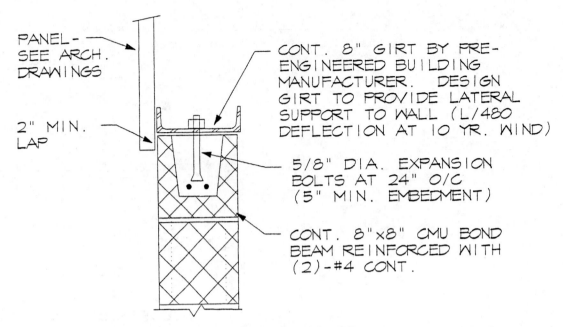

PANEL-
SEE ARCH.
DRAWINGS

2" MIN.
LAP

CONT. 8" GIRT BY PRE-
ENGINEERED BUILDING
MANUFACTURER. DESIGN
GIRT TO PROVIDE LATERAL
SUPPORT TO WALL (L/480
DEFLECTION AT 10 YR. WIND)

5/8" DIA. EXPANSION
BOLTS AT 24" O/C
(5" MIN. EMBEDMENT)

CONT. 8"x8" CMU BOND
BEAM REINFORCED WITH
(2)-#4 CONT.

<u>SINGLE WYTHE CMU WALL</u>

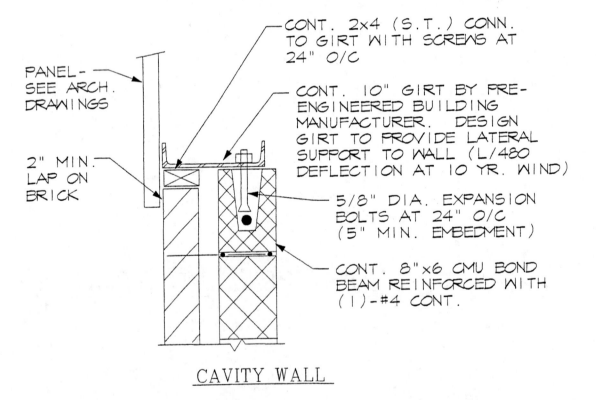

PANEL-
SEE ARCH.
DRAWINGS

2" MIN.
LAP ON
BRICK

CONT. 2x4 (S.T.) CONN.
TO GIRT WITH SCREWS AT
24" O/C

CONT. 10" GIRT BY PRE-
ENGINEERED BUILDING
MANUFACTURER. DESIGN
GIRT TO PROVIDE LATERAL
SUPPORT TO WALL (L/480
DEFLECTION AT 10 YR. WIND)

5/8" DIA. EXPANSION
BOLTS AT 24" O/C
(5" MIN. EMBEDMENT)

CONT. 8"x6 CMU BOND
BEAM REINFORCED WITH
(1)-#4 CONT.

<u>CAVITY WALL</u>

<u>TYP. WALL TO GIRT CONNECTION DETAILS</u>

NOT TO SCALE   (DETAIL T4-WGIRT)

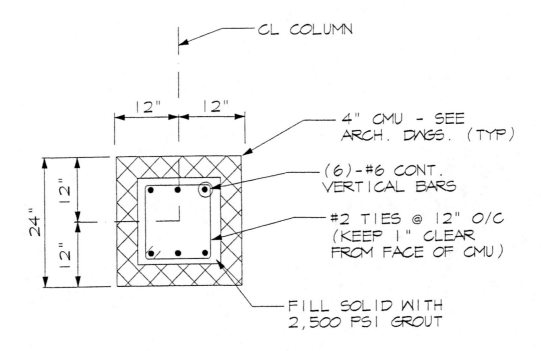

CL COLUMN

12"   12"

4" CMU - SEE
ARCH. DWGS. (TYP)

(6)-#6 CONT.
VERTICAL BARS

#2 TIES @ 12" O/C
(KEEP 1" CLEAR
FROM FACE OF CMU)

24"   12"   12"

FILL SOLID WITH
2,500 PSI GROUT

TYPICAL MASONRY COLUMN DETAIL

NOT TO SCALE   (DETAIL T4-CCOL)

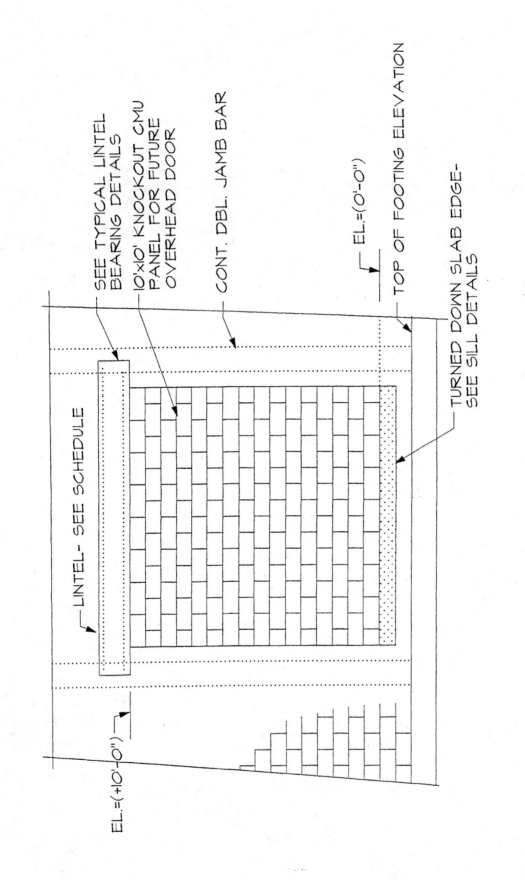

SEE TYPICAL LINTEL BEARING DETAILS

10'x10' KNOCKOUT CMU PANEL FOR FUTURE OVERHEAD DOOR

CONT. DBL. JAMB BAR

EL.=(0'-0")

TOP OF FOOTING ELEVATION

TURNED DOWN SLAB EDGE-SEE SILL DETAILS

LINTEL- SEE SCHEDULE

EL.=(+10'-0")

TYP. KNOCKOUT PANEL ELEVATION DETAIL
(DETAIL T4—WKNOC)

NOT TO SCALE

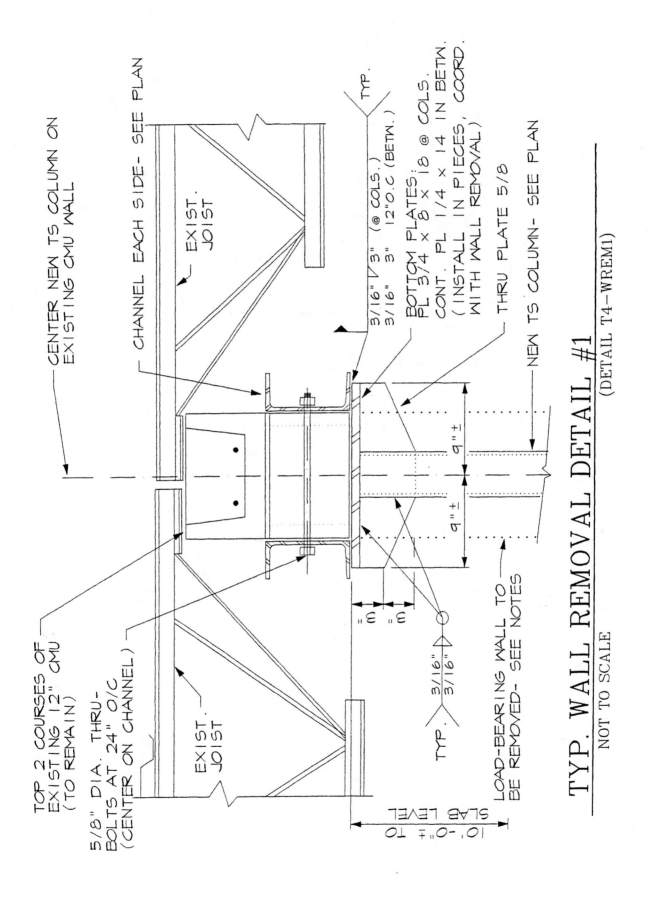

CENTER NEW TS COLUMN ON EXISTING CMU WALL

CHANNEL EACH SIDE- SEE PLAN

EXIST. JOIST

TOP 2 COURSES OF EXISTING 12" CMU (TO REMAIN)

5/8" DIA. THRU-BOLTS AT 24" O/C (CENTER ON CHANNEL)

EXIST. JOIST

TYP.

3/16" $\vee$ 3" (@ COLS.)
3/16" 3" 12" O.C (BETW.)

BOTTOM PLATES:
PL 3/4 × 8 × 18 @ COLS.
CONT. PL 1/4 × 14 IN BETW.
(INSTALL IN PIECES, COORD.
WITH WALL REMOVAL)

THRU PLATE 5/8

NEW TS COLUMN- SEE PLAN

9" ±

9" ±

3"

3"

TYP. 3/16"
      3/16"

LOAD-BEARING WALL TO BE REMOVED- SEE NOTES

SLAB LEVEL
10'-0" ± TO

TYP. WALL REMOVAL DETAIL #1

NOT TO SCALE

(DETAIL T4-WREM1)

201

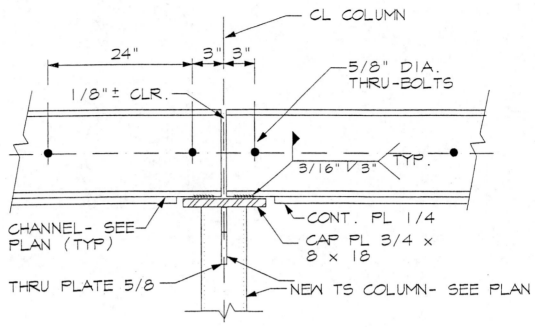

CL COLUMN

24"   3"   3"

1/8"± CLR.

5/8" DIA.
THRU-BOLTS

3/16" ⌵ 3"   TYP.

CHANNEL- SEE
PLAN (TYP)

CONT. PL 1/4

CAP PL 3/4 ×
8 × 18

THRU PLATE 5/8

NEW TS COLUMN- SEE PLAN

CHANNEL BEARING AT NEW TS COLUMNS

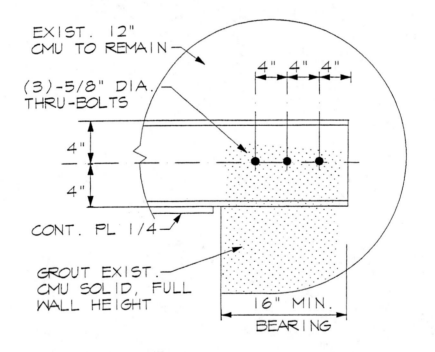

EXIST. 12"
CMU TO REMAIN

(3)-5/8" DIA.
THRU-BOLTS

4"   4"   4"

4"

4"

CONT. PL 1/4

GROUT EXIST.
CMU SOLID, FULL
WALL HEIGHT

16" MIN.
BEARING

CHANNEL BEARING AT EXISTING CMU WALL

TYP. WALL REMOVAL DETAIL #2

NOT TO SCALE   (DETAIL T4-WREM2)

NOTE: PROVIDE LATERAL BRACING FOR ALL INTERIOR PARTITION WALLS WHERE LENGTH BETW. BONDED CROSS WALLS IS MORE THAN 10' FOR 4" WALLS, 12' FOR 6" WALLS, & 14' FOR 8" WALLS.

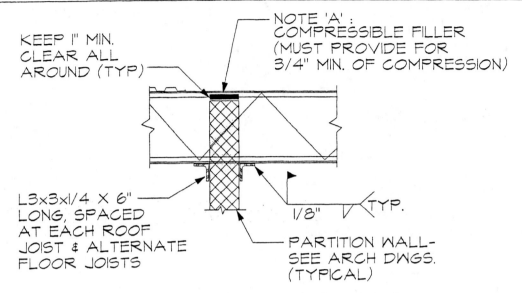

KEEP 1" MIN. CLEAR ALL AROUND (TYP)

NOTE 'A':
COMPRESSIBLE FILLER
(MUST PROVIDE FOR
3/4" MIN. OF COMPRESSION)

L3x3x1/4 X 6" LONG, SPACED AT EACH ROOF JOIST & ALTERNATE FLOOR JOISTS

1/8" TYP.

PARTITION WALL-
SEE ARCH DWGS.
(TYPICAL)

WALL PERPENDICULAR TO JOISTS

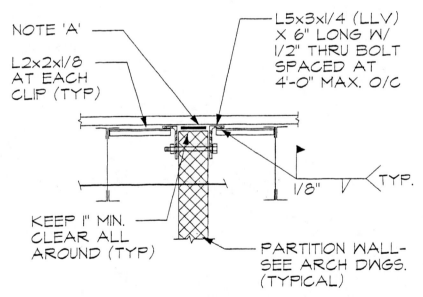

NOTE 'A'

L2x2x1/8 AT EACH CLIP (TYP)

L5x3x1/4 (LLV) X 6" LONG W/ 1/2" THRU BOLT SPACED AT 4'-0" MAX. O/C

1/8" TYP.

KEEP 1" MIN. CLEAR ALL AROUND (TYP)

PARTITION WALL-
SEE ARCH DWGS.
(TYPICAL)

WALL PARALLEL TO JOISTS

TYP. PARTITION BRACING DETAIL #1
NOT TO SCALE   (DETAIL T4-PWB1)

NOTE: PROVIDE LATERAL BRACING FOR ALL INTERIOR PARTITION WALLS WHERE LENGTH BETW. BONDED CROSS WALLS IS MORE THAN 10' FOR 4" WALLS, 12' FOR 6" WALLS, & 14' FOR 8" WALLS.

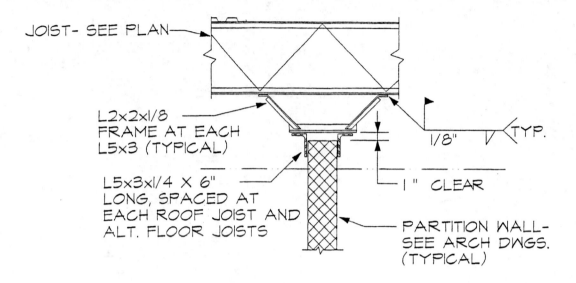

JOIST- SEE PLAN

L2x2x1/8 FRAME AT EACH L5x3 (TYPICAL)

L5x3x1/4 X 6" LONG, SPACED AT EACH ROOF JOIST AND ALT. FLOOR JOISTS

1/8" TYP.

1" CLEAR

PARTITION WALL- SEE ARCH DWGS. (TYPICAL)

WALL PERPENDICULAR TO JOISTS

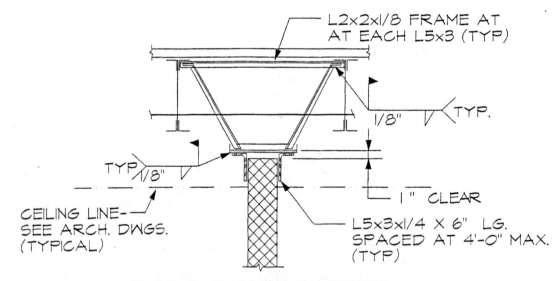

L2x2x1/8 FRAME AT AT EACH L5x3 (TYP)

1/8" TYP.

TYP 1/8"

CEILING LINE- SEE ARCH. DWGS. (TYPICAL)

1" CLEAR

L5x3x1/4 X 6" LG. SPACED AT 4'-0" MAX. (TYP)

WALL PARALLEL TO JOISTS

TYP. PARTITION BRACING DETAIL #2

NOT TO SCALE   (DETAIL T4-PWB2)

Commentary—Masonry Details

MASONRY WALLS

TYPICAL LOW-LIFT WALL CONSTRUCTION (DETAIL T4-WLOW)

This detail shows the typical low-lift masonry procedure. Low-lift construction involves constructing the masonry in relatively short vertical segments, allowing the lower segment to "set" prior to constructing the next segment. The reinforcing is centered in the wall.

TYPICAL HIGH-LIFT WALL CONSTRUCTION (DETAIL T4-WHIGH)

This detail shows the typical high-lift masonry procedure. High-lift construction involves constructing the masonry in much larger vertical segments. One disadvantage is that the procedure requires strict quality control. Clean out openings are required at the base to inspect the reinforcing. Again, the reinforcing is centered in the wall.

TYPICAL CMU CONTROL JOINT DETAIL & TYPICAL BRICK EXPANSION JOINT DETAIL (DETAIL T4-WJT)

Two details are shown. One shows a cmu control joint and the other a brick expansion joint. The locations of these joints are typically shown on the wall elevations on the architectural drawings.

DETAIL AT WALL INTERSECTION (DETAIL T4-WTIE)

The intersections of load-bearing masonry walls can be tied together in several ways. One way is to interlock and bond the intersecting wall directly. Another way is to provide a mechanical tie as shown in the detail.

TYPICAL BOND BEAM REINFORCING DETAILS (DETAIL T4-BBM1)

Bond beams are typically made with U-shaped masonry units. They contain reinforcing bars and are filled with grout. Bond beams are typically located directly below the bearing points of floor and roof framing members. They safely distribute the floor and roof loads to the masonry wall. The reinforcing is run continuous. This detail shows how to run the continuous reinforcing at corners and intersections. It also includes a section view.

LINTELS

BOND BEAM LINTEL DETAILS (DETAIL T4-LNTL1)

This detail shows a typical bond beam lintel section and schedule. The lintel schedule indicated is intended for nonbearing wall applications and would apply only for lintels not otherwise shown. Lintels in load-bearing walls should be shown in a separate lintel schedule.

TYPICAL BOND BEAM LINTEL BEARING DETAILS (DETAIL T4-LNTL2)

This detail shows two different bond beam lintel bearing conditions. One detail shows the bearing condition at a control joint location. The other detail shows the more typical condition.

TYPICAL BRICK SHELF ANGLE BEARING DETAILS (DETAIL T4-SHELF)

This detail shows three different brick shelf angle lintel bearing conditions. One condition is shown at the bearing condition at an offset brick expansion joint. Another condition shown is the bearing condition at a vertical brick expansion joint. The third condition shown is the more typical bearing.

STEEL BAR JOIST FRAMING

STEEL BAR JOIST FRAMING DETAIL #1 (DETAIL T4-JOIS1)

This detail shows a typical perimeter roof joist bearing condition at a 12″ masonry wall. The roof joist uplift and anchorage should be carefully investigated for the actual loading condition. A bearing plate is provided at each roof joist. The Steel Joist Institute requires that the face of the plate be no more than ½″ from the face of the wall. It is recommended that the joist bearing plate anchors be extended a minimum of 16″ into the masonry wall. In this detail, the uplift is transferred to the top 16″ of the masonry wall through the headed studs. This force is resisted by the vertical wall reinforcing. The bond beam transfers the force from the headed studs by spanning between the vertical reinforcing bars. It is recommended that the entire course below the bond beam be grouted solid. This increases the effective size of the bond beam under uplift loading and also provides more wall mass to resist wind uplift.

STEEL BAR JOIST FRAMING DETAIL #2
(DETAIL T4-JOIS2)

This detail is similar to the previous one, but it shows a brick cavity wall. The parapet shown is not a structural requirement; however, it does provide more mass to resist the wind uplift.

STEEL BAR JOIST FRAMING DETAIL #3
(DETAIL T4-JOIS3)

This detail is similar to the previous two, but it shows an 8″ perimeter masonry wall.

STEEL BAR JOIST FRAMING DETAIL #4
(DETAIL T4-JOIS4)

Masonry walls function well as shearwalls; however, the diaphragm forces must be properly transmitted to the walls. The wall top must also be properly braced. Although it would be easier to place a joist immediately adjacent to the wall, this would neither provide diaphragm transfer nor brace the top of the wall. In this detail, the continuous angle transfers the force from the metal deck diaphragm through the expansion bolts. The wall top is now braced. Expansion bolts are preferable to anchor bolts from a constructibility viewpoint. The top three courses are noted as being grouted solid to facilitate the expansion bolting, which is much more effective in solid masonry.

STEEL BAR JOIST FRAMING DETAIL #5
(DETAIL T4-JOIS5)

Two details are shown. One details a typical cmu anchor bolt. The disadvantage of anchor bolts compared to expansion bolts is that they are pre-set while the wall is being constructed. In many cases, presetting the bolts is not practical. The other detail is similar to the previous one, but it shows a perimeter cavity wall where the perimeter angle is connected with anchor bolts in lieu of expansion bolts. The continuous angle is turned down, which is typically advantageous where there is not a parapet wall. In this case, the angle could also be turned up, which would eliminate coping the bridging angles.

STEEL BAR JOIST FRAMING DETAIL #6
(DETAIL T4-JOIS6)

This detail shows a floor joist bearing condition at a perimeter wall. Since the floor joists are spaced fairly closely, a continuous bearing plate is provided. The vertical reinforcing is offset at the plate as shown. Floor joists do not have wind uplift; therefore, the plate anchors need minimal embedment.

STEEL BAR JOIST FRAMING DETAIL #7
(DETAIL T4-JOIS7)

This detail is similar to the previous one, but an interior wall condition is shown. The wall stops below the floor joists and a rail is provided.

STEEL BAR JOIST FRAMING DETAIL #8
(DETAIL T4-JOIS8)

This detail is similar to the previous one, but joists are bearing on the interior wall from both sides.

STEEL BAR JOIST FRAMING DETAIL #9
(DETAIL T4-JOIS9)

This detail is similar to detail T4-JOIS7, but the joists run parallel to the wall. The joist bearing beyond is at a masonry coursing elevation; therefore, running the wall to the top of joist elevation (an additional 2½") requires a cut block course below the bond beam course. Another option would be to provide a 10½" deep bond beam course. This could be accomplished by cutting a 16" deep U-shaped unit.

STEEL BAR JOIST FRAMING DETAIL #10
(DETAIL T4-JOI10)

This detail is similar to the previous one, but the condition at an existing masonry wall is shown. A joist is placed next to the existing wall so that minimal vertical load is transferred to the existing wall. The clip angle is provided because the existing wall is being used as a shearwall. The delineation between new and existing construction is clearly shown.

STEEL BAR JOIST FRAMING DETAIL #11
(DETAIL T4-JOI11)

This detail is similar to the previous one, but a new masonry wall condition is shown. The continuous angle is turned up.

STEEL BAR JOIST FRAMING DETAIL #12
(DETAIL T4-JOI12)

This detail is similar to the previous two, but an existing masonry wall and floor joist condition are shown.

Miscellaneous Details

TYPICAL WALL TO GIRT CONNECTION DETAILS (DETAIL T4-WGIRT)

Industrial-related buildings typically have masonry skirt walls, which run 8 to 10′ above the floor slab level. At that point, the wall system changes to metal siding or metal panels. These two details show wall connections between the metal building girt and two different masonry wall systems.

TYPICAL MASONRY COLUMN DETAIL (DETAIL T4-CCOL)

Reinforced masonry can be successfully used as columns. This detail shows a plan detail of a 24 in^2 column.

TYPICAL KNOCKOUT PANEL ELEVATION DETAIL (DETAIL T4-WKNOC)

Many shell buildings are constructed without knowing if overhead doors will be required by the future tenant. This detail shows an elevation of a knock-out panel so that an overhead door can be added without causing major modifications to the wall system.

TYPICAL WALL REMOVAL DETAIL #1 (DETAIL T4-WREM1)

Renovation projects frequently involve the removal of load-bearing masonry walls. This detail shows one way to remove a load-bearing masonry wall and transfer that load to a new steel framing system, which consists of steel channels on each side of the wall. The existing load is transferred to these channels by the thru-bolts and continuous bottom plate.

TYPICAL WALL REMOVAL DETAIL #2 (DETAIL T4-WREM2)

This detail is related to the previous one and shows the side views of how the new steel framing system is connected.

TYPICAL PARTITION BRACING DETAIL #1 (DETAIL T4-PWB1)

The tops of interior non-load-bearing partition walls need to be properly braced. These two details show bracing conditions for walls that run continuous to the deck level.

TYPICAL PARTITION BRACING DETAIL #2
(DETAIL T4-PWB2)

This detail is similar to the previous one, but the partition walls stop just above the ceiling level.

Steel

(CSI Division 5)

Discussion—Steel

TYPES OF DETAILS INCLUDED

STEEL FRAMING DETAILS ARE PROVIDED IN THIS CHAPTER FOR THE FOLLOWING:

▶ *Shear Connections*

▶ *Moment Connections*

▶ *Cantilevered Roof Beam Connections*

▶ *Column Splices*

▶ *Beams Bearing on Masonry*

▶ *Miscellaneous Connections*

▶ *Braced Bays and Wall Elevations*

▶ *Joists and Joist Girders*

▶ *Composite Construction*

▶ *Overhead Crane Framing*

▶ *Cold-formed Metal Framing*

▶ *Miscellaneous Framing*

SHEAR CONNECTIONS

Steel members must be checked for *shear,* a force that is normally a maximum at the supports. Shear connections transfer shear forces from beams to their supports, typically steel columns. Steel beams are usually wide-flange (WF) sections. Steel columns may be wide-flange, tubular, or pipe sections.

There are many different types of shear connections. The American Institute of Steel Construction (AISC) provides details and recommendations for a wide variety of conditions. For wide-flange beams connected to wide-flange columns, perhaps the most frequently used shear connection is to shop weld double angles to the beam ends. The outstanding legs of the beam end angles are field bolted to the columns.

Steel fabricators normally review the construction drawings and then prepare *shop drawings,* which show all steel framing members, connections, material strengths, and so on. The shop drawings are submitted to the engineer for review prior to steel fabrication.

Field bolting is normally more practical and economical than field welding. Whether to shop bolt or shop weld is usually left up to the steel fabricator. However, be very careful when detailing welds. Although it is always the fabricator's prerogative to use shop welding in lieu of field welding, it can be costly if a field weld is mistakenly indicated where a shop weld would actually be better suited. Field welding is definitely more difficult and expensive. However, in some cases, it cannot be avoided.

In many regions, the actual design and detailing of shear connections are left up to the steel fabricator. The reasoning is that steel fabricators are normally accustomed to fabricating shear connections in a certain way that makes them more competitive. To limit the type of shear connection to a specific type of detail may place undue limitations on some fabricators. The information necessary for the fabricator to design the connections must be shown on the drawings. A schematic type of shear connection should be shown. The connection shear force requirement must also be indicated. This capacity is normally given in kips (1000 lb).

Many engineers require the fabricators to base the shear connection design on the maximum capacity allowed by AISC. In AISC's *Manual of Steel Construction,* the maximum allowable uniform load capacity of all steel beams is given. This capacity varies depending on the span. However, there are several reasons that this practice should be avoided. One reason is that the allowable loads are based on uniform loading conditions and do not consider concentrated loads. For concentrated loads, the AISC values may not be adequate and the required shear force must be indicated. If not indicated, the connection may be undersized. A second reason to avoid this practice is that the AISC values are based on "maximum" capacities and may be overly conservative for a majority of shear connection conditions, thus increasing the connection cost. A third reason is that this practice requires the steel fabricator to spend much more time determining the beam shear force requirements. This additional time will increase the project cost.

The most practical and economical way is to indicate the beam shear forces on the plan or in a schedule. The engineer can easily determine the beam end shear forces while doing the calculations during the beam design process. This ensures that all conditions, such as at concentrated loads, are properly addressed. It also results in a more economical connection system cost. It should be noted that in some regions the connection design is not left up to the steel fabricator, but is provided by the engineer. The steel fabricator still provides shop drawings.

MOMENT CONNECTIONS

Steel members must be checked for *moment*, which is related to beam flexure. A moment can be thought of as a "force times a lever arm" and is typically denoted in the units of foot-kips. Simple span beams do not rely on flexural restraint at the supports. It is assumed that the beam end can rotate with respect to its support, and the moment due to flexure will be a maximum somewhere along the span and zero at the supports. Simple span beams with uniform loads have the maximum moment at midspan.

Simple span beams are generally very economical. Only shear connections are required. However, for certain types of structures, some flexural restraint is required at the supports. This is true for structures that rely on the beam to column connection to provide rigid frame action. Rigid frames frequently serve as part of the lateral load resisting system. For beams in these frames, moment forces occur at the beam ends.

Moment connections transfer the moment end forces from the beams to their supports, typically steel columns. Shear also exists at the supports so a shear connection is also required. There are many different types of moment connections. The American Institute of Steel Construction provides details and recommendations for a wide variety of conditions.

Moment transfer is normally accomplished by providing *flange continuity*. For wide-flange beams connected to wide-flange columns, perhaps the most frequently used moment connection is to field weld the beam flanges directly to the column flanges.

As with shear connections, in many regions the actual design and detailing of the moment connection are left up to the steel fabricator. Limiting the type of moment connection to a specific type of detail may place undue limitations on certain fabricators. The information necessary for the fabricator to design the connections must be shown on the drawings. A schematic type of moment connection should be provided. The connection moment force requirement must also be indicated. The most practical and economical way is to indicate the moments on plan or in schedule. Since moment connections are more complicated than shear connections and frequently require field welding, it is better to avoid them whenever possible.

CANTILEVERED ROOF BEAM SYSTEMS

These systems are primarily used along the interior framing lines at roof levels. The system involves running the roof beams over the columns and cantilevering a short distance beyond. The cantilever placements are arranged so that shorter, simple span beams are created in alternate bays. The beam moments are significantly lower than a comparable system that uses simple span beams between columns. This moment reduction provides a more efficient and economical framing system. The AISC *Manual of Steel Construction* provides coefficients to help determine these moments.

The system is also advantageous from the steel erector's point of view. Beams can rest directly on top of the columns. Simple shear connections are provided at the cantilevered beam to beam connections. Due to the beam continuity over the interior supports, the bottom flange of the beam needs to be braced. The cantilevered system is ideally suited for roofs because the columns terminate. For that reason, it is not normally used for floor framing systems.

COLUMN SPLICES

In most cases, the fabricated steel members are delivered to the site via flatbed trucks. This limits the maximum member lengths; hence, splices may be required. For many low-rise buildings, column lengths are such that splices are not required. However, in other cases, a column splice is the solution. The AISC provides details and recommendations for a wide variety of column splice conditions.

BEAMS BEARING ON MASONRY

When steel beams bear on masonry walls, the beam reaction must be safely distributed to the wall via a bearing plate. The bearing plates are preset in the masonry during the wall construction. The plate must be properly anchored to the masonry.

The beam can be either field welded or field bolted to the plate. However, field welding for this type of connection is not desirable for several reasons. Welding provides an extremely rigid connection. Since masonry is a very brittle material, welding the beams can cause masonry cracks to develop due to any slight movement of the steel beam. This is a special concern if the beam spans are relatively long and exposed to drastic temperature variations (especially temperature drops) during construction. For this reason, it is preferable to field bolt the beam to the plate. Field bolting is also more economical than field welding.

MISCELLANEOUS CONNECTIONS

A variety of other potential connections may be required, such as column weak-axis braces, knee braces, and girts. Wide-flange columns have a strong axis and a weak axis. The column axial capacity depends on its unbraced length. The longer the unbraced length, the lower the column capacity; hence, weak-axis braces are an effective means to significantly increase the column axial capacity. Knee braces can be used to provide rigid frame action and alleviate the difficulty involved with moment connections. Girts are wall framing members that are typically Z or C shapes. They can be light-gauge metal or structural steel channel sections. These members need to be connected properly to their supports.

BRACED BAYS AND WALLS

Braced bay systems are a very efficient means of resisting lateral loads, such as from wind and earthquakes. This type of system is generally more economical than rigid frames for a variety of reasons. Building *drift*, or side sway, is minimized. This increases the column capacities. The framing generally involves simple spans; therefore, moment connections are not required.

Braced bays rely on the lateral forces to be transferred to the diagonal braces. Diagonals usually consist of small-diameter rods, steel angles, or tubular sections. Rods and angles are designed only for tensile forces (they are too slender to resist compressive forces). Threaded rods can be easily tightened and straightened; however, angles are more difficult to keep straight. Tubular sections are advantageous since they can be designed to resist both tensile and compressive forces. This results in a much higher overall capacity per braced bay and minimizes the required number of braced bays. Regarding details, braced bay elevations and diagonal brace end connections should be shown. If the connection is not fully detailed, the connection force requirement must also be indicated. This force is normally shown on the braced bay elevations.

Perimeter walls can be very economically framed with steel systems. Metal wall panels supported by steel girts are a system frequently used for warehouses and other industrial buildings. A short masonry skirt wall can be used for the first 8 to 12'. The masonry skirt wall provides better durability and increased protection from forklifts.

JOISTS AND JOIST GIRDERS

Open web steel bar joists have been in use for over 70 years. Steel joists are basically steel trusses used as secondary members and spaced closely together. This is a very efficient way to frame floor and roof systems. Joists are primarily intended to be used with simple spans and uniform loading conditions. However, they can also be used with other loadings, such as for concentrated loads, uplift loads, end moments, and so forth. Joist spacing depends on several factors, including the spanning capacity of the deck system and the allowable load capacity of the joists. For floor framing systems using 3 to 3½" noncomposite concrete floor decks, joists are normally spaced from 24 to 36" on center. A 2½" concrete deck system is considered the absolute minimum; however, floor vibrations may be excessive. To minimize floor vibrations, it is recommended to space the joists further apart and use a thicker deck system. For roof framing systems using 1½" metal roof decks, joist spacings typically range from 4 to 6'.

Joists are plant fabricated under strict quality control procedures. Double angles are currently used for the top and bottom chords. Round steel bars are used for the web diagonals. The Steel Joist Institute (SJI) and a variety of joist manufacturers publish catalogs that describe available joist sizes, span limitations, allowable loadings, deflection limitations, and so on. Several different series are available. K-series joists range from 8 to 30" in depth and can span up to 60'. LH-series (long span) joists range from 18 to 48" in depth and can span up to 96'. DLH-series (deep long span) joists range from 52 to 72" in depth and can span up to 144'.

Joist girders are primary members designed as simple spans to support concentrated loads from equally spaced bar joists. The concentrated loads are considered to act at the panel points of the joist girders. The standard configuration of a joist girder is with parallel chords, which typically consist of double angles. The vertical and diagonal webs also typically consist of double angles. Joist girders are a very economical framing system, especially for larger projects; however, for smaller projects, it may be more economical to use simple span steel floor beams and the cantilevered roof beam system.

COMPOSITE CONSTRUCTION

Composite systems are primarily used for floor systems where there are higher live loads and/or more stringent vibration requirements than typically encountered. The system can also span further than noncomposite decks. It relies on the metal decking to bond with the concrete slab, forming a composite system. The metal decking surface has a pattern of deformations, allowing this bond to take place. The deck acts as the bottom (tension) reinforcing. Welded wire fabric is also provided in the concrete slab to serve as top reinforcing. For decks with extreme loadings, such as for forklift trucks, reinforcing bars should be considered in lieu of welded wire fabric.

Since composite deck systems can span further than noncomposite systems, heavier loads are supported by the secondary framing members. Due to this heavy loading requirement, wide-flange steel beams are frequently used as *stringers*. Shear studs are frequently added on top of the steel beams so that the stringer beams also act compositely with the deck system. This effectively increases the properties of the stringer, increasing the allowable load capacity. The shear studs are typically field welded after the metal decking is installed.

OVERHEAD CRANE SYSTEMS

Overhead cranes are used in many industrial and warehouse buildings. Crane capacities vary from ¼ ton on up. The design of the crane equipment is provided by the crane manufacturer; however, the design of the crane support system is provided by the design professional.

Two special cases regarding cranes are frequently encountered. One case is adding crane systems within existing buildings. New runway girders, columns, and foundations are typically required. For low-capacity cranes, it may be possible to avoid new foundations by properly distributing the column reactions to the existing floor slab via extra large column base plates.

Another special case is with projects that involve pre-engineered metal buildings. Although the design of the main building is provided by the building manufacturer, the design of the overhead crane support system usually is not. Heavy crane systems should have their own vertical support system (independent columns); however, the crane support columns will still be braced back to the metal building columns. Therefore, the bracing forces and crane loading information necessary for the metal building manufacturer to design the main building columns must be shown on the drawings.

COLD-FORMED METAL FRAMING

Cold-formed metal framing members are produced from steel sheets and plant rolled into a variety of shapes. C-shaped members are most frequently used. The webs can be punched with openings to facilitate field-installed work, such as electrical conduit and plumbing.

One major problem currently associated with cold-formed metal framing is design responsibility. Many metal subcontractors find themselves unwilling participants in the design process because the complete design is not shown on the drawings. The member size, flange width, metal gauge (thickness), member spacing, and connection requirements must be indicated.

Unfortunately, many design professionals specify minimal information pertaining to the cold-formed framing, leaving it up to the contractor to figure out. This may lead to serious problems. For instance, if the curtain wall system is merely denoted as "6″ metal studs at 24″ on center," an improper gauge or flange width may be provided. This practice also increases the overall construction cost. Subcontractors will allocate price contingencies to allow for unknowns. Therefore, it is recommended that the design of all structural cold-formed members be performed by the design professional. The complete member specification and connection requirements should be indicated.

MISCELLANEOUS FRAMING

A variety of other details are also presented to address pipe hangers, stair hangers, roof deck openings, roof screen walls, strengthening existing members, roof deck attachments, and supporting mechanical rooftop units.

GENERAL COMMENT

Practices vary from region to region and the details that follow have generally been taken from specific applications. Any member sizes or connections shown were intended for that particular application. Accordingly, the applicability of all details should be investigated and all member sizes and connections should be determined prior to use.

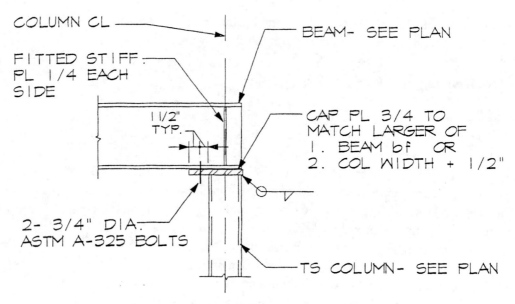

COLUMN CL

BEAM- SEE PLAN

FITTED STIFF.
PL 1/4 EACH
SIDE

1 1/2"
TYP.

CAP PL 3/4 TO
MATCH LARGER OF
1. BEAM bf OR
2. COL WIDTH + 1/2"

2- 3/4" DIA.
ASTM A-325 BOLTS

TS COLUMN- SEE PLAN

BEAM FRAMES OVER TS COLUMN

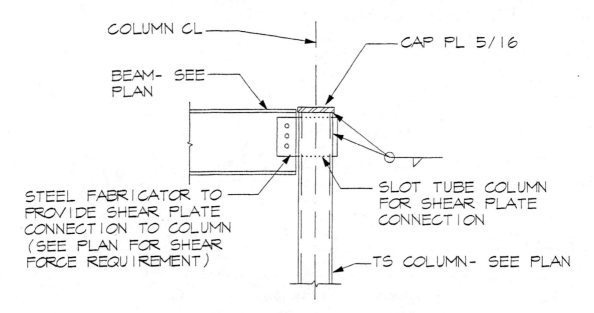

COLUMN CL

CAP PL 5/16

BEAM- SEE
PLAN

STEEL FABRICATOR TO
PROVIDE SHEAR PLATE
CONNECTION TO COLUMN
(SEE PLAN FOR SHEAR
FORCE REQUIREMENT)

SLOT TUBE COLUMN
FOR SHEAR PLATE
CONNECTION

TS COLUMN- SEE PLAN

BEAM FRAMES INTO TS COLUMN

BEAM SHEAR CONNECTION DETAIL #1

NOT TO SCALE   (DETAIL T5-CSH1)

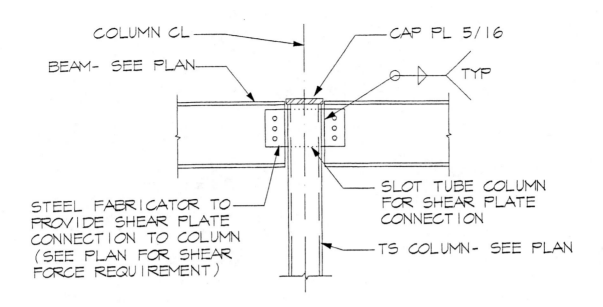

COLUMN CL

CAP PL 5/16

BEAM- SEE PLAN

TYP

STEEL FABRICATOR TO PROVIDE SHEAR PLATE CONNECTION TO COLUMN (SEE PLAN FOR SHEAR FORCE REQUIREMENT)

SLOT TUBE COLUMN FOR SHEAR PLATE CONNECTION

TS COLUMN- SEE PLAN

BEAM FRAMES ON BOTH SIDES OF TS COLUMN

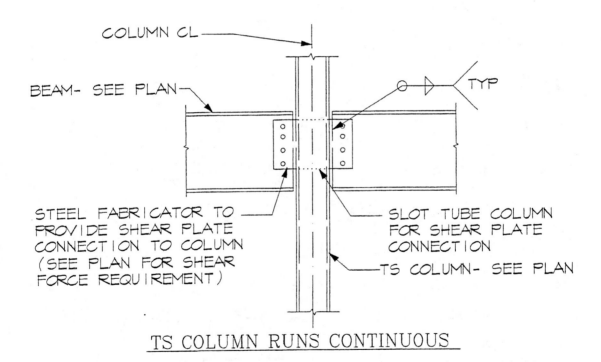

COLUMN CL

BEAM- SEE PLAN

TYP

STEEL FABRICATOR TO PROVIDE SHEAR PLATE CONNECTION TO COLUMN (SEE PLAN FOR SHEAR FORCE REQUIREMENT)

SLOT TUBE COLUMN FOR SHEAR PLATE CONNECTION

TS COLUMN- SEE PLAN

TS COLUMN RUNS CONTINUOUS

BEAM SHEAR CONNECTION DETAIL #2
NOT TO SCALE   (DETAIL T5-CSH2)

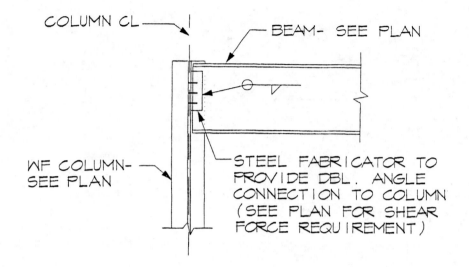

COLUMN CL

BEAM- SEE PLAN

WF COLUMN-
SEE PLAN

STEEL FABRICATOR TO
PROVIDE DBL. ANGLE
CONNECTION TO COLUMN
(SEE PLAN FOR SHEAR
FORCE REQUIREMENT)

BEAM FRAMES INTO WEB OF WF COLUMN

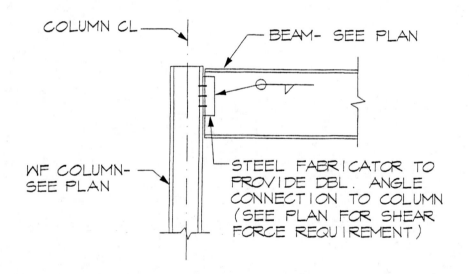

COLUMN CL

BEAM- SEE PLAN

WF COLUMN-
SEE PLAN

STEEL FABRICATOR TO
PROVIDE DBL. ANGLE
CONNECTION TO COLUMN
(SEE PLAN FOR SHEAR
FORCE REQUIREMENT)

BEAM FRAMES INTO FLANGE OF WF COLUMN

BEAM SHEAR CONNECTION DETAIL #3
NOT TO SCALE   (DETAIL T5-CSH3)

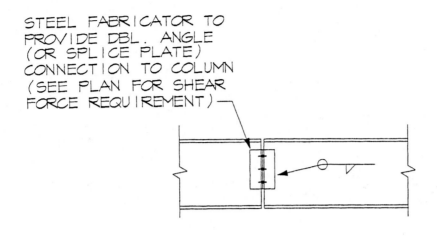

STEEL FABRICATOR TO
PROVIDE DBL. ANGLE
(OR SPLICE PLATE)
CONNECTION TO COLUMN
(SEE PLAN FOR SHEAR
FORCE REQUIREMENT)

<u>BEAM TO BEAM SHEAR SPLICE</u>

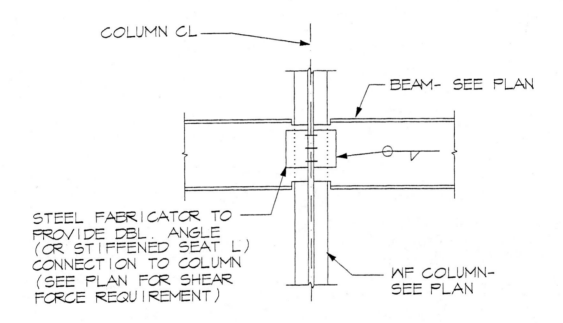

COLUMN CL

BEAM- SEE PLAN

STEEL FABRICATOR TO
PROVIDE DBL. ANGLE
(OR STIFFENED SEAT L)
CONNECTION TO COLUMN
(SEE PLAN FOR SHEAR
FORCE REQUIREMENT)

WF COLUMN-
SEE PLAN

<u>BEAM FRAMES BOTH SIDES INTO WEB OF WF COLUMN</u>

BEAM SHEAR CONNECTION DETAIL #4
NOT TO SCALE   (DETAIL T5—CSH4)

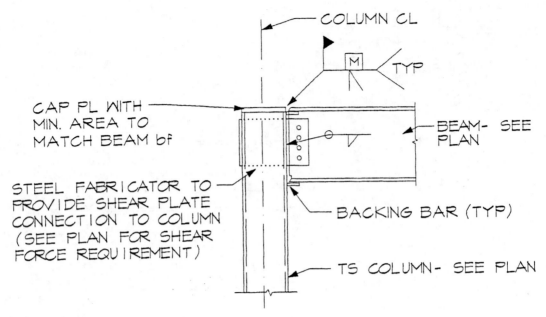

CAP PL WITH MIN. AREA TO MATCH BEAM bf

STEEL FABRICATOR TO PROVIDE SHEAR PLATE CONNECTION TO COLUMN (SEE PLAN FOR SHEAR FORCE REQUIREMENT)

COLUMN CL

M

TYP

BEAM- SEE PLAN

BACKING BAR (TYP)

TS COLUMN- SEE PLAN

<u>BEAM FRAMES ONE SIDE OF TS COLUMN</u>

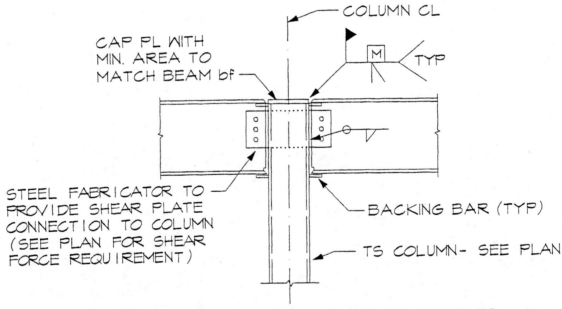

CAP PL WITH MIN. AREA TO MATCH BEAM bf

COLUMN CL

M

TYP

STEEL FABRICATOR TO PROVIDE SHEAR PLATE CONNECTION TO COLUMN (SEE PLAN FOR SHEAR FORCE REQUIREMENT)

BACKING BAR (TYP)

TS COLUMN- SEE PLAN

<u>BEAM FRAMES BOTH SIDES OF TS COLUMN</u>

<u>BEAM MOMENT CONNECTION DETAIL #1</u>

NOT TO SCALE

(DETAIL T5-CM01)

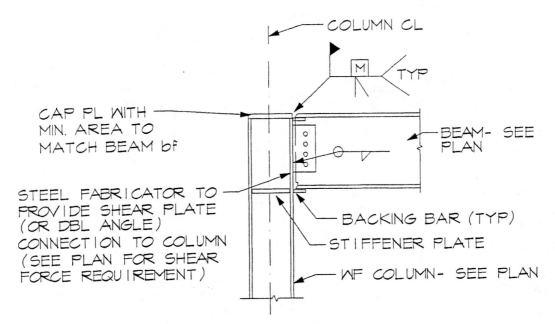

COLUMN CL

TYP

CAP PL WITH
MIN. AREA TO
MATCH BEAM bf

BEAM- SEE
PLAN

STEEL FABRICATOR TO
PROVIDE SHEAR PLATE
(OR DBL ANGLE)
CONNECTION TO COLUMN
(SEE PLAN FOR SHEAR
FORCE REQUIREMENT)

BACKING BAR (TYP)
STIFFENER PLATE
WF COLUMN- SEE PLAN

BEAM FRAMES ONE SIDE OF WF COLUMN

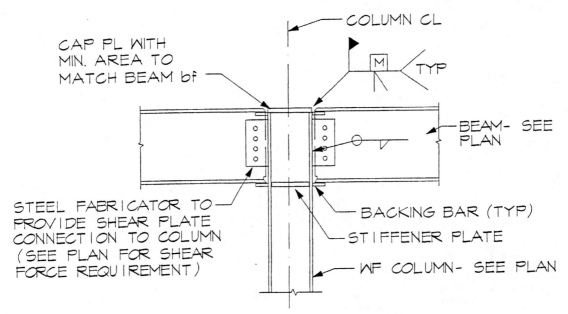

COLUMN CL

TYP

CAP PL WITH
MIN. AREA TO
MATCH BEAM bf

BEAM- SEE
PLAN

STEEL FABRICATOR TO
PROVIDE SHEAR PLATE
CONNECTION TO COLUMN
(SEE PLAN FOR SHEAR
FORCE REQUIREMENT)

BACKING BAR (TYP)
STIFFENER PLATE
WF COLUMN- SEE PLAN

BEAM FRAMES BOTH SIDES OF WF COLUMN

BEAM MOMENT CONNECTION DETAIL #2
NOT TO SCALE   (DETAIL T5—CM02)

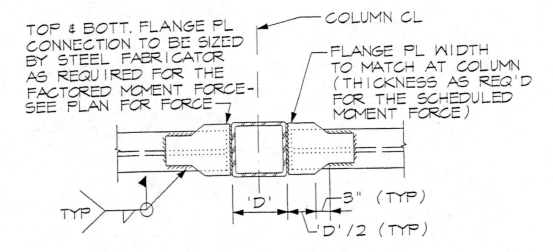

TOP & BOTT. FLANGE PL CONNECTION TO BE SIZED BY STEEL FABRICATOR AS REQUIRED FOR THE FACTORED MOMENT FORCE- SEE PLAN FOR FORCE

COLUMN CL

FLANGE PL WIDTH TO MATCH AT COLUMN (THICKNESS AS REQ'D FOR THE SCHEDULED MOMENT FORCE)

'D'

3" (TYP)

'D'/2 (TYP)

TYP

PLAN AT TOP FLANGE
(DETAIL AT BOTTOM FLANGE IS SIM.)

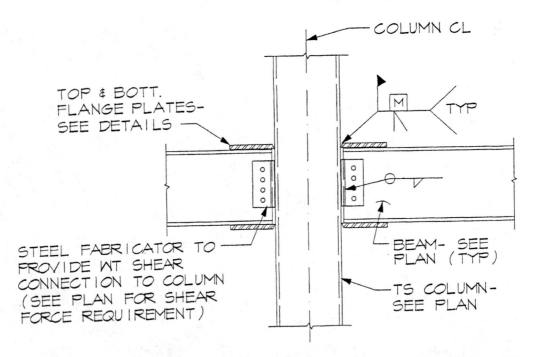

TOP & BOTT. FLANGE PLATES- SEE DETAILS

COLUMN CL

M TYP

STEEL FABRICATOR TO PROVIDE WT SHEAR CONNECTION TO COLUMN (SEE PLAN FOR SHEAR FORCE REQUIREMENT)

BEAM- SEE PLAN (TYP)

TS COLUMN- SEE PLAN

BEAM FRAMES BOTH SIDES OF TS COLUMN

BEAM MOMENT CONNECTION DETAIL #3
NOT TO SCALE

(DETAIL T5-CM03)

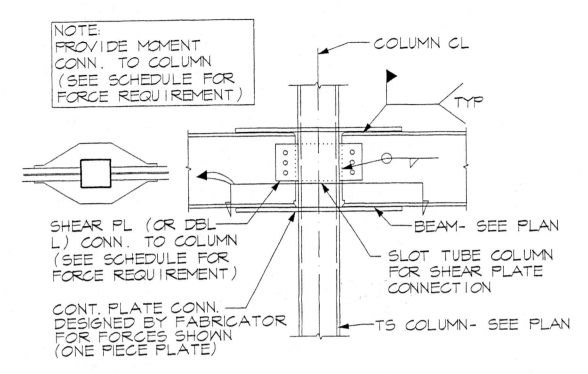

NOTE:
PROVIDE MOMENT
CONN. TO COLUMN
(SEE SCHEDULE FOR
FORCE REQUIREMENT)

COLUMN CL

TYP

SHEAR PL (OR DBL
L) CONN. TO COLUMN
(SEE SCHEDULE FOR
FORCE REQUIREMENT)

CONT. PLATE CONN.
DESIGNED BY FABRICATOR
FOR FORCES SHOWN
(ONE PIECE PLATE)

BEAM- SEE PLAN

SLOT TUBE COLUMN
FOR SHEAR PLATE
CONNECTION

TS COLUMN- SEE PLAN

BEAM FRAMES BOTH SIDES OF TS COLUMN

BEAM MOMENT CONNECTION DETAIL #4

NOT TO SCALE   (DETAIL T5-CM04)

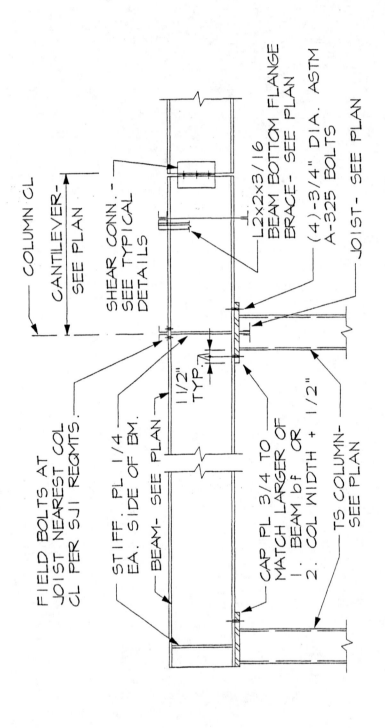

COLUMN CL

CANTILEVER- SEE PLAN

SHEAR CONN.- SEE TYPICAL DETAILS

FIELD BOLTS AT JOIST NEAREST COL CL PER SJI REQMTS.

STIFF. PL 1/4 EA. SIDE OF BM.

BEAM- SEE PLAN

1 1/2" TYP.

CAP PL 3/4 TO MATCH LARGER OF
1. BEAM bf OR
2. COL WIDTH + 1/2"

TS COLUMN- SEE PLAN

L2×2×3/16 BEAM BOTTOM FLANGE BRACE- SEE PLAN

(4)-3/4" DIA. ASTM A-325 BOLTS

JOIST- SEE PLAN

END TS COLUMN   INTERIOR TS COLUMN

TYPICAL CANTILEVER ROOF BEAM DETAILS

(DETAIL T5-CBM1)

NOT TO SCALE

226

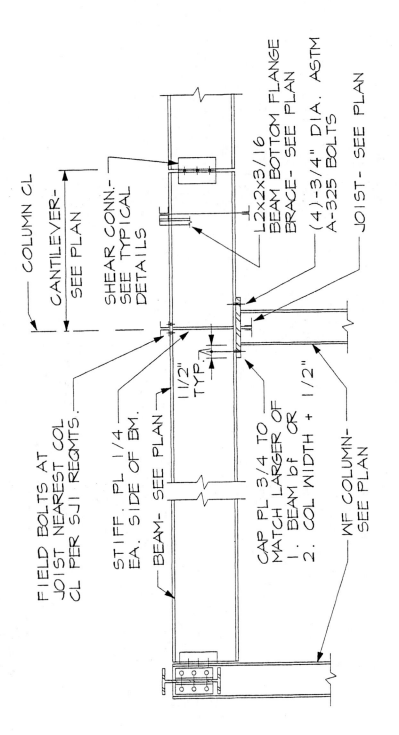

FIELD BOLTS AT JOIST NEAREST COL CL PER SJI REQMTS.

COLUMN CL

CANTILEVER- SEE PLAN

SHEAR CONN.- SEE TYPICAL DETAILS

STIFF. PL 1/4 EA. SIDE OF BM.

BEAM- SEE PLAN

1 1/2" TYP.

CAP PL 3/4 TO MATCH LARGER OF
1. BEAM bf OR
2. COL WIDTH + 1/2"

L2x2x3/16 BEAM BOTTOM FLANGE BRACE- SEE PLAN

(4)-3/4" DIA. ASTM A-325 BOLTS

JOIST- SEE PLAN

WF COLUMN- SEE PLAN

END WF COLUMN   INTERIOR WF COLUMN

TYPICAL CANTILEVER ROOF BEAM DETAILS

(DETAIL T5-CBM2)

NOT TO SCALE

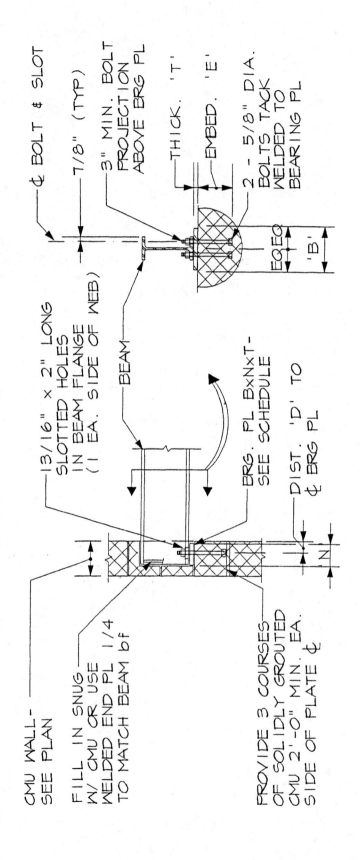

NOTE: DETAIL IS SHOWN FOR BEAMS PERPENDICULAR TO WALL. PROVIDE SIMILAR DETAIL FOR BEAMS PARALLEL TO WALL.

CMU WALL - SEE PLAN

13/16" x 2" LONG SLOTTED HOLES IN BEAM FLANGE (1 EA. SIDE OF WEB)

FILL IN SNUG W/ CMU OR USE WELDED END PL 1/4 TO MATCH BEAM bf

BEAM

¢ BOLT & SLOT

7/8" (TYP)

3" MIN. BOLT PROJECTION ABOVE BRG PL

THICK. 'T'

EMBED. 'E'

2 - 5/8" DIA. BOLTS TACK WELDED TO BEARING PL

BRG. PL BxNxT- SEE SCHEDULE

DIST. 'D' TO ¢ BRG PL

PROVIDE 3 COURSES OF SOLIDLY GROUTED CMU 2'-0" MIN. EA. SIDE OF PLATE ¢

EQ EQ

'B'

N

'FIXED' CONNECTION

TYPICAL BEAM BEARING ON MASONRY DETAILS

(DETAIL T5-CMU1)

NOT TO SCALE

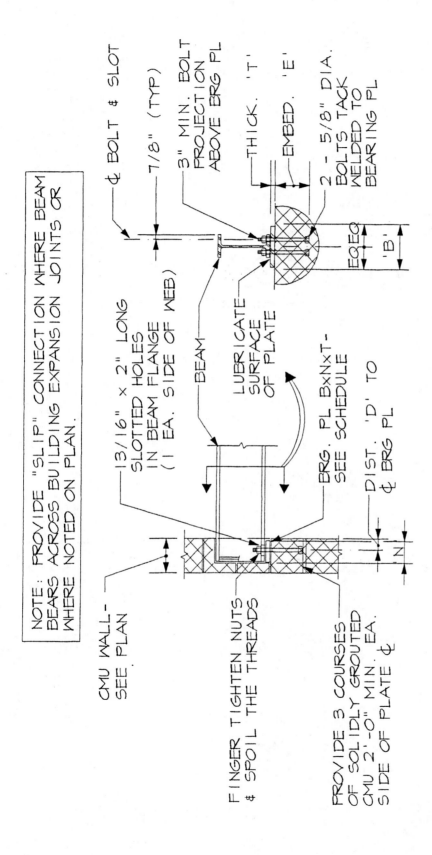

NOTE: PROVIDE "SLIP" CONNECTION WHERE BEAM BEARS ACROSS BUILDING EXPANSION JOINTS OR WHERE NOTED ON PLAN.

₵ BOLT & SLOT

7/8" (TYP)

3" MIN. BOLT PROJECTION ABOVE BRG PL

THICK. 'T'

EMBED. 'E'

2 - 5/8" DIA. BOLTS TACK WELDED TO BEARING PL

13/16" x 2" LONG SLOTTED HOLES IN BEAM FLANGE (1 EA. SIDE OF WEB)

BEAM

LUBRICATE SURFACE OF PLATE

CMU WALL - SEE PLAN

FINGER TIGHTEN NUTS & SPOIL THE THREADS

PROVIDE 3 COURSES OF SOLIDLY GROUTED CMU 2'-0" MIN. EA. SIDE OF PLATE ₵

BRG. PL BxNxT - SEE SCHEDULE

DIST. 'D' TO ₵ BRG PL

'SLIP' CONNECTION

TYPICAL BEAM BEARING ON MASONRY DETAILS

(DETAIL T5-CMU2)

NOT TO SCALE

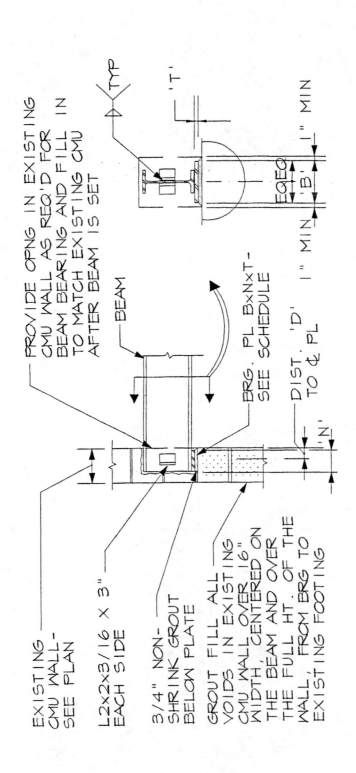

EXISTING CMU WALL– SEE PLAN

PROVIDE OPNG IN EXISTING CMU WALL AS REQ'D FOR BEAM BEARING AND FILL IN TO MATCH EXISTING CMU AFTER BEAM IS SET

BEAM

L2x2x3/16 × 3" EACH SIDE

3/4" NON– SHRINK GROUT BELOW PLATE

BRG. PL BxNxT– SEE SCHEDULE

DIST. 'D' TO ₵ PL

GROUT FILL ALL VOIDS IN EXISTING CMU WALL OVER 16" WIDTH, CENTERED ON THE BEAM AND OVER THE FULL HT. OF THE WALL, FROM BRG TO EXISTING FOOTING

'N'

'T'

EQ EQ

'B' 1" MIN

1" MIN

TYP

TYPICAL BEAM BEARING ON EXISTING MASONRY DETAILS

(DETAIL T5–CMU3)

NOT TO SCALE

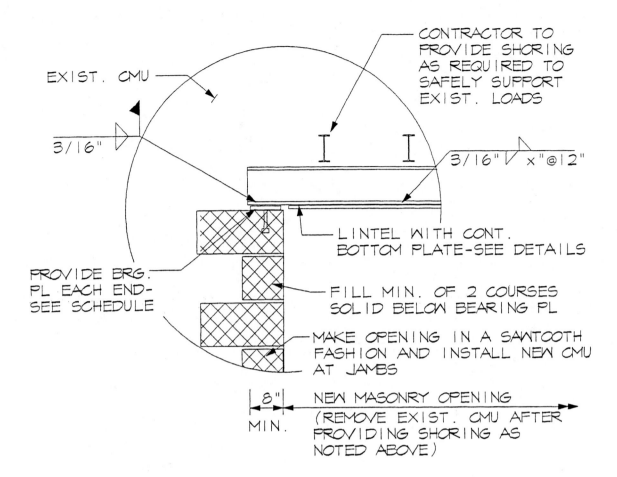

EXIST. CMU

CONTRACTOR TO PROVIDE SHORING AS REQUIRED TO SAFELY SUPPORT EXIST. LOADS

3/16"

3/16" x"@12"

LINTEL WITH CONT. BOTTOM PLATE-SEE DETAILS

PROVIDE BRG. PL EACH END- SEE SCHEDULE

FILL MIN. OF 2 COURSES SOLID BELOW BEARING PL

MAKE OPENING IN A SAWTOOTH FASHION AND INSTALL NEW CMU AT JAMBS

8" MIN.

NEW MASONRY OPENING (REMOVE EXIST. CMU AFTER PROVIDING SHORING AS NOTED ABOVE)

LINTEL BRG. ON EXISTING MASONRY

NOT TO SCALE

(DETAIL T5-CMU4)

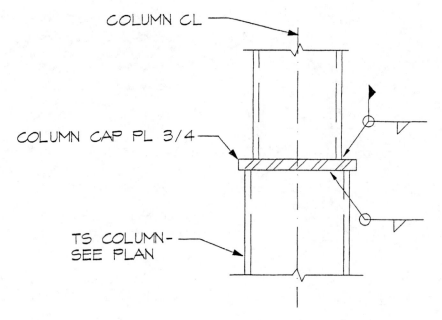

COLUMN CL

COLUMN CAP PL 3/4

TS COLUMN-
SEE PLAN

TS COLUMN SPLICE, FILLET WELDED

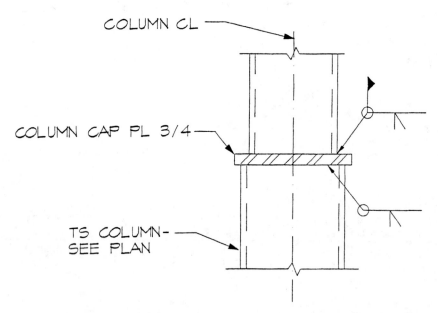

COLUMN CL

COLUMN CAP PL 3/4

TS COLUMN-
SEE PLAN

TS COLUMN SPLICE, PENETRATION WELDED

TYP. TS COLUMN SPLICE DETAILS

NOT TO SCALE   (DETAIL T5—COL1)

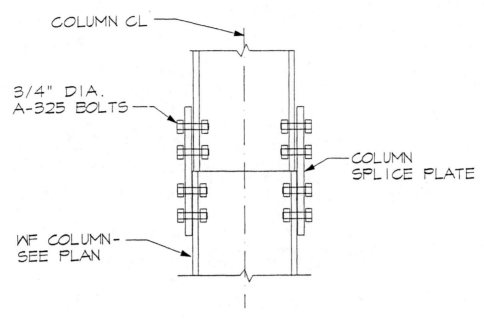

COLUMN CL

3/4" DIA.
A-325 BOLTS

COLUMN
SPLICE PLATE

WF COLUMN-
SEE PLAN

WF COLUMN SPLICE, BOLTED

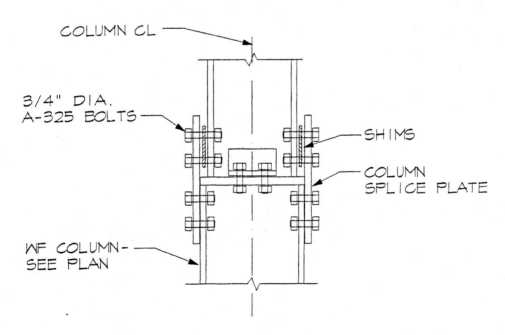

COLUMN CL

3/4" DIA.
A-325 BOLTS

SHIMS

COLUMN
SPLICE PLATE

WF COLUMN-
SEE PLAN

WF COLUMN SPLICE, BUTT PLATE

TYP. WF COLUMN SPLICE DETAILS

NOT TO SCALE   (DETAIL T5-COL2)

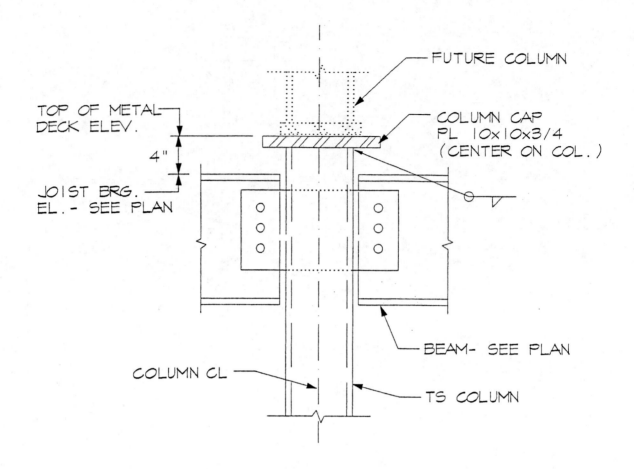

FUTURE COLUMN

TOP OF METAL
DECK ELEV.

COLUMN CAP
PL 10x10x3/4
(CENTER ON COL.)

4"

JOIST BRG.
EL. - SEE PLAN

BEAM- SEE PLAN

COLUMN CL

TS COLUMN

TYP. FUTURE COLUMN CONNECTION

NOT TO SCALE   (DETAIL T5—COL3)

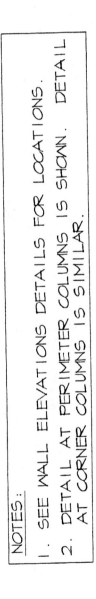

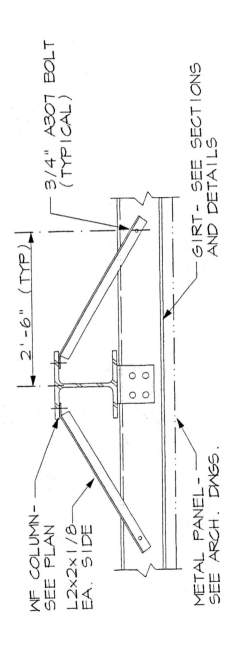

NOTES:

1. SEE WALL ELEVATIONS DETAILS FOR LOCATIONS.

2. DETAIL AT PERIMETER COLUMNS IS SHOWN. DETAIL AT CORNER COLUMNS IS SIMILAR.

3/4" A307 BOLT (TYPICAL)

2'-6" (TYP)

GIRT- SEE SECTIONS AND DETAILS

WF COLUMN- SEE PLAN

L2x2x1/8 EA. SIDE

METAL PANEL- SEE ARCH. DWGS.

TYPICAL COLUMN BRACE DETAIL

NOT TO SCALE

(DETAIL T5-BRCE1)

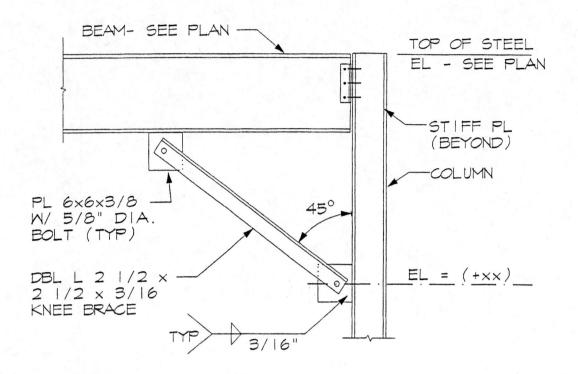

BEAM- SEE PLAN

TOP OF STEEL
EL - SEE PLAN

STIFF PL
(BEYOND)

COLUMN

45°

PL 6x6x3/8
W/ 5/8" DIA.
BOLT (TYP)

DBL L 2 1/2 ×
2 1/2 × 3/16
KNEE BRACE

EL = (+xx)

TYP   3/16"

TYP. KNEE BRACE DETAIL

NOT TO SCALE   (DETAIL T5—BRCE2)

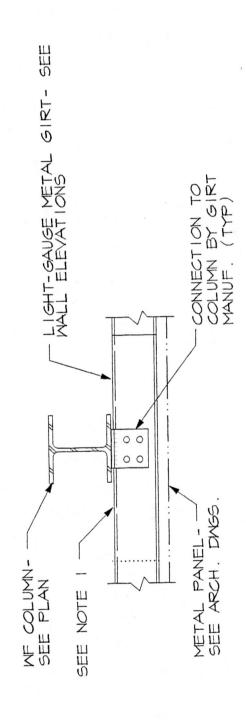

NOTES:
1. LAP GIRTS AT COLUMN AS REQUIRED BY THE MANUFACTURER.
2. PROVIDE GIRT WEB STIFFENERS AS REQUIRED BY THE MANUF.

LIGHT-GAUGE METAL GIRT- SEE WALL ELEVATIONS

CONNECTION TO COLUMN BY GIRT MANUF. (TYP)

WF COLUMN- SEE PLAN

SEE NOTE 1

METAL PANEL- SEE ARCH. DWGS.

TYPICAL LIGHT-GAUGE METAL GIRT CONNECTION DETAIL #1

(DETAIL T5-GIRT1)

NOT TO SCALE

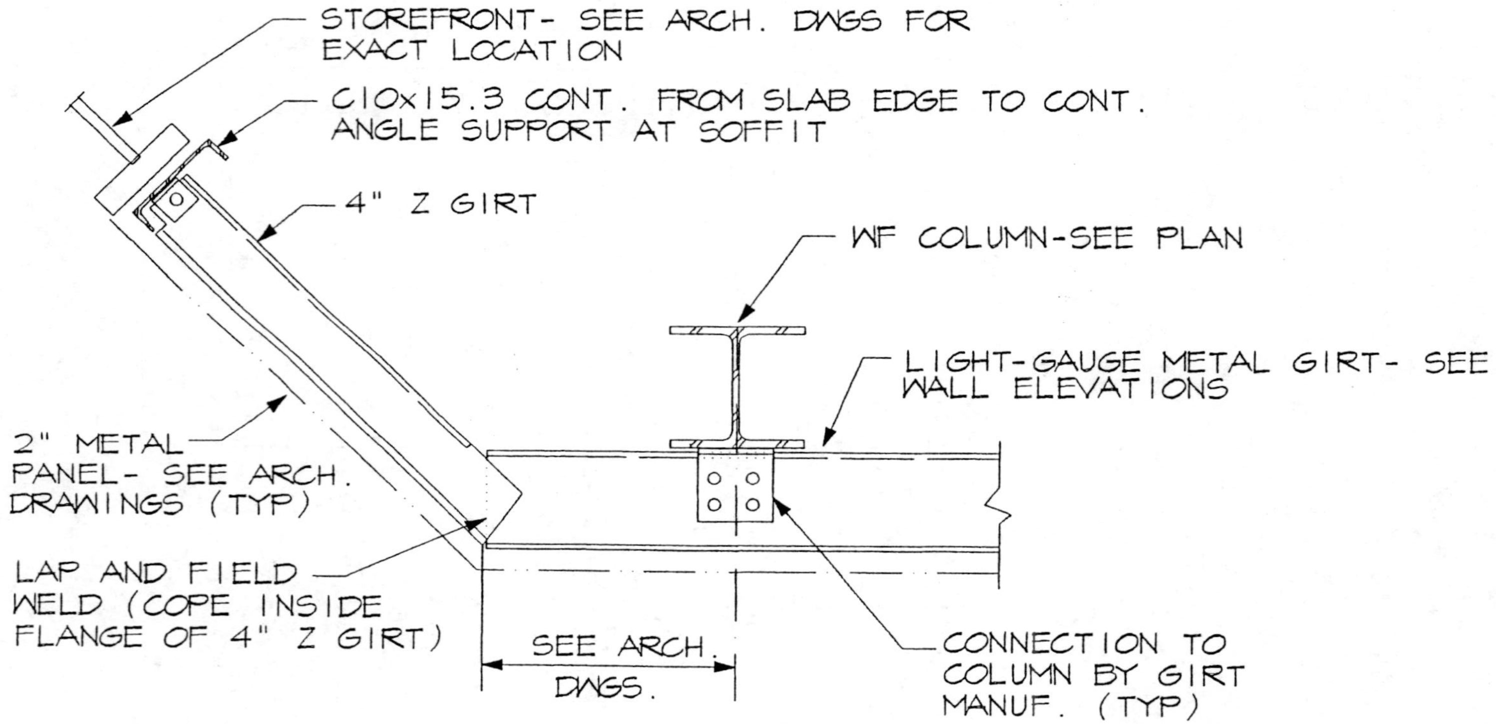

STOREFRONT- SEE ARCH. DWGS FOR
EXACT LOCATION

C10x15.3 CONT. FROM SLAB EDGE TO CONT.
ANGLE SUPPORT AT SOFFIT

4" Z GIRT

WF COLUMN-SEE PLAN

LIGHT-GAUGE METAL GIRT- SEE
WALL ELEVATIONS

CONNECTION TO
COLUMN BY GIRT
MANUF. (TYP)

SEE ARCH.
DWGS.

2" METAL
PANEL- SEE ARCH.
DRAWINGS (TYP)

LAP AND FIELD
WELD (COPE INSIDE
FLANGE OF 4" Z GIRT)

DETAIL AT BUILDING RECESSES

TYPICAL LIGHT-GAUGE METAL GIRT CONNECTION DETAIL #2

(DETAIL T5-GIRT2)

NOT TO SCALE

238

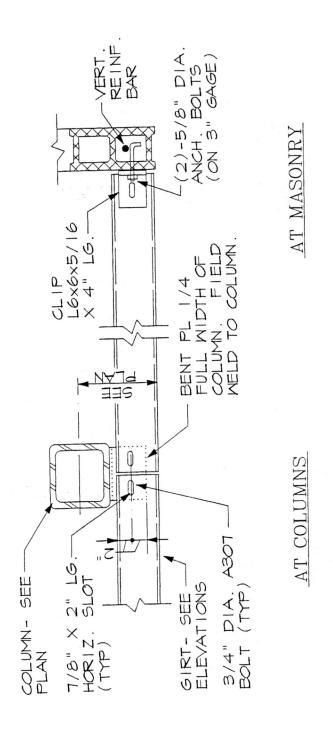

COLUMN- SEE
PLAN

7/8" X 2" LG.
HORIZ. SLOT
(TYP)

GIRT- SEE
ELEVATIONS

3/4" DIA. A307
BOLT (TYP)

CLIP
L6x6x5/16
X 4" LG.

SEE PLAN

BENT PL 1/4
FULL WIDTH OF
COLUMN. FIELD
WELD TO COLUMN.

VERT.
REINF.
BAR

(2)-5/8" DIA.
ANCH. BOLTS
(ON 3" GAGE)

AT MASONRY

AT COLUMNS

TYPICAL CHANNEL GIRT CONNECTION DETAILS
(DETAIL T5-GIRT3)

NOT TO SCALE

239

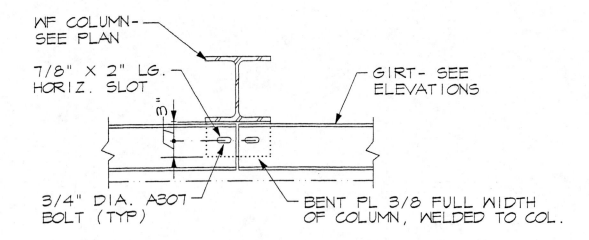

WF COLUMN-
SEE PLAN

7/8" X 2" LG.
HORIZ. SLOT

GIRT- SEE
ELEVATIONS

3/4" DIA. A307
BOLT (TYP)

BENT PL 3/8 FULL WIDTH
OF COLUMN, WELDED TO COL.

<u>AT TYP. GIRTS</u>

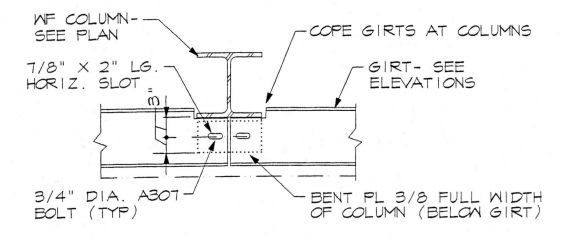

WF COLUMN-
SEE PLAN

7/8" X 2" LG.
HORIZ. SLOT

COPE GIRTS AT COLUMNS

GIRT- SEE
ELEVATIONS

3/4" DIA. A307
BOLT (TYP)

BENT PL 3/8 FULL WIDTH
OF COLUMN (BELOW GIRT)

<u>AT COPED GIRTS</u>

TYP. GIRT CONNECTION DETAILS

NOT TO SCALE   (DETAIL T5-GIRT4)

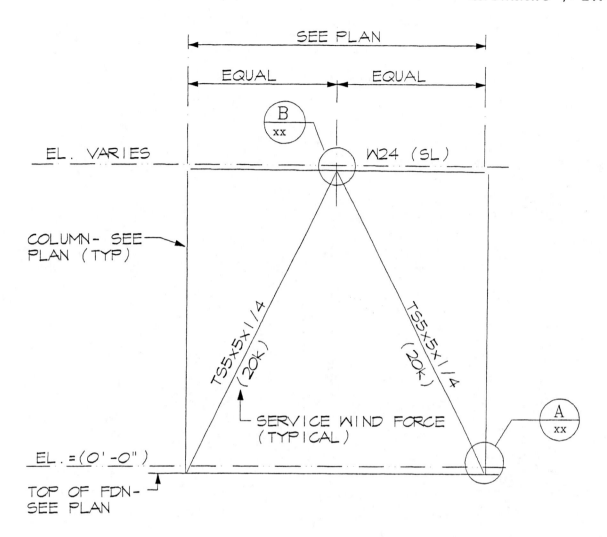

SEE PLAN

EQUAL   EQUAL

B
xx

EL. VARIES   W24 (SL)

COLUMN- SEE
PLAN (TYP)

TS5x5x1/4
(20k)

TS5x5x1/4
(20k)

A
xx

SERVICE WIND FORCE
(TYPICAL)

EL.=(0'-0")

TOP OF FDN-
SEE PLAN

TYP. BRACED BAY ELEVATION 1

NOT TO SCALE   (DETAIL T5-BAYE1)

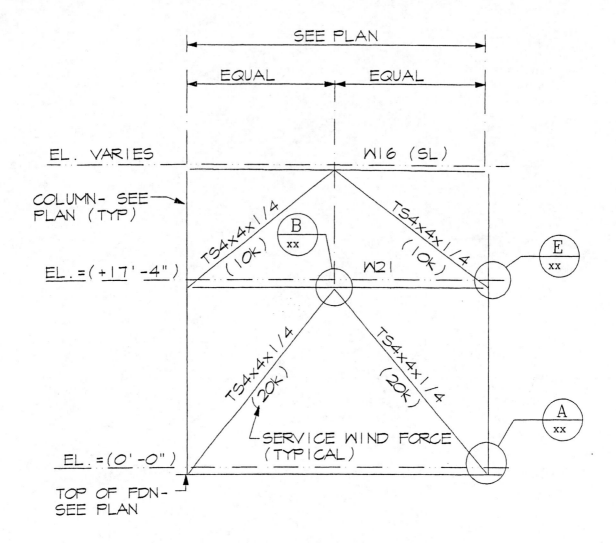

TYP. BRACED BAY ELEVATION 2

NOT TO SCALE   (DETAIL T5—BAYE2)

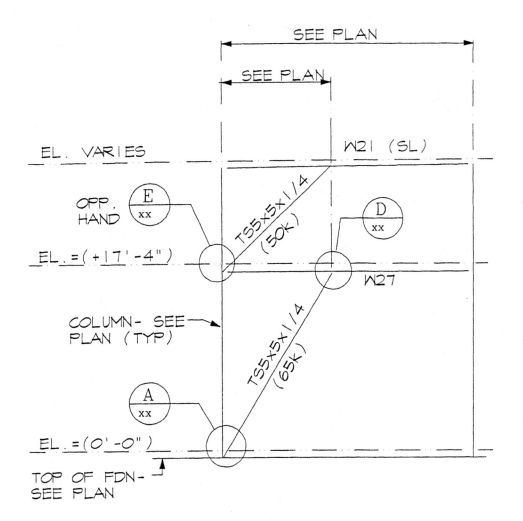

SEE PLAN

SEE PLAN

EL. VARIES

W21 (SL)

OPP.
HAND

$\begin{array}{c} E \\ \hline xx \end{array}$

TS5x5x1/4
(50k)

$\begin{array}{c} D \\ \hline xx \end{array}$

EL.=(+17'-4")

W27

COLUMN- SEE
PLAN (TYP)

TS5x5x1/4
(65k)

$\begin{array}{c} A \\ \hline xx \end{array}$

EL.=(0'-0")

TOP OF FDN-
SEE PLAN

TYP. BRACED BAY ELEVATION 3
NOT TO SCALE   (DETAIL T5—BAYE3)

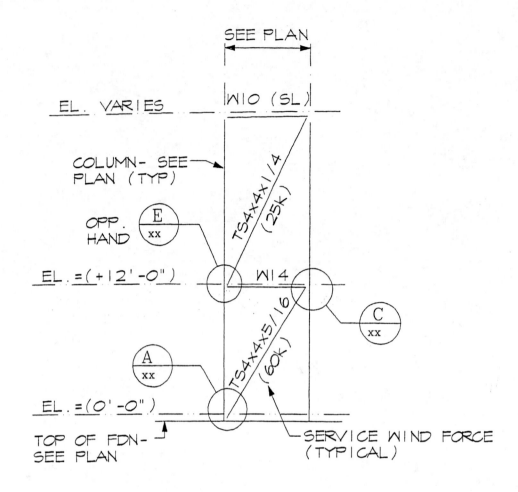

TYP. BRACED BAY ELEVATION 4

NOT TO SCALE   (DETAIL T5—BAYE4)

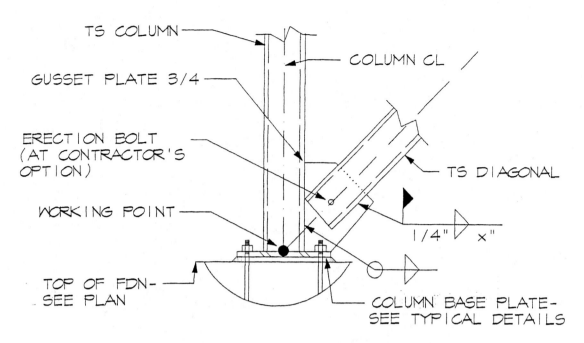

TS COLUMN

GUSSET PLATE 3/4

ERECTION BOLT
(AT CONTRACTOR'S
OPTION)

WORKING POINT

TOP OF FDN-
SEE PLAN

COLUMN CL

TS DIAGONAL

1/4"   x"

COLUMN BASE PLATE-
SEE TYPICAL DETAILS

DETAIL A, AT TS COLUMN

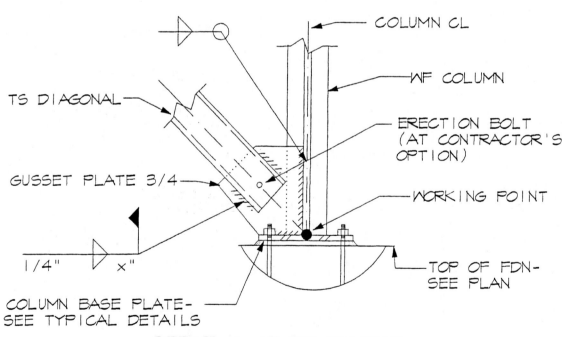

COLUMN CL

WF COLUMN

ERECTION BOLT
(AT CONTRACTOR'S
OPTION)

WORKING POINT

TOP OF FDN-
SEE PLAN

TS DIAGONAL

GUSSET PLATE 3/4

1/4"   x"

COLUMN BASE PLATE-
SEE TYPICAL DETAILS

DETAIL A, AT WF COLUMN

BRACED BAY DETAIL A

NOT TO SCALE   (DETAIL T5-BAYD1)

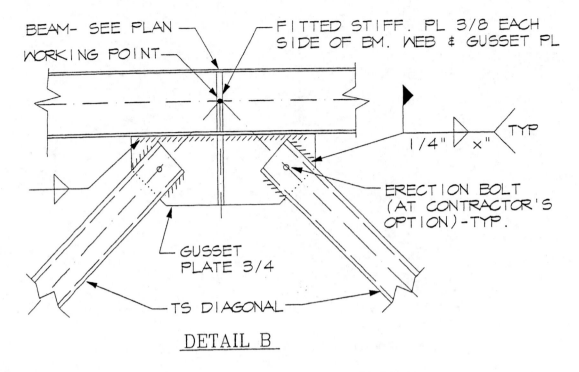

BEAM- SEE PLAN

WORKING POINT

FITTED STIFF. PL 3/8 EACH
SIDE OF BM. WEB & GUSSET PL

1/4" x" TYP

ERECTION BOLT
(AT CONTRACTOR'S
OPTION)-TYP.

GUSSET
PLATE 3/4

TS DIAGONAL

DETAIL B

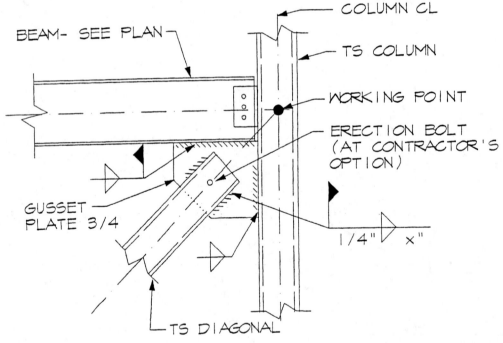

BEAM- SEE PLAN

COLUMN CL

TS COLUMN

WORKING POINT

ERECTION BOLT
(AT CONTRACTOR'S
OPTION)

1/4" x"

GUSSET
PLATE 3/4

TS DIAGONAL

DETAIL C

BRACED BAY DETAILS B AND C

NOT TO SCALE

(DETAIL T5—BAYD2)

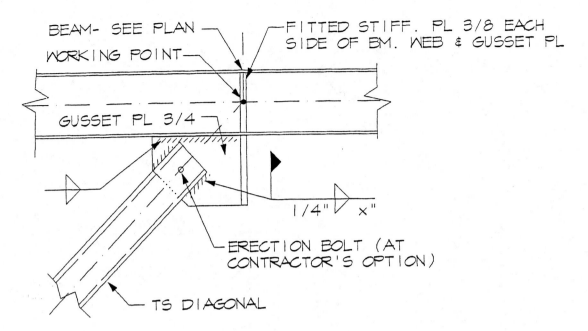

BEAM- SEE PLAN

WORKING POINT

FITTED STIFF. PL 3/8 EACH SIDE OF BM. WEB & GUSSET PL

GUSSET PL 3/4

1/4" ⊿ x"

ERECTION BOLT (AT CONTRACTOR'S OPTION)

TS DIAGONAL

DETAIL D

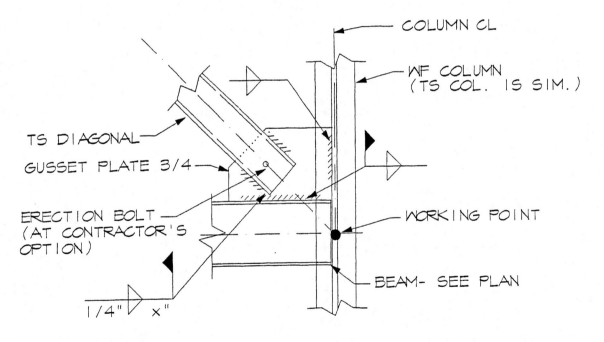

COLUMN CL

WF COLUMN (TS COL. IS SIM.)

TS DIAGONAL

GUSSET PLATE 3/4

ERECTION BOLT (AT CONTRACTOR'S OPTION)

WORKING POINT

1/4" ⊿ x"

BEAM- SEE PLAN

DETAIL E

BRACED BAY DETAILS D AND E

NOT TO SCALE   (DETAIL T5-BAYD3)

NOTE: PROVIDE BOLTED
CONNECTION BETWEEN
PL 7/8" AND THE 3/4"
GUSSET PER THE FORCES
INDICATED ON THE BRACED
BAY ELEVATIONS

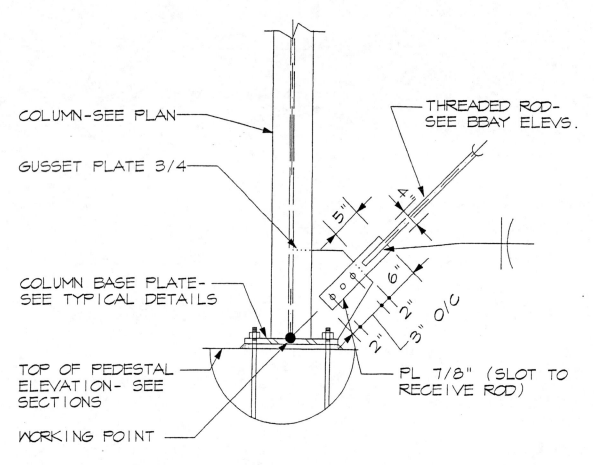

COLUMN-SEE PLAN

THREADED ROD-
SEE BBAY ELEVS.

GUSSET PLATE 3/4

COLUMN BASE PLATE-
SEE TYPICAL DETAILS

TOP OF PEDESTAL
ELEVATION- SEE
SECTIONS

WORKING POINT

PL 7/8" (SLOT TO
RECEIVE ROD)

5" 4" 6" 2" 2" 3" O/C

DETAIL A, USING THREADED RODS

BRACED BAY DETAIL F

NOT TO SCALE   (DETAIL T5-BAYD4)

NOTE: PROVIDE BOLTED
CONNECTION BETWEEN
PL 7/8" AND THE 3/4"
GUSSET PER THE FORCES
INDICATED ON THE BRACED
BAY ELEVATIONS

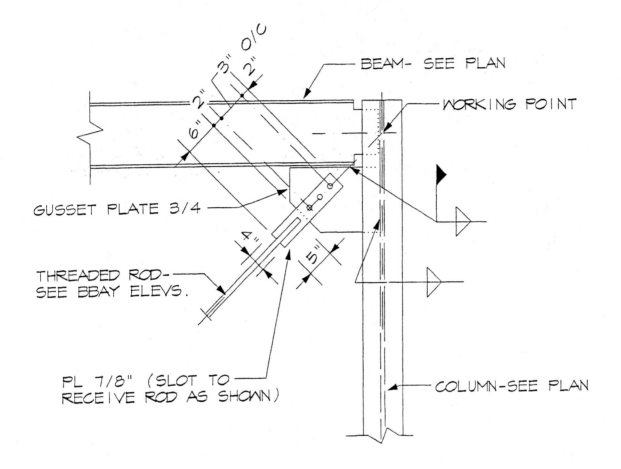

BEAM- SEE PLAN

WORKING POINT

GUSSET PLATE 3/4

THREADED ROD-
SEE BBAY ELEVS.

PL 7/8" (SLOT TO
RECEIVE ROD AS SHOWN)

COLUMN-SEE PLAN

DETAIL C, USING THREADED RODS

BRACED BAY DETAIL G

NOT TO SCALE   (DETAIL T5-BAYD5)

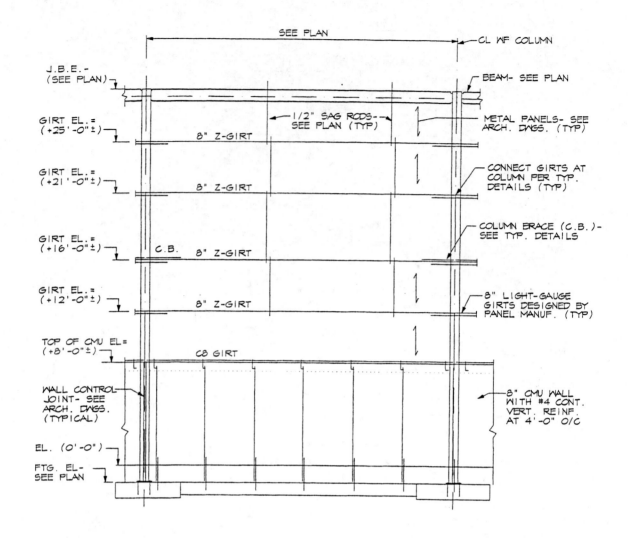

SEE PLAN

CL WF COLUMN

J.B.E.-
(SEE PLAN)

BEAM- SEE PLAN

1/2" SAG RODS-
SEE PLAN (TYP)

METAL PANELS- SEE
ARCH. DWGS. (TYP)

GIRT EL.=
(+25'-0"±)

8" Z-GIRT

GIRT EL.=
(+21'-0"±)

8" Z-GIRT

CONNECT GIRTS AT
COLUMN PER TYP.
DETAILS (TYP)

GIRT EL.=
(+16'-0"±)

C.B.   8" Z-GIRT

COLUMN BRACE (C.B.)-
SEE TYP. DETAILS

GIRT EL.=
(+12'-0"±)

8" Z-GIRT

8" LIGHT-GAUGE
GIRTS DESIGNED BY
PANEL MANUF. (TYP)

TOP OF CMU EL=
(+8'-0"±)

C8 GIRT

WALL CONTROL-
JOINT- SEE
ARCH. DWGS.
(TYPICAL)

8" CMU WALL
WITH #4 CONT.
VERT. REINF.
AT 4'-0" O/C

EL. (0'-0")

FTG. EL-
SEE PLAN

TYP. PERIMETER WALL ELEVATION 1
NOT TO SCALE

(DETAIL T5-WALL1)

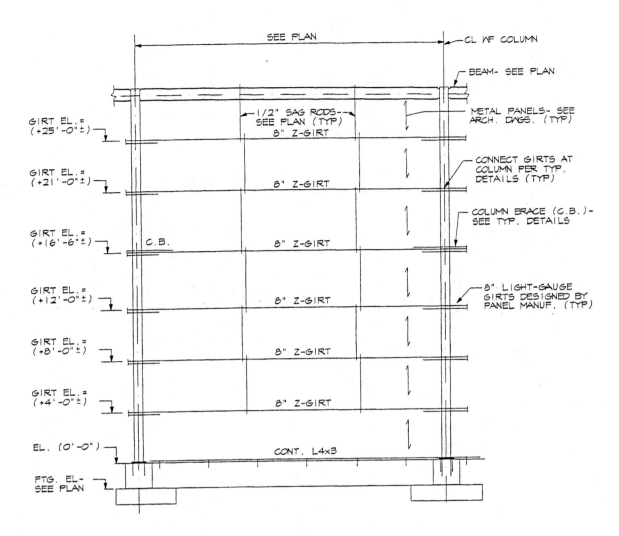

SEE PLAN

CL WF COLUMN

BEAM- SEE PLAN

1/2" SAG RODS-
SEE PLAN (TYP)

8" Z-GIRT

METAL PANELS- SEE
ARCH. DWGS. (TYP)

GIRT EL.=
(+25'-0"±)

GIRT EL.=
(+21'-0"±)

8" Z-GIRT

CONNECT GIRTS AT
COLUMN PER TYP.
DETAILS (TYP)

GIRT EL.=
(+16'-6"±)

C.B.

8" Z-GIRT

COLUMN BRACE (C.B.)-
SEE TYP. DETAILS

GIRT EL.=
(+12'-0"±)

8" Z-GIRT

8" LIGHT-GAUGE
GIRTS DESIGNED BY
PANEL MANUF. (TYP)

GIRT EL.=
(+8'-0"±)

8" Z-GIRT

GIRT EL.=
(+4'-0"±)

8" Z-GIRT

EL. (0'-0")

CONT. L4x3

FTG. EL-
SEE PLAN

TYP. PERIMETER WALL ELEVATION 2
NOT TO SCALE
(DETAIL T5-WALL2)

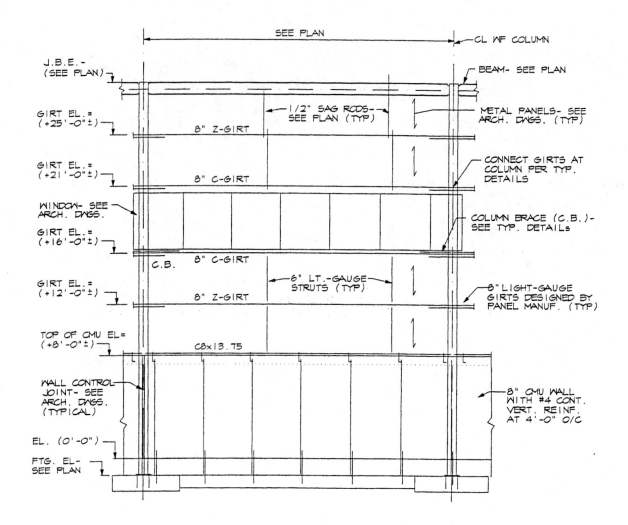

SEE PLAN

CL WF COLUMN

J.B.E.- (SEE PLAN)

BEAM- SEE PLAN

1/2" SAG RODS- SEE PLAN (TYP)

METAL PANELS- SEE ARCH. DWGS. (TYP)

GIRT EL.= (+25'-0"±)

8" Z-GIRT

GIRT EL.= (+21'-0"±)

8" C-GIRT

CONNECT GIRTS AT COLUMN PER TYP. DETAILS

WINDOW- SEE ARCH. DWGS.

GIRT EL.= (+16'-0"±)

COLUMN BRACE (C.B.)- SEE TYP. DETAILs

C.B.   8" C-GIRT

GIRT EL.= (+12'-0"±)

8" Z-GIRT

6" LT.-GAUGE STRUTS (TYP)

8" LIGHT-GAUGE GIRTS DESIGNED BY PANEL MANUF. (TYP)

TOP OF CMU EL= (+8'-0"±)

C8×13.75

WALL CONTROL JOINT- SEE ARCH. DWGS. (TYPICAL)

8" CMU WALL WITH #4 CONT. VERT. REINF. AT 4'-0" O/C

EL. (0'-0")

FTG. EL- SEE PLAN

TYP. PERIMETER WALL ELEVATION 3
NOT TO SCALE   (DETAIL T5—WALL3)

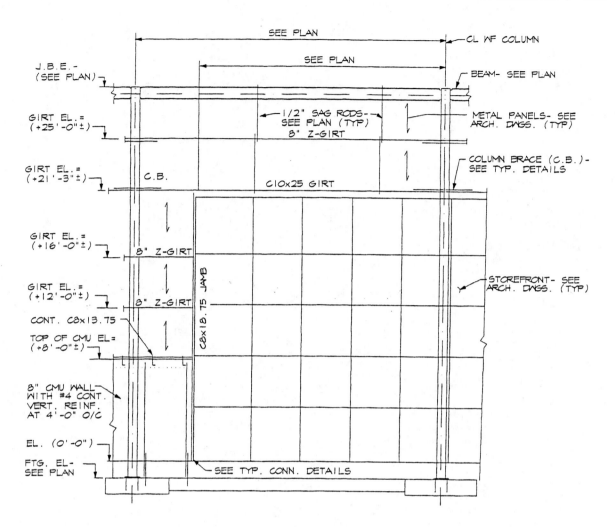

J.B.E.- (SEE PLAN)

GIRT EL.= (+25'-0"±)

GIRT EL.= (+21'-3"±)

C.B.

GIRT EL.= (+16'-0"±)

8" Z-GIRT

GIRT EL.= (+12'-0"±)

8" Z-GIRT

CONT. C8x13.75

TOP OF CMU EL= (+8'-0"±)

8" CMU WALL WITH #4 CONT. VERT. REINF. AT 4'-0" O/C

EL. (0'-0")

FTG. EL- SEE PLAN

SEE PLAN

SEE PLAN

CL WF COLUMN

BEAM- SEE PLAN

1/2" SAG RODS- SEE PLAN (TYP)
8" Z-GIRT

METAL PANELS- SEE ARCH. DWGS. (TYP)

COLUMN BRACE (C.B.)- SEE TYP. DETAILS

C10x25 GIRT

C8x18.75 JAMB

STOREFRONT- SEE ARCH. DWGS. (TYP)

SEE TYP. CONN. DETAILS

TYP. PERIMETER WALL ELEVATION 4
NOT TO SCALE   (DETAIL T5-WALL4)

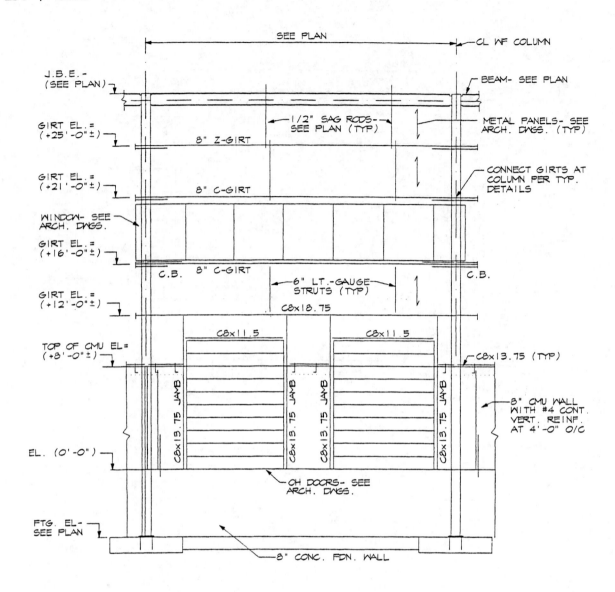

TYP. PERIMETER WALL ELEVATION 5
NOT TO SCALE   (DETAIL T5—WALL5)

NOTES:
1. AT JOIST NEAREST (OR ON) COLUMN CL CONNECT JOISTS WITH FIELD BOLTS PER S.J.I. REQUIREMENTS.
2. OFFSET JOISTS IF BEAM FLANGE IS LESS THAN 5" AND PROVIDE MINIMUM OF 2 1/2" JOIST BEARING.

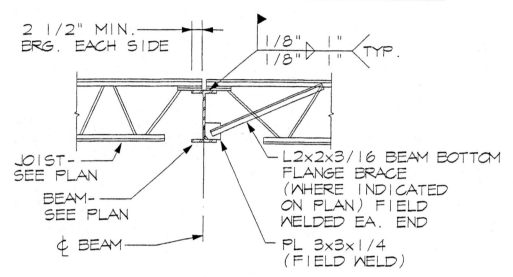

2 1/2" MIN. BRG. EACH SIDE

1/8" 1"
1/8" 1" TYP.

JOIST-SEE PLAN

BEAM-SEE PLAN

℄ BEAM

L2x2x3/16 BEAM BOTTOM FLANGE BRACE (WHERE INDICATED ON PLAN) FIELD WELDED EA. END

PL 3x3x1/4 (FIELD WELD)

AT INTERIOR BEAMS

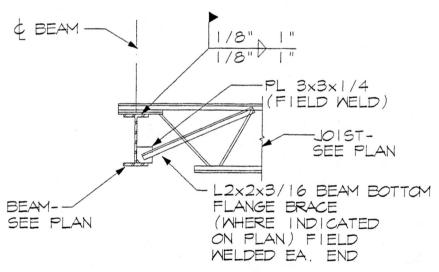

℄ BEAM

1/8" 1"
1/8" 1"

PL 3x3x1/4 (FIELD WELD)

JOIST-SEE PLAN

L2x2x3/16 BEAM BOTTOM FLANGE BRACE (WHERE INDICATED ON PLAN) FIELD WELDED EA. END

BEAM-SEE PLAN

AT PERIMETER BEAMS

TYP. K-JOIST CONNECTION DETAILS

NOT TO SCALE   (DETAIL T5-J1)

NOTES:
1. AT JOIST NEAREST (OR ON) COLUMN CL, CONNECT JOISTS WITH FIELD BOLTS PER S.J.I. REQUIREMENTS.
2. OFFSET JOISTS IF BEAM FLANGE IS LESS THAN 8" AND PROVIDE MINIMUM OF 4" JOIST BEARING.

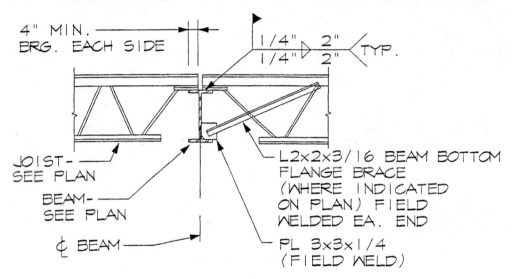

4" MIN. BRG. EACH SIDE

1/4" 2"
1/4" 2" TYP.

JOIST- SEE PLAN

BEAM- SEE PLAN

℄ BEAM

L2x2x3/16 BEAM BOTTOM FLANGE BRACE (WHERE INDICATED ON PLAN) FIELD WELDED EA. END

PL 3x3x1/4 (FIELD WELD)

AT INTERIOR BEAMS

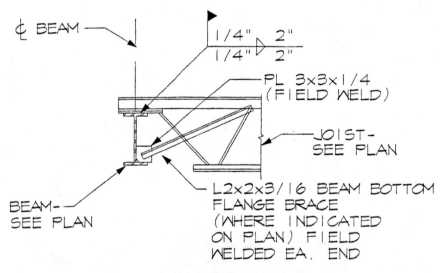

℄ BEAM

1/4" 2"
1/4" 2"

PL 3x3x1/4 (FIELD WELD)

JOIST- SEE PLAN

L2x2x3/16 BEAM BOTTOM FLANGE BRACE (WHERE INDICATED ON PLAN) FIELD WELDED EA. END

BEAM- SEE PLAN

AT PERIMETER BEAMS

TYP. LH–JOIST CONNECTION DETAILS

NOT TO SCALE   (DETAIL T5–J2)

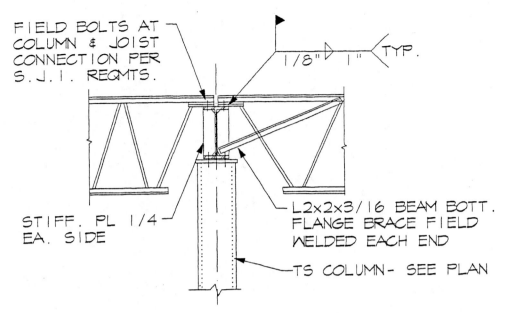

FIELD BOLTS AT
COLUMN & JOIST
CONNECTION PER
S.J.I. REQMTS.

1/8" ⟋ 1" TYP.

STIFF. PL 1/4
EA. SIDE

L2x2x3/16 BEAM BOTT.
FLANGE BRACE FIELD
WELDED EACH END

TS COLUMN- SEE PLAN

DETAIL AT CANTILEVERED BEAM SYSTEM

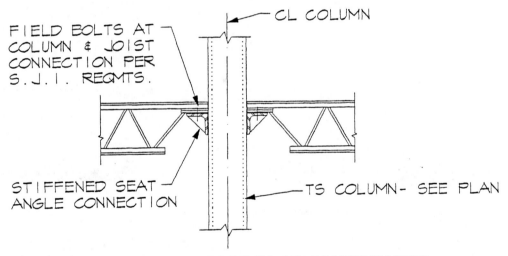

CL COLUMN

FIELD BOLTS AT
COLUMN & JOIST
CONNECTION PER
S.J.I. REQMTS.

STIFFENED SEAT
ANGLE CONNECTION

TS COLUMN- SEE PLAN

DETAIL WHERE COLUMN IS CONTINUOUS

TYP. CONNECTION DETAILS AT COL. CL

NOT TO SCALE   (DETAIL T5-J3)

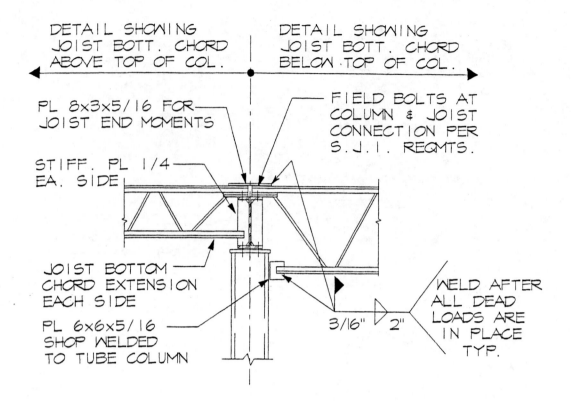

DETAIL SHOWING
JOIST BOTT. CHORD
ABOVE TOP OF COL.

DETAIL SHOWING
JOIST BOTT. CHORD
BELOW TOP OF COL.

PL 8x3x5/16 FOR
JOIST END MOMENTS

FIELD BOLTS AT
COLUMN & JOIST
CONNECTION PER
S.J.I. REQMTS.

STIFF. PL 1/4
EA. SIDE

JOIST BOTTOM
CHORD EXTENSION
EACH SIDE

PL 6x6x5/16
SHOP WELDED
TO TUBE COLUMN

3/16"   2"

WELD AFTER
ALL DEAD
LOADS ARE
IN PLACE
TYP.

TYP. CONNECTION DETAIL AT COL. CL

NOT TO SCALE   (DETAIL T5-J4)

NOTES:
1. CONCENTRATED LOAD LOCATED AT JOIST PANEL POINT LOCATION - NO ADDITIONAL ANGLES REQUIRED.
2. CONCENTRATED LOAD (100 LBS. OR HEAVIER) NOT LOCATED AT JOIST PANEL POINT LOCATION - PROVIDE ⌐L1×1×1/8 TO PANEL POINT AS SHOWN.

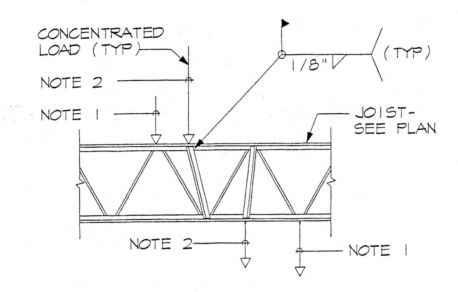

TYP. CONCENTRATED LOAD DETAIL
NOT TO SCALE   (DETAIL T5-J5)

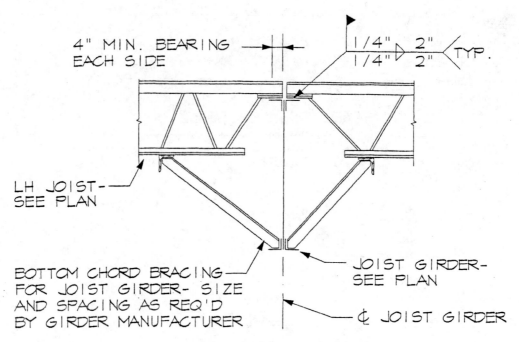

4" MIN. BEARING
EACH SIDE

1/4" 2"
1/4" 2" TYP.

LH JOIST-
SEE PLAN

BOTTOM CHORD BRACING
FOR JOIST GIRDER- SIZE
AND SPACING AS REQ'D
BY GIRDER MANUFACTURER

JOIST GIRDER-
SEE PLAN

₵ JOIST GIRDER

LH–SERIES JOISTS

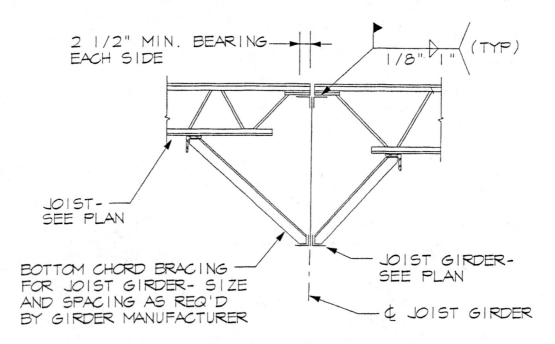

2 1/2" MIN. BEARING
EACH SIDE

1/8" 1" (TYP)

JOIST-
SEE PLAN

BOTTOM CHORD BRACING
FOR JOIST GIRDER- SIZE
AND SPACING AS REQ'D
BY GIRDER MANUFACTURER

JOIST GIRDER-
SEE PLAN

₵ JOIST GIRDER

K– SERIES JOISTS

TYP. JOIST TO GIRDER CONNECTION

NOT TO SCALE   (DETAIL T5–JG1)

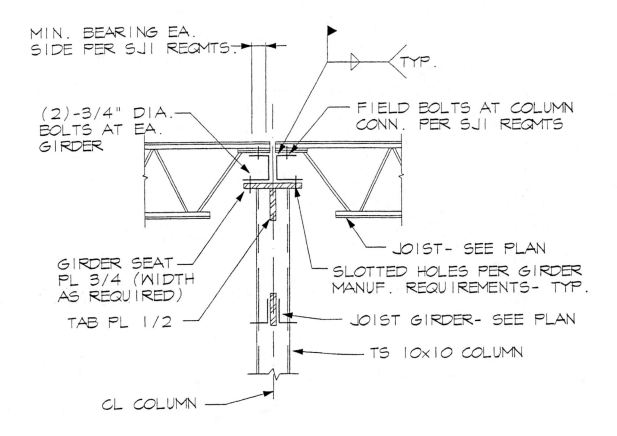

MIN. BEARING EA. SIDE PER SJI REQMTS.

TYP.

(2)-3/4" DIA. BOLTS AT EA. GIRDER

FIELD BOLTS AT COLUMN CONN. PER SJI REQMTS

GIRDER SEAT PL 3/4 (WIDTH AS REQUIRED)

TAB PL 1/2

JOIST- SEE PLAN

SLOTTED HOLES PER GIRDER MANUF. REQUIREMENTS- TYP.

JOIST GIRDER- SEE PLAN

TS 10x10 COLUMN

CL COLUMN

TYP. DETAIL AT INTERIOR COLUMN CL

NOT TO SCALE

(DETAIL T5-JG2)

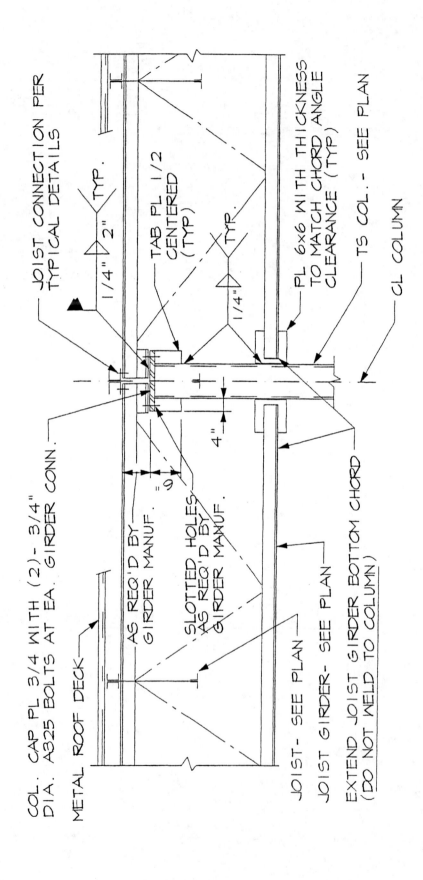

COL. CAP PL 3/4 WITH (2)- 3/4" DIA. A325 BOLTS AT EA. GIRDER CONN.

METAL ROOF DECK

JOIST CONNECTION PER TYPICAL DETAILS

1/4" 2" TYP.

TAB PL 1/2 CENTERED (TYP)

1/4" TYP.

PL 6x6 WITH THICKNESS TO MATCH CHORD ANGLE CLEARANCE (TYP)

TS COL.- SEE PLAN

CL COLUMN

4"

6" = AS REQ'D BY GIRDER MANUF.

SLOTTED HOLES AS REQ'D BY GIRDER MANUF.

JOIST- SEE PLAN

JOIST GIRDER- SEE PLAN

EXTEND JOIST GIRDER BOTTOM CHORD (DO NOT WELD TO COLUMN)

TYPICAL JOIST GIRDER CONNECTION DETAIL AT INTERIOR

NOT TO SCALE
(DETAIL T5-JG3)

262

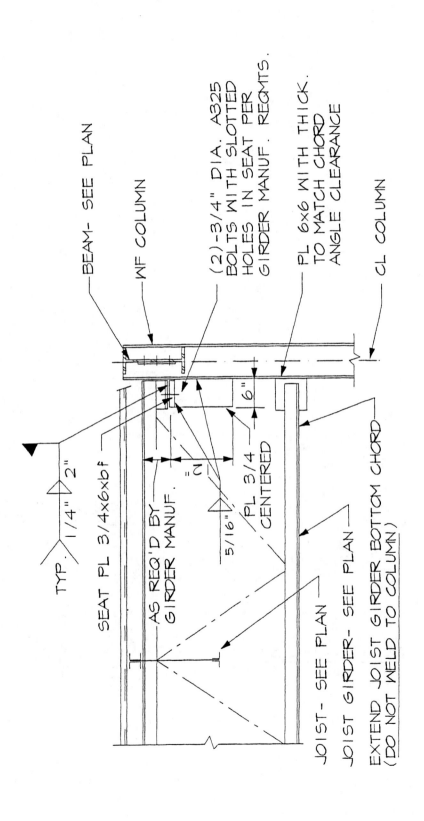

TYP. $\frac{1/4"}{}$ 2"

SEAT PL 3/4x6xbf

AS REQ'D BY GIRDER MANUF.

2"

5/16

PL 3/4 CENTERED

BEAM- SEE PLAN

WF COLUMN

(2)-3/4" DIA. A325 BOLTS WITH SLOTTED HOLES IN SEAT PER GIRDER MANUF. REQMTS.

PL 6x6 WITH THICK. TO MATCH CHORD ANGLE CLEARANCE

CL COLUMN

6"

JOIST- SEE PLAN

JOIST GIRDER- SEE PLAN

EXTEND JOIST GIRDER BOTTOM CHORD (DO NOT WELD TO COLUMN)

TYPICAL JOIST GIRDER CONNECTION DETAIL AT PERIMETER

DETAIL AT PERIMETER WF COLUMN AND WF BEAM

(DETAIL T5-JG4)

NOT TO SCALE

263

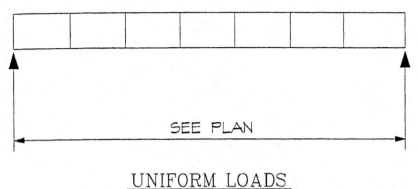

W1 = 23 PSF DL + 20 PSF LL
W2 = 15 PSF NET WIND UPLIFT

SEE PLAN

UNIFORM LOADS

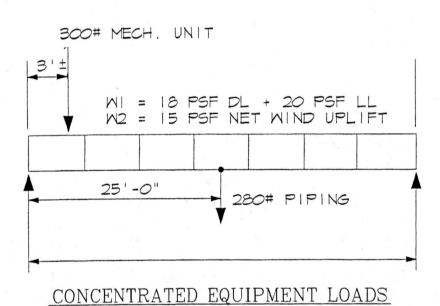

300# MECH. UNIT

3' ±

W1 = 18 PSF DL + 20 PSF LL
W2 = 15 PSF NET WIND UPLIFT

25'-0"

280# PIPING

CONCENTRATED EQUIPMENT LOADS

SPECIAL JOIST LOADING DIAGRAM 1

NOT TO SCALE   (DETAIL T5–JL1)

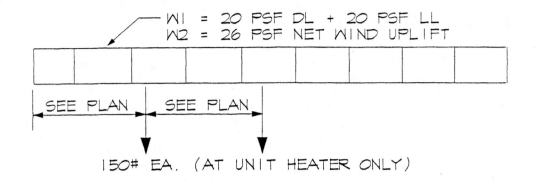

W1 = 20 PSF DL + 20 PSF LL
W2 = 26 PSF NET WIND UPLIFT

SEE PLAN   SEE PLAN

150# EA. (AT UNIT HEATER ONLY)

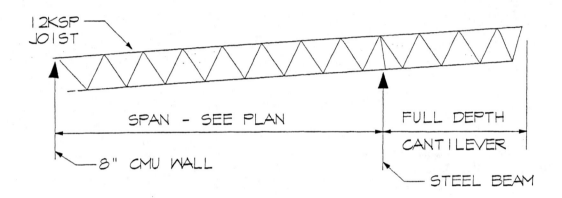

12KSP JOIST

SPAN - SEE PLAN   FULL DEPTH

8" CMU WALL   CANTILEVER

                                   STEEL BEAM

SPECIAL JOIST LOADING DIAGRAM 2
NOT TO SCALE   (DETAIL T5-JL2)

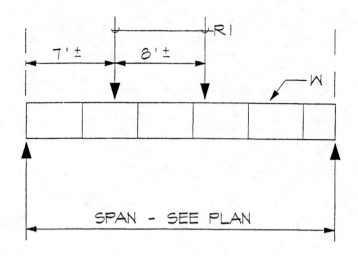

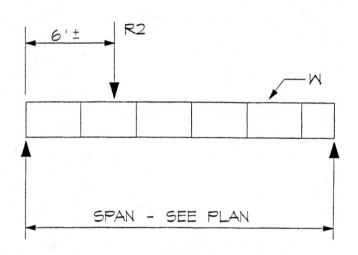

SPECIAL JOIST LOADING DIAGRAM 3

NOT TO SCALE   (DETAIL T5–JL3)

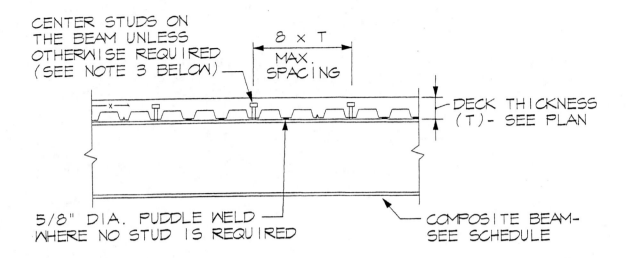

CENTER STUDS ON THE BEAM UNLESS OTHERWISE REQUIRED (SEE NOTE 3 BELOW)

8 × T MAX. SPACING

DECK THICKNESS (T) - SEE PLAN

5/8" DIA. PUDDLE WELD WHERE NO STUD IS REQUIRED

COMPOSITE BEAM- SEE SCHEDULE

DECK IS PERPENDICULAR OR SKEWED TO BEAM

NOTES:

1. THE MIN. NUMBER OF STUDS FOR EACH BEAM IS SHOWN IN THE COMPOSITE BEAM SCHEDULE.

2. SPACE STUDS AS EVENLY AS POSSIBLE IN AVAILABLE DECK FLUTES. WHERE STUD SPACING EXCEEDS THE MAX. SPACING ALLOWED, PROVIDE ADDITIONAL STUDS TO SATISFY THE SPACING REQUIREMENTS.

3. WHERE THE NUMBER OF STUDS EXCEEDS THE NUMBER OF FLUTES, PROVIDE TWO STUDS IN EVERY OTHER FLUTE, STARTING AT EACH END OF THE BEAM. THE TRANSVERSE SPACING BETWEEN TWO STUDS IN A SINGLE FLUTE SHALL BE 4 × STUD DIAMETERS (MIN.).

4. SEE THE COMPOSITE BEAM SCHEDULE FOR ADDITIONAL REQUIREMENTS. TURN THE NATURAL BEAM CAMBER UP.

TYP. COMPOSITE BEAM ELEVATION 1

NOT TO SCALE

(DETAIL T5-COMP1)

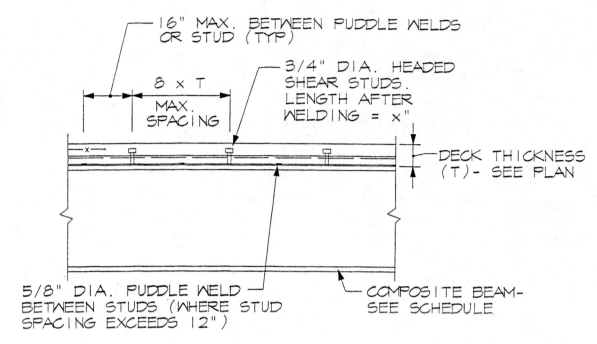

16" MAX. BETWEEN PUDDLE WELDS
OR STUD (TYP)

3/4" DIA. HEADED
SHEAR STUDS.
LENGTH AFTER
WELDING = x"

8 × T
MAX.
SPACING

DECK THICKNESS
(T)- SEE PLAN

5/8" DIA. PUDDLE WELD
BETWEEN STUDS (WHERE STUD
SPACING EXCEEDS 12")

COMPOSITE BEAM-
SEE SCHEDULE

DECK IS PARALLEL TO BEAM

COMPOSITE BEAM SCHEDULE

MARK	BEAM	MAX. END REACTION	MIDSPAN CAMBER	NUMBER OF STUDS	REMARKS
CB-1	W10x15	11 KIPS	—	12	INTERIOR
CB-2	W10x15	8 KIPS	—	8	INTERIOR
CB-3	W10x26	13 KIPS	—	12	INTERIOR
CB-4	W10x15	8 KIPS	—	3	CANTILEVER
CB-5	W10x26	13 KIPS	—	5	CANTILEVER
CB-6	W10x12	7 KIPS	—	3	CANTILEVER
CB-7	W18x76	86 KIPS	0.6"	22	GIRDER
CB-8	W16x40	36 KIPS	—	16	GIRDER

NOTE: NATURAL CAMBER OF BEAM SHALL BE TURNED UP.

TYP. COMPOSITE BEAM ELEVATION 2

NOT TO SCALE   (DETAIL T5—COMP2)

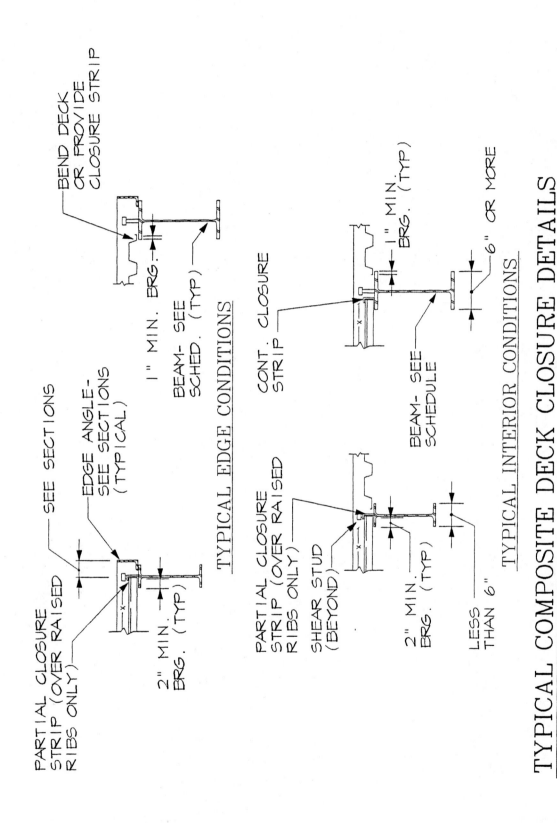

PARTIAL CLOSURE STRIP (OVER RAISED RIBS ONLY)

SEE SECTIONS

EDGE ANGLE- SEE SECTIONS (TYPICAL)

BEND DECK OR PROVIDE CLOSURE STRIP

1" MIN. BRG.

BEAM- SEE SCHED. (TYP)

2" MIN. BRG. (TYP)

TYPICAL EDGE CONDITIONS

PARTIAL CLOSURE STRIP (OVER RAISED RIBS ONLY)

SHEAR STUD (BEYOND)

CONT. CLOSURE STRIP

2" MIN. BRG. (TYP)

1" MIN. BRG. (TYP)

BEAM- SEE SCHEDULE

LESS THAN 6"

6" OR MORE

TYPICAL INTERIOR CONDITIONS

TYPICAL COMPOSITE DECK CLOSURE DETAILS

(DETAIL T5-COMP3)

NOT TO SCALE

269

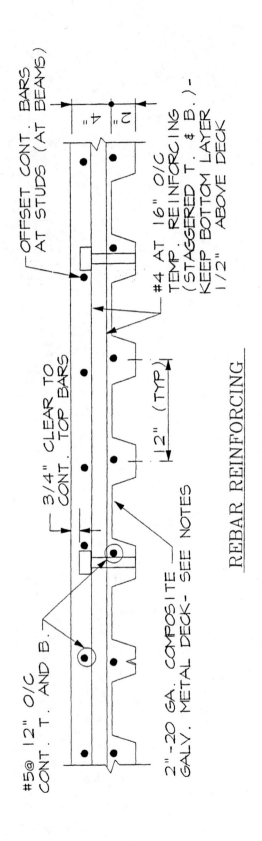

OFFSET CONT. BARS AT STUDS (AT BEAMS)

#4 AT 16" O/C TEMP. REINFORCING (STAGGERED T. & B.) - KEEP BOTTOM LAYER 1/2" ABOVE DECK

3/4" CLEAR TO CONT. TOP BARS

2" (TYP)

#5@ 12" O/C CONT. T. AND B.

2"-20 GA. COMPOSITE GALV. METAL DECK- SEE NOTES

REBAR REINFORCING

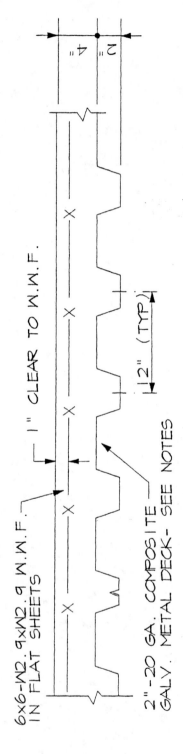

1" CLEAR TO W.W.F.

6x6-W2.9xW2.9 W.W.F. IN FLAT SHEETS

2" (TYP)

2"-20 GA. COMPOSITE GALV. METAL DECK- SEE NOTES

WELDED WIRE FABRIC REINFORCING

TYPICAL COMPOSITE DECK REINFORCING DETAILS

(DETAIL T5-COMP4)

NOT TO SCALE

270

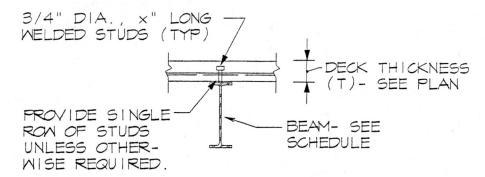

NOTE: SLAB REINFORCING IS NOT SHOWN FOR CLARITY.

3/4" DIA., x" LONG WELDED STUDS (TYP)

DECK THICKNESS (T) - SEE PLAN

PROVIDE SINGLE ROW OF STUDS UNLESS OTHER- WISE REQUIRED.

BEAM- SEE SCHEDULE

DECK IS PERPENDICULAR OR SKEWED TO BEAM

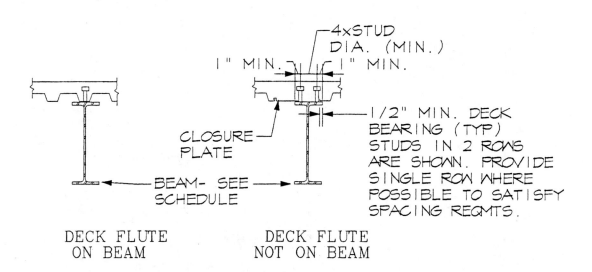

4xSTUD DIA. (MIN.)

1" MIN.   1" MIN.

1/2" MIN. DECK BEARING (TYP) STUDS IN 2 ROWS ARE SHOWN. PROVIDE SINGLE ROW WHERE POSSIBLE TO SATISFY SPACING REQMTS.

CLOSURE PLATE

BEAM- SEE SCHEDULE

DECK FLUTE ON BEAM

DECK FLUTE NOT ON BEAM

DECK IS PARALLEL TO BEAM

COMPOSITE BEAM SECTION DETAILS

NOT TO SCALE   (DETAIL T5-COMP5)

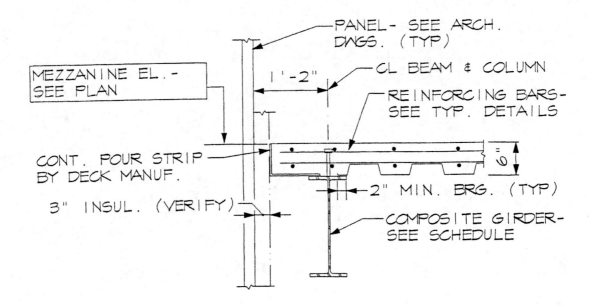

PANEL- SEE ARCH. DWGS. (TYP)

CL BEAM & COLUMN

REINFORCING BARS- SEE TYP. DETAILS

1'-2"

MEZZANINE EL.- SEE PLAN

CONT. POUR STRIP BY DECK MANUF.

3" INSUL. (VERIFY)

6"

2" MIN. BRG. (TYP)

COMPOSITE GIRDER- SEE SCHEDULE

PERIMETER GIRDER CONDITION

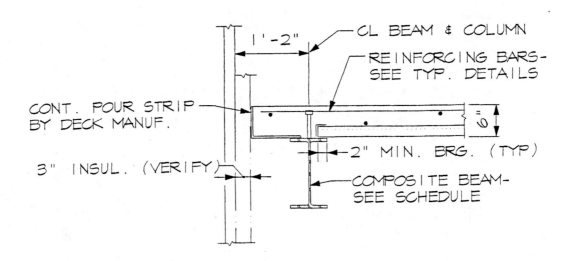

CL BEAM & COLUMN

REINFORCING BARS- SEE TYP. DETAILS

1'-2"

CONT. POUR STRIP BY DECK MANUF.

3" INSUL. (VERIFY)

6"

2" MIN. BRG. (TYP)

COMPOSITE BEAM- SEE SCHEDULE

PERIMETER BEAM CONDITION

COMPOSITE BEAM SECTION DETAILS

NOT TO SCALE

(DETAIL T5-COMP6)

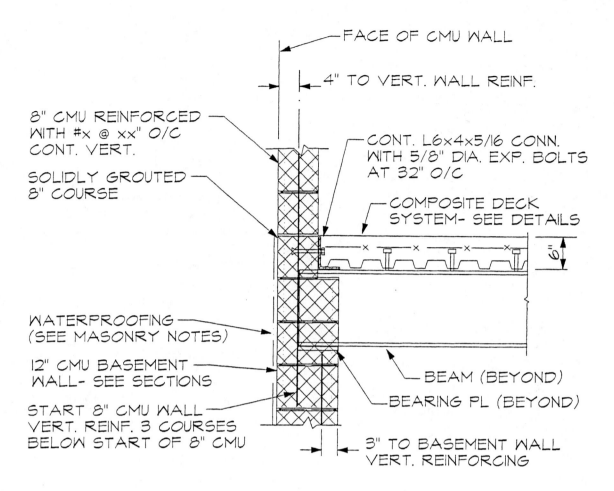

FACE OF CMU WALL

4" TO VERT. WALL REINF.

8" CMU REINFORCED WITH #x @ xx" O/C CONT. VERT.

SOLIDLY GROUTED 8" COURSE

CONT. L6x4x5/16 CONN. WITH 5/8" DIA. EXP. BOLTS AT 32" O/C

COMPOSITE DECK SYSTEM- SEE DETAILS

6"

WATERPROOFING (SEE MASONRY NOTES)

12" CMU BASEMENT WALL- SEE SECTIONS

START 8" CMU WALL VERT. REINF. 3 COURSES BELOW START OF 8" CMU

BEAM (BEYOND)

BEARING PL (BEYOND)

3" TO BASEMENT WALL VERT. REINFORCING

COMPOSITE BEAM SECTION DETAILS

NOT TO SCALE   (DETAIL T5-COMP7)

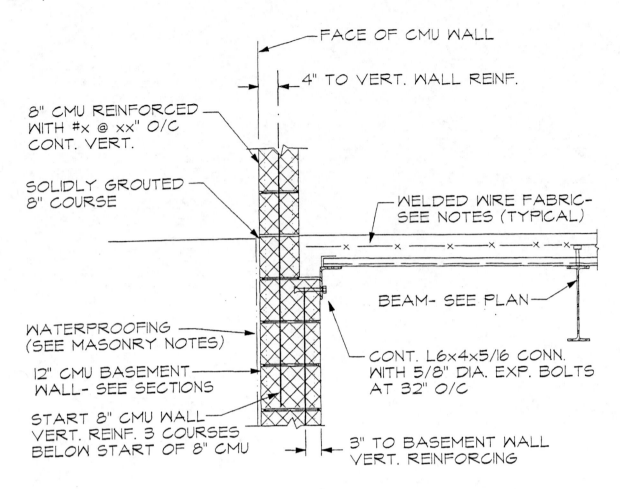

FACE OF CMU WALL

4" TO VERT. WALL REINF.

8" CMU REINFORCED WITH #x @ xx" O/C CONT. VERT.

SOLIDLY GROUTED 8" COURSE

WELDED WIRE FABRIC- SEE NOTES (TYPICAL)

BEAM- SEE PLAN

WATERPROOFING (SEE MASONRY NOTES)

12" CMU BASEMENT WALL- SEE SECTIONS

START 8" CMU WALL VERT. REINF. 3 COURSES BELOW START OF 8" CMU

CONT. L6x4x5/16 CONN. WITH 5/8" DIA. EXP. BOLTS AT 32" O/C

3" TO BASEMENT WALL VERT. REINFORCING

COMPOSITE BEAM SECTION DETAILS

NOT TO SCALE

(DETAIL T5-COMP8)

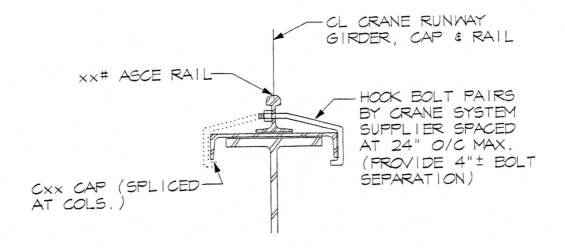

TYPICAL CRANE RAIL CONNECTION
NOT TO SCALE

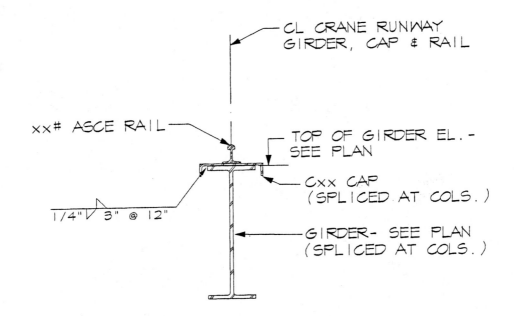

TYPICAL RUNWAY GIRDER DETAIL
NOT TO SCALE

(DETAIL T5—CRAN1)

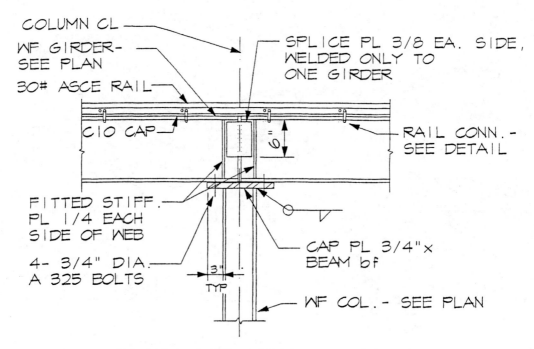

COLUMN CL

WF GIRDER-
SEE PLAN

30# ASCE RAIL

C10 CAP

SPLICE PL 3/8 EA. SIDE,
WELDED ONLY TO
ONE GIRDER

RAIL CONN. -
SEE DETAIL

FITTED STIFF.
PL 1/4 EACH
SIDE OF WEB

4- 3/4" DIA.
A 325 BOLTS

CAP PL 3/4"x
BEAM bf

WF COL. - SEE PLAN

3"
TYP

6"

INTERIOR FRAMING CONDITION

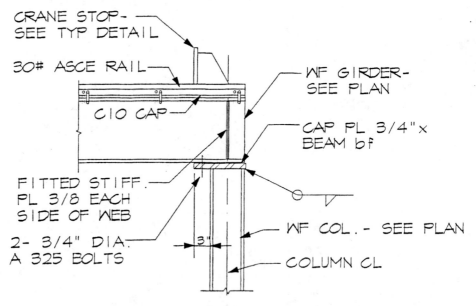

CRANE STOP-
SEE TYP DETAIL

30# ASCE RAIL

C10 CAP

WF GIRDER-
SEE PLAN

CAP PL 3/4"x
BEAM bf

FITTED STIFF.
PL 3/8 EACH
SIDE OF WEB

2- 3/4" DIA.
A 325 BOLTS

WF COL. - SEE PLAN

COLUMN CL

3"

END FRAMING CONDITION

TYPICAL GIRDER CONNECTION DETAILS

NOT TO SCALE   (DETAIL T5-CRAN2)

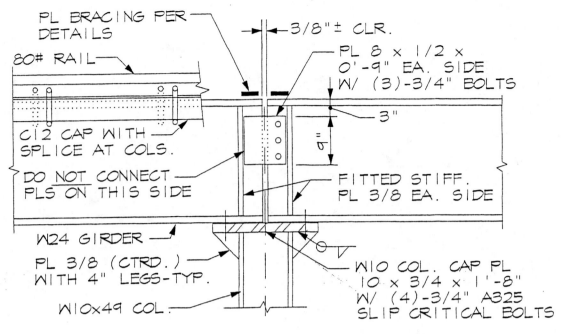

PL BRACING PER DETAILS

80# RAIL

C12 CAP WITH SPLICE AT COLS.

DO NOT CONNECT PLS ON THIS SIDE

W24 GIRDER

PL 3/8 (CTRD.) WITH 4" LEGS-TYP.

W10x49 COL.

3/8"± CLR.

PL 8 × 1/2 × 0'-9" EA. SIDE W/ (3)-3/4" BOLTS

3"

9"

FITTED STIFF. PL 3/8 EA. SIDE

W10 COL. CAP PL 10 × 3/4 × 1'-8" W/ (4)-3/4" A325 SLIP CRITICAL BOLTS

INTERIOR FRAMING CONDITION

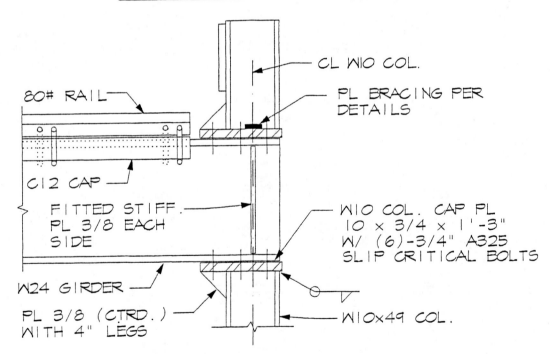

80# RAIL

C12 CAP

FITTED STIFF. PL 3/8 EACH SIDE

W24 GIRDER

PL 3/8 (CTRD.) WITH 4" LEGS

CL W10 COL.

PL BRACING PER DETAILS

W10 COL. CAP PL 10 × 3/4 × 1'-3" W/ (6)-3/4" A325 SLIP CRITICAL BOLTS

W10x49 COL.

END FRAMING CONDITION

TYPICAL GIRDER CONNECTION DETAILS

NOT TO SCALE   (DETAIL T5-CRAN3)

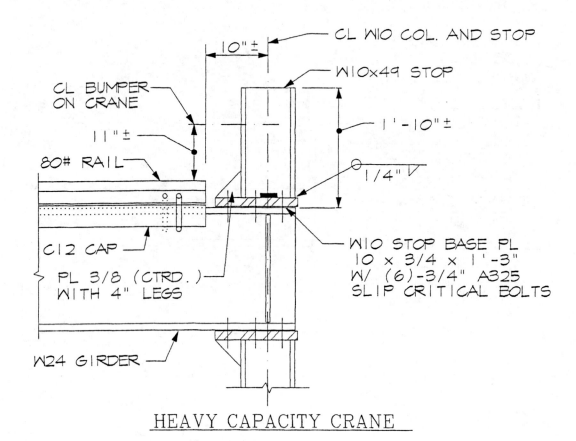

CL W10 COL. AND STOP

W10x49 STOP

10"±

1'-10"±

CL BUMPER
ON CRANE

11"±

1/4"

80# RAIL

W10 STOP BASE PL
10 × 3/4 × 1'-3"
W/ (6)-3/4" A325
SLIP CRITICAL BOLTS

C12 CAP

PL 3/8 (CTRD.)
WITH 4" LEGS

W24 GIRDER

HEAVY CAPACITY CRANE

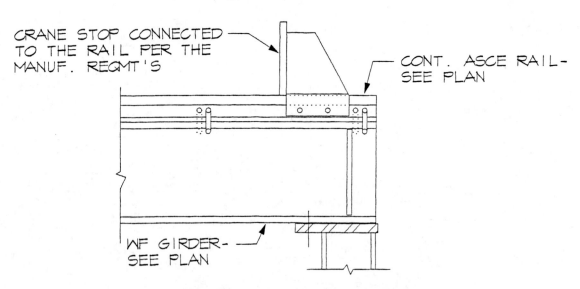

CRANE STOP CONNECTED
TO THE RAIL PER THE
MANUF. REQMT'S

CONT. ASCE RAIL-
SEE PLAN

WF GIRDER-
SEE PLAN

LIGHT CAPACITY CRANE

TYPICAL CRANE STOP DETAILS

NOT TO SCALE   (DETAIL T5-CRAN4)

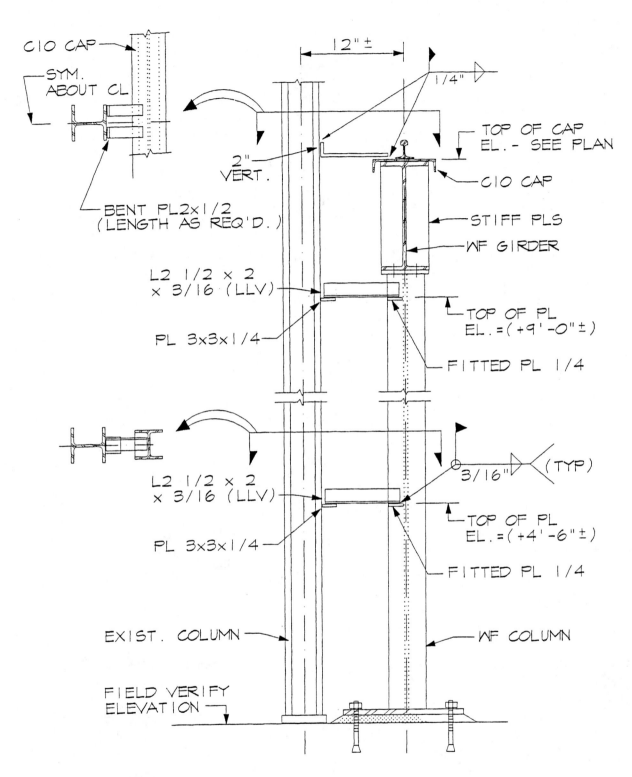

C10 CAP

SYM. ABOUT CL

BENT PL2x1/2 (LENGTH AS REQ'D.)

12"±

2" VERT.

1/4"

TOP OF CAP EL. - SEE PLAN

C10 CAP

STIFF PLS

WF GIRDER

L2 1/2 x 2 x 3/16 (LLV)

PL 3x3x1/4

TOP OF PL EL. = (+9'-0"±)

FITTED PL 1/4

L2 1/2 x 2 x 3/16 (LLV)

3/16" (TYP)

PL 3x3x1/4

TOP OF PL EL. = (+4'-6"±)

FITTED PL 1/4

EXIST. COLUMN

WF COLUMN

FIELD VERIFY ELEVATION

TYP. COLUMN TIE TO EXIST. COLUMN
NOT TO SCALE   (DETAIL T5-CRAN5)

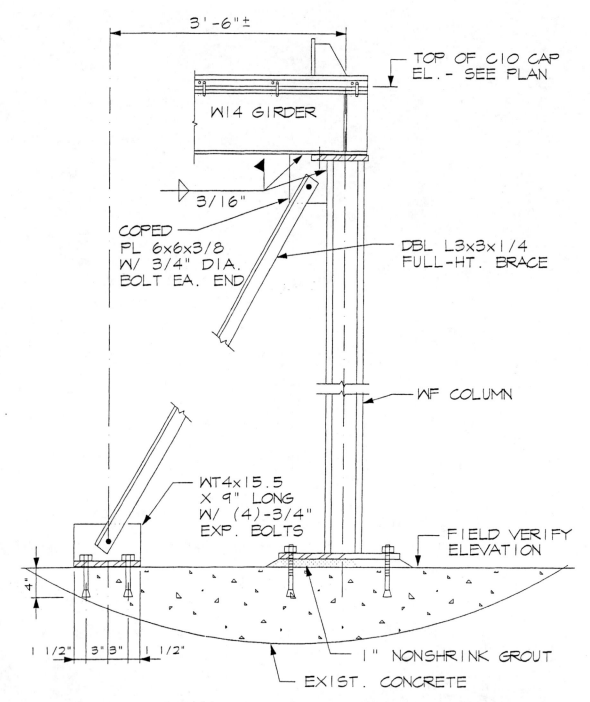

3'-6"±

TOP OF C10 CAP
EL.- SEE PLAN

W14 GIRDER

3/16"

COPED
PL 6x6x3/8
W/ 3/4" DIA.
BOLT EA. END

DBL L3x3x1/4
FULL-HT. BRACE

WF COLUMN

WT4x15.5
X 9" LONG
W/ (4)-3/4"
EXP. BOLTS

FIELD VERIFY
ELEVATION

4"

1 1/2" | 3" | 3" | 1 1/2"

1" NONSHRINK GROUT

EXIST. CONCRETE

TYPICAL COLUMN END BRACE DETAIL
NOT TO SCALE   (DETAIL T5-CRAN6)

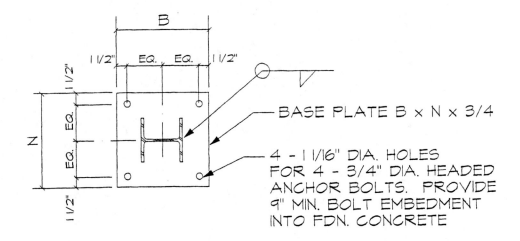

BASE PLATE B x N x 3/4

4 - 1 1/16" DIA. HOLES
FOR 4 - 3/4" DIA. HEADED
ANCHOR BOLTS. PROVIDE
9" MIN. BOLT EMBEDMENT
INTO FDN. CONCRETE

COLUMN ON NEW FOUNDATION

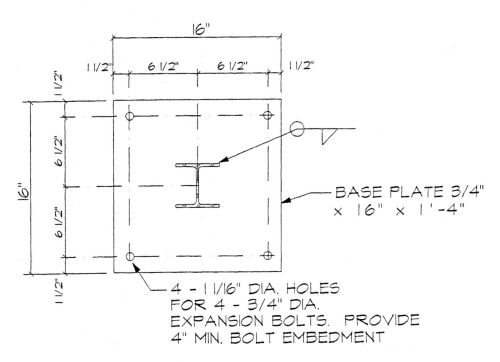

BASE PLATE 3/4"
x 16" x 1'-4"

4 - 1 1/16" DIA. HOLES
FOR 4 - 3/4" DIA.
EXPANSION BOLTS. PROVIDE
4" MIN. BOLT EMBEDMENT

COLUMN ON EXISTING SLAB

TYPICAL COLUMN BASE PL DETAILS

NOT TO SCALE   (DETAIL T5-CRAN7)

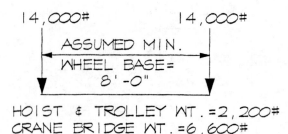

14,000#   14,000#

ASSUMED MIN.
WHEEL BASE=
8'-0"

HOIST & TROLLEY WT.=2,200#
CRANE BRIDGE WT.=6,600#

ASSUMED CRANE WHEEL LOADINGS
(10 TON CRANE, 2 WHEELS PER END TRUCK)

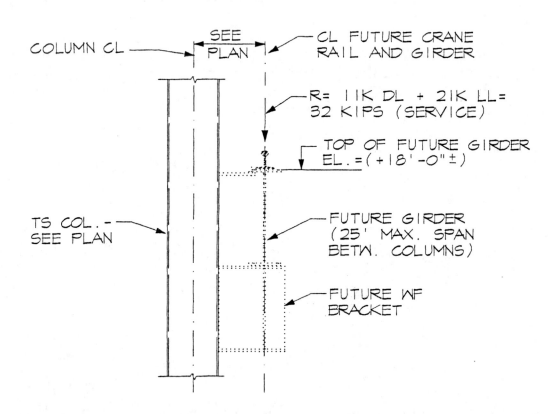

COLUMN CL ——

SEE
PLAN

— CL FUTURE CRANE
RAIL AND GIRDER

— R= 11K DL + 21K LL=
32 KIPS (SERVICE)

— TOP OF FUTURE GIRDER
EL.=(+18'-0"±)

TS COL.-
SEE PLAN

— FUTURE GIRDER
(25' MAX. SPAN
BETW. COLUMNS)

— FUTURE WF
BRACKET

ASSUMED FUTURE CRANE CONDITION
NOT TO SCALE   (DETAIL T5-CRAN8)

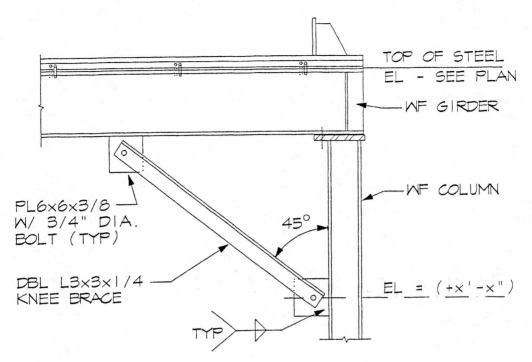

TOP OF STEEL
EL - SEE PLAN

WF GIRDER

WF COLUMN

45°

PL6x6x3/8
W/ 3/4" DIA.
BOLT (TYP)

DBL L3x3x1/4
KNEE BRACE

EL = (+x'-x")

TYP

TYPICAL KNEE BRACE DETAIL

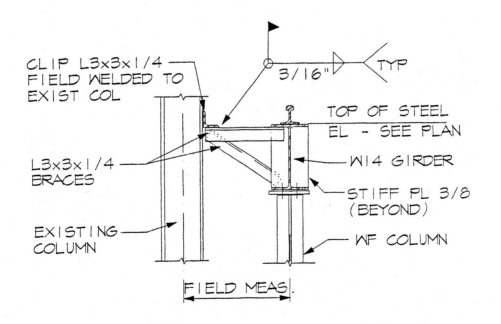

3/16" TYP

CLIP L3x3x1/4
FIELD WELDED TO
EXIST COL

L3x3x1/4
BRACES

EXISTING
COLUMN

FIELD MEAS.

TOP OF STEEL
EL - SEE PLAN

W14 GIRDER

STIFF PL 3/8
(BEYOND)

WF COLUMN

TYPICAL GIRDER BRACE DETAIL

MISC. CRANE FRAMING DETAILS

NOT TO SCALE   (DETAIL T5-CRAN9)

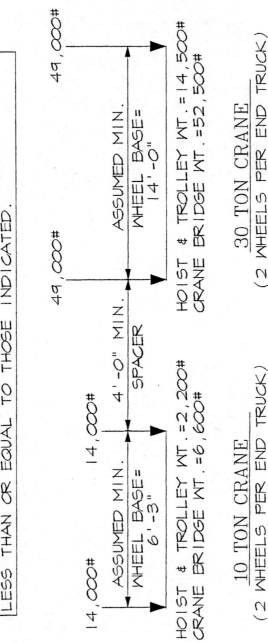

NOTE: WHEEL LOADINGS SHOWN DO NOT INCLUDE IMPACT. CONTRACTOR TO PROVIDE CRANE SYSTEM WITH WHEEL LOADS LESS THAN OR EQUAL TO THOSE INDICATED.

14,000#   14,000#

ASSUMED MIN.
WHEEL BASE=
6'-3"

HOIST & TROLLEY WT.=2,200#
CRANE BRIDGE WT.=6,600#

10 TON CRANE
(2 WHEELS PER END TRUCK)

4'-0" MIN.
SPACER

49,000#   49,000#

ASSUMED MIN.
WHEEL BASE=
14'-0"

HOIST & TROLLEY WT.=14,500#
CRANE BRIDGE WT.=52,500#

30 TON CRANE
(2 WHEELS PER END TRUCK)

ASSUMED CRANE WHEEL LOADING DIAGRAMS
(DETAIL T5—CRA10)

NOT TO SCALE

284

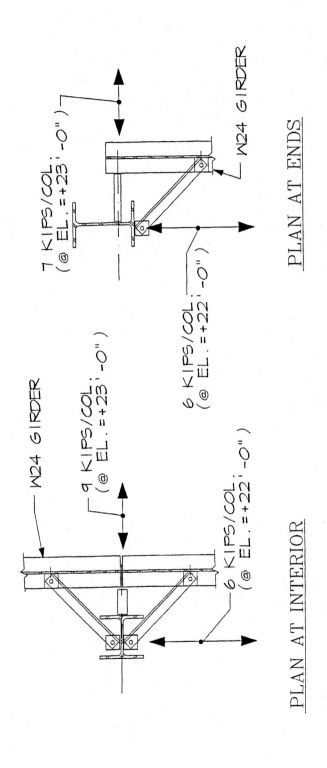

W24 GIRDER

9 KIPS/COL.
(@ EL.=+23'-0")

6 KIPS/COL.
(@ EL.=+22'-0")

PLAN AT INTERIOR

7 KIPS/COL.
(@ EL.=+23'-0")

W24 GIRDER

6 KIPS/COL.
(@ EL.=+22'-0")

PLAN AT ENDS

TYP. CRANE LOADINGS TO MAIN BUILDING COLUMNS
(DETAIL T5-CRA11)

NOT TO SCALE

285

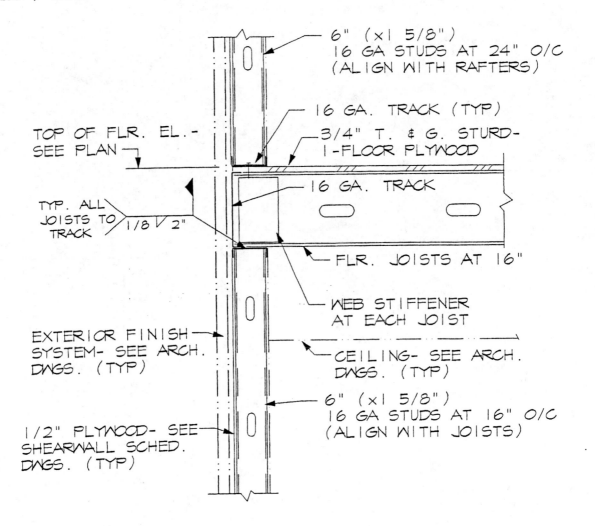

6" (x1 5/8")
16 GA STUDS AT 24" O/C
(ALIGN WITH RAFTERS)

16 GA. TRACK (TYP)

3/4" T. & G. STURD-
I-FLOOR PLYWOOD

TOP OF FLR. EL.-
SEE PLAN

16 GA. TRACK

TYP. ALL
JOISTS TO
TRACK   1/8 2"

FLR. JOISTS AT 16"

WEB STIFFENER
AT EACH JOIST

EXTERIOR FINISH
SYSTEM- SEE ARCH.
DWGS. (TYP)

CEILING- SEE ARCH.
DWGS. (TYP)

6" (x1 5/8")
16 GA STUDS AT 16" O/C
(ALIGN WITH JOISTS)

1/2" PLYWOOD- SEE
SHEARWALL SCHED.
DWGS. (TYP)

TYPICAL FLOOR JOIST BEARING DETAIL
NOT TO SCALE   (DETAIL T5-CF1)

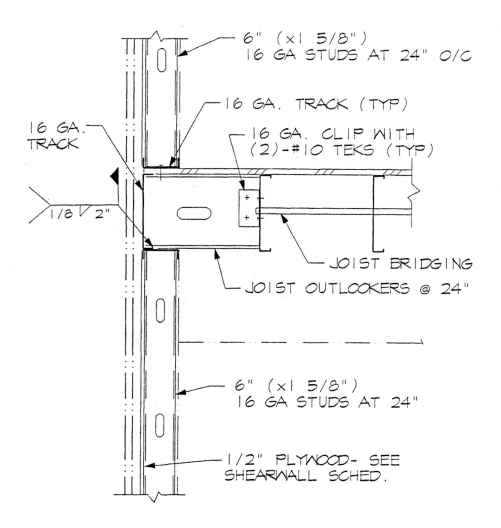

6" (x1 5/8")
16 GA STUDS AT 24" O/C

16 GA. TRACK (TYP)

16 GA. CLIP WITH
(2)-#10 TEKS (TYP)

16 GA.
TRACK

1/8 2"

JOIST BRIDGING

JOIST OUTLOOKERS @ 24"

6" (x1 5/8")
16 GA STUDS AT 24"

1/2" PLYWOOD- SEE
SHEARWALL SCHED.

TYP. FLOOR JOIST NONBEARING DETAIL

NOT TO SCALE   (DETAIL T5-CF2)

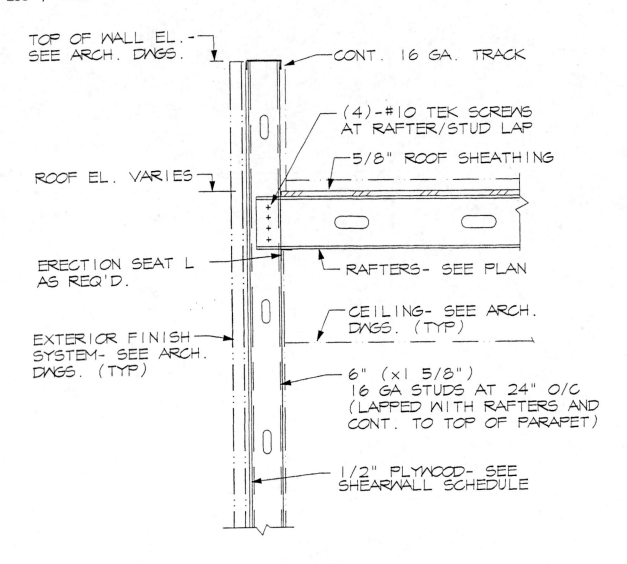

TOP OF WALL EL.-
SEE ARCH. DWGS.

CONT. 16 GA. TRACK

(4)-#10 TEK SCREWS
AT RAFTER/STUD LAP

5/8" ROOF SHEATHING

ROOF EL. VARIES

ERECTION SEAT L
AS REQ'D.

RAFTERS- SEE PLAN

CEILING- SEE ARCH.
DWGS. (TYP)

EXTERIOR FINISH
SYSTEM- SEE ARCH.
DWGS. (TYP)

6" (x1 5/8")
16 GA STUDS AT 24" O/C
(LAPPED WITH RAFTERS AND
CONT. TO TOP OF PARAPET)

1/2" PLYWOOD- SEE
SHEARWALL SCHEDULE

TYPICAL ROOF JOIST BEARING DETAIL

NOT TO SCALE

(DETAIL T5-CF3)

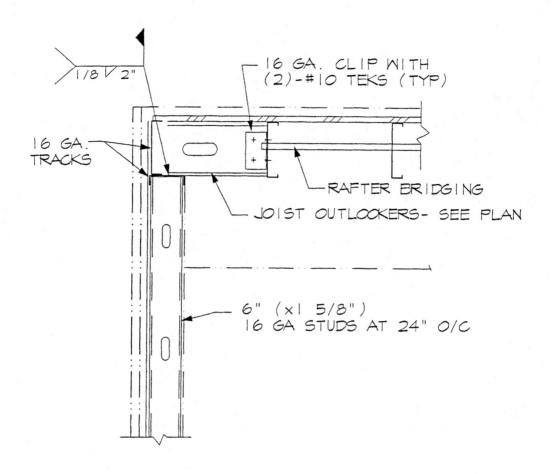

16 GA. CLIP WITH
(2)-#10 TEKS (TYP)

16 GA.
TRACKS

RAFTER BRIDGING

JOIST OUTLOOKERS- SEE PLAN

6" (x1 5/8")
16 GA STUDS AT 24" O/C

TYP. ROOF JOIST NONBEARING DETAIL
NOT TO SCALE   (DETAIL T5-CF4)

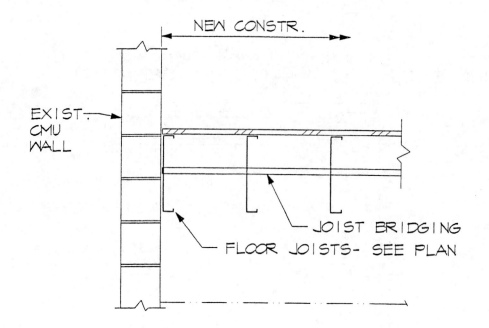

NEW CONSTR.

EXIST. CMU WALL

JOIST BRIDGING

FLOOR JOISTS- SEE PLAN

FLOOR DETAIL AT EXISTING MASONRY WALL

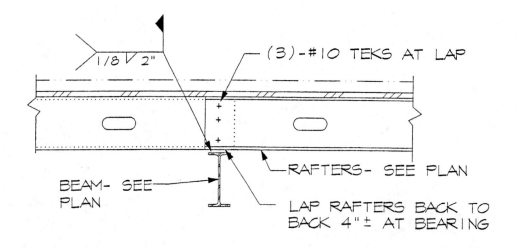

1/8 V 2"

(3)-#10 TEKS AT LAP

RAFTERS- SEE PLAN

BEAM- SEE PLAN

LAP RAFTERS BACK TO BACK 4"± AT BEARING

ROOF DETAIL AT INTERIOR STEEL BEAMS

MISCELLANEOUS FRAMING DETAILS

NOT TO SCALE   (DETAIL T5-CF5)

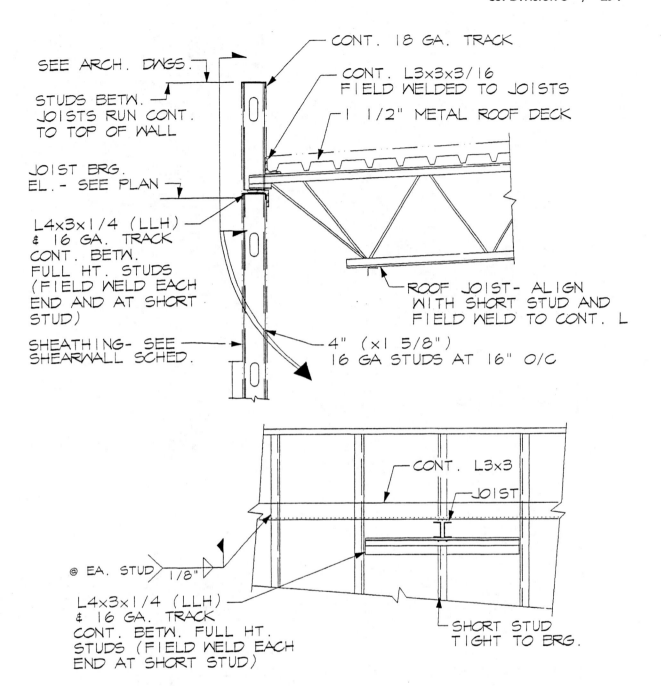

SEE ARCH. DWGS.

STUDS BETW. JOISTS RUN CONT. TO TOP OF WALL

JOIST BRG. EL.- SEE PLAN

L4x3x1/4 (LLH) & 16 GA. TRACK CONT. BETW. FULL HT. STUDS (FIELD WELD EACH END AND AT SHORT STUD)

SHEATHING- SEE SHEARWALL SCHED.

CONT. 18 GA. TRACK

CONT. L3x3x3/16 FIELD WELDED TO JOISTS

1 1/2" METAL ROOF DECK

ROOF JOIST- ALIGN WITH SHORT STUD AND FIELD WELD TO CONT. L

4" (x1 5/8") 16 GA STUDS AT 16" O/C

CONT. L3x3

JOIST

@ EA. STUD 1/8"

L4x3x1/4 (LLH) & 16 GA. TRACK CONT. BETW. FULL HT. STUDS (FIELD WELD EACH END AT SHORT STUD)

SHORT STUD TIGHT TO BRG.

TYP. ROOF JOIST BEARING DETAIL 1

NOT TO SCALE (DETAIL T5-CF6)

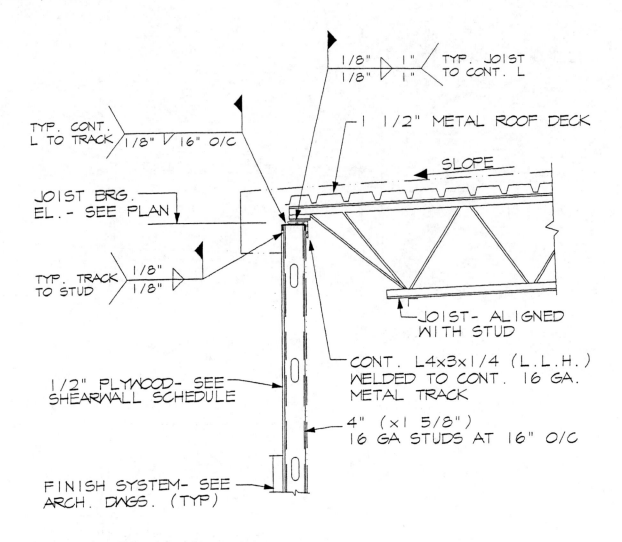

TYP. JOIST TO CONT. L

1 1/2" METAL ROOF DECK

SLOPE

TYP. CONT. L TO TRACK — 1/8" 16" O/C

JOIST BRG. EL. - SEE PLAN

TYP. TRACK TO STUD — 1/8" 1/8"

JOIST- ALIGNED WITH STUD

CONT. L4x3x1/4 (L.L.H.) WELDED TO CONT. 16 GA. METAL TRACK

1/2" PLYWOOD- SEE SHEARWALL SCHEDULE

4" (x1 5/8") 16 GA STUDS AT 16" O/C

FINISH SYSTEM- SEE ARCH. DWGS. (TYP)

TYP. ROOF JOIST BEARING DETAIL 2

NOT TO SCALE (DETAIL T5—CF7)

NOTE: LOCATE JOISTS DIRECTLY OVER
STUDS. OFFSET JOISTS AND BEAR 3" MIN.
ON CONT. L4x3

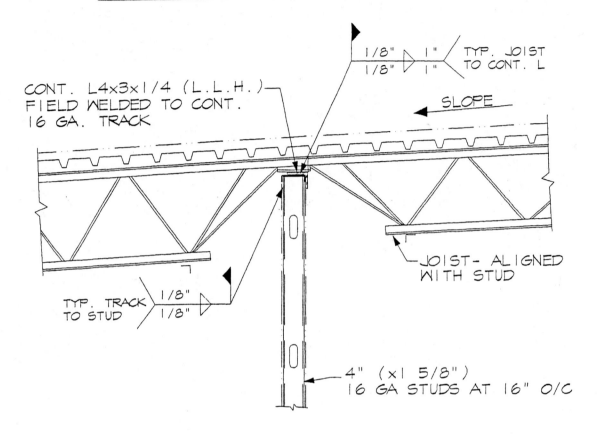

1/8" ⊳ 1"
1/8" ⊳ 1"   TYP. JOIST
            TO CONT. L

CONT. L4x3x1/4 (L.L.H.)
FIELD WELDED TO CONT.
16 GA. TRACK

SLOPE

JOIST- ALIGNED
WITH STUD

TYP. TRACK   1/8"
TO STUD   1/8"

4" (x1 5/8")
16 GA STUDS AT 16" O/C

TYP. INTERIOR JOIST BEARING DETAIL

NOT TO SCALE   (DETAIL T5-CF8)

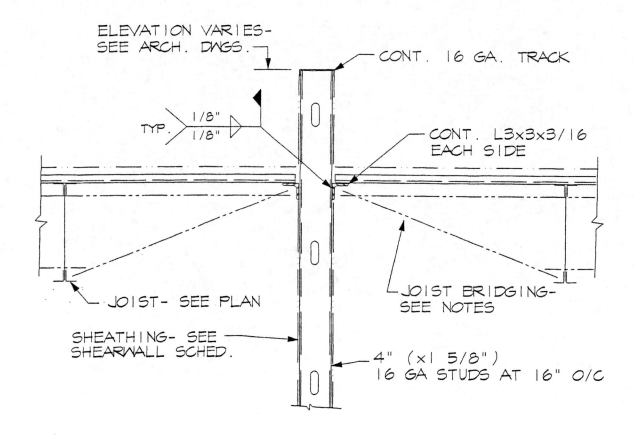

ELEVATION VARIES-
SEE ARCH. DWGS.

CONT. 16 GA. TRACK

TYP. 1/8"
1/8"

CONT. L3x3x3/16
EACH SIDE

JOIST- SEE PLAN

JOIST BRIDGING-
SEE NOTES

SHEATHING- SEE
SHEARWALL SCHED.

4" (x1 5/8")
16 GA STUDS AT 16" O/C

MISC. INTERIOR FRAMING DETAIL 1

NOT TO SCALE   (DETAIL T5-CF9)

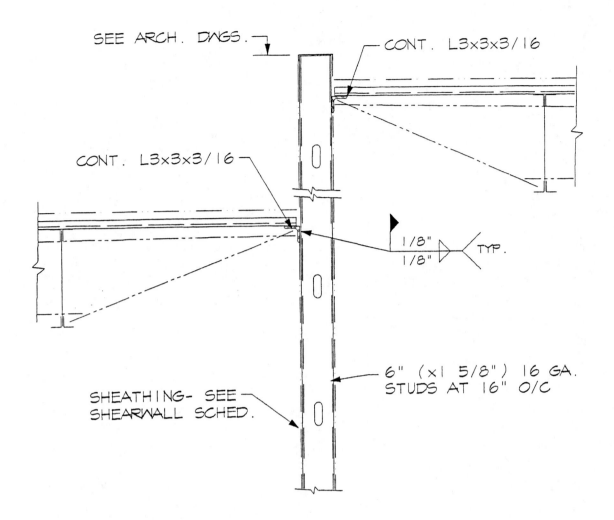

SEE ARCH. DWGS.

CONT. L3x3x3/16

CONT. L3x3x3/16

1/8"
1/8"
TYP.

6" (x1 5/8") 16 GA.
STUDS AT 16" O/C

SHEATHING- SEE
SHEARWALL SCHED.

MISC. INTERIOR FRAMING DETAIL 2

NOT TO SCALE   (DETAIL T5—CF10)

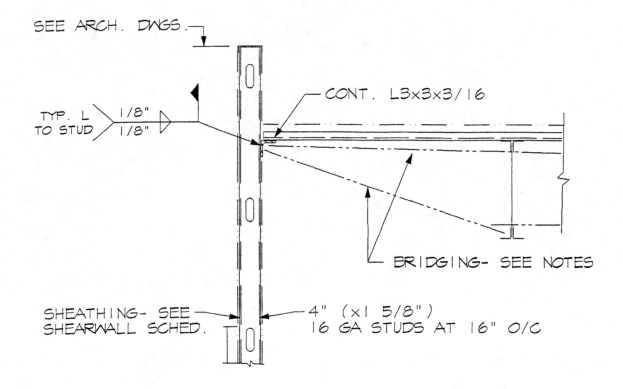

SEE ARCH. DWGS.

CONT. L3x3x3/16

TYP. L TO STUD 1/8" 1/8"

BRIDGING- SEE NOTES

SHEATHING- SEE SHEARWALL SCHED.

4" (x1 5/8") 16 GA STUDS AT 16" O/C

TYP. PERIMETER NONBEARING DETAIL

NOT TO SCALE   (DETAIL T5–CF11)

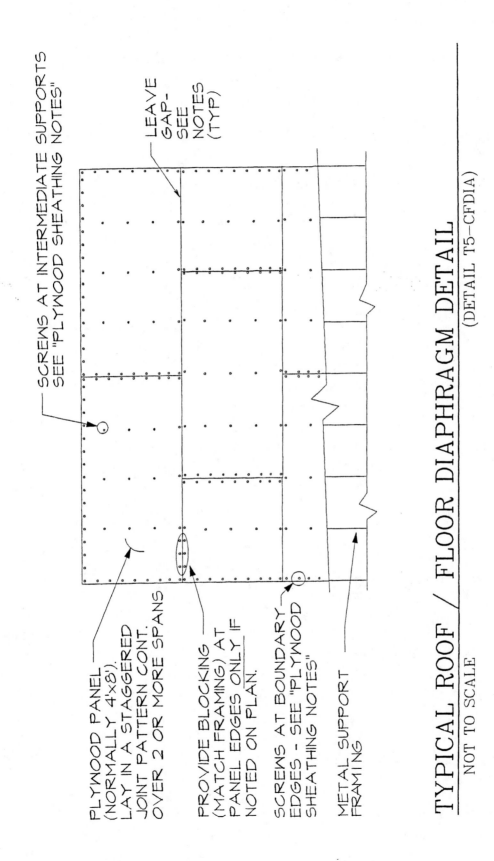

SCREWS AT INTERMEDIATE SUPPORTS SEE "PLYWOOD SHEATHING NOTES"

LEAVE GAP- SEE NOTES (TYP)

PLYWOOD PANEL (NORMALLY 4'x8'). LAY IN A STAGGERED JOINT PATTERN CONT. OVER 2 OR MORE SPANS

PROVIDE BLOCKING (MATCH FRAMING) AT PANEL EDGES ONLY IF NOTED ON PLAN.

SCREWS AT BOUNDARY EDGES - SEE "PLYWOOD SHEATHING NOTES"

METAL SUPPORT FRAMING

TYPICAL ROOF / FLOOR DIAPHRAGM DETAIL
(DETAIL T5-CFDIA)

NOT TO SCALE

297

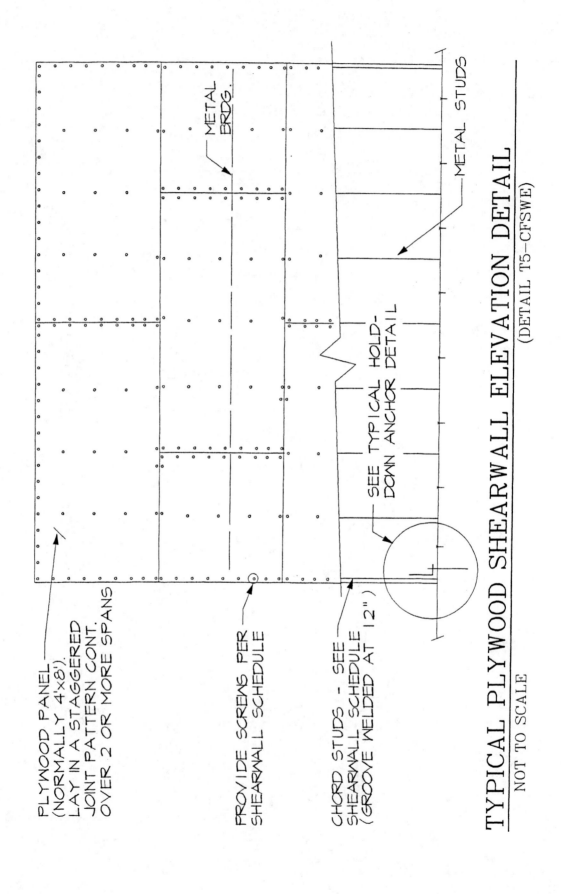

PLYWOOD PANEL
(NORMALLY 4'×8').
LAY IN A STAGGERED
JOINT PATTERN CONT.
OVER 2 OR MORE SPANS

PROVIDE SCREWS PER
SHEARWALL SCHEDULE

CHORD STUDS - SEE
SHEARWALL SCHEDULE
(GROOVE WELDED AT 12")

SEE TYPICAL HOLD-
DOWN ANCHOR DETAIL

METAL
BRDG.

METAL STUDS

TYPICAL PLYWOOD SHEARWALL ELEVATION DETAIL

(DETAIL T5-CFSWE)

NOT TO SCALE

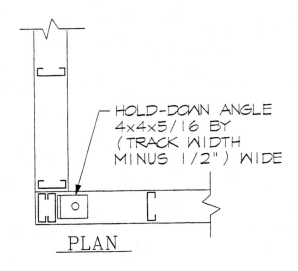

PLAN

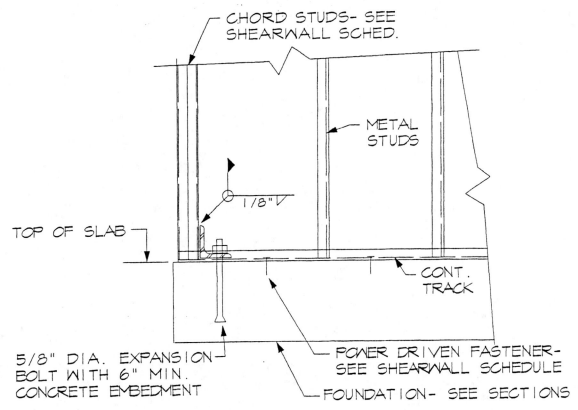

CHORD STUDS- SEE
SHEARWALL SCHED.

HOLD-DOWN ANGLE
4x4x5/16 BY
(TRACK WIDTH
MINUS 1/2") WIDE

METAL
STUDS

TOP OF SLAB

1/8"

CONT.
TRACK

5/8" DIA. EXPANSION
BOLT WITH 6" MIN.
CONCRETE EMBEDMENT

POWER DRIVEN FASTENER-
SEE SHEARWALL SCHEDULE

FOUNDATION- SEE SECTIONS

ELEVATION AT FOUNDATION LEVEL

TYP. HOLD-DOWN ANCHOR DETAILS
NOT TO SCALE   (DETAIL T5-CFHDA)

COLD-FORMED METAL HEADER SCHEDULE

MARK	SECTION	DESCRIPTION	GAUGE	JAMB STUDS	
				JACK	FULL-HT.
H-1		(2)- 6"(×1 5/8")	16 GA.	SINGLE	SINGLE
H-2		(2)- 8"(×1 5/8")	16 GA.	SINGLE	DOUBLE

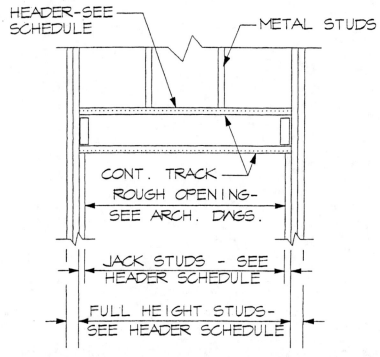

TYPICAL FRAMED OPENING DETAIL
NOT TO SCALE   (DETAIL T5-CFHDR)

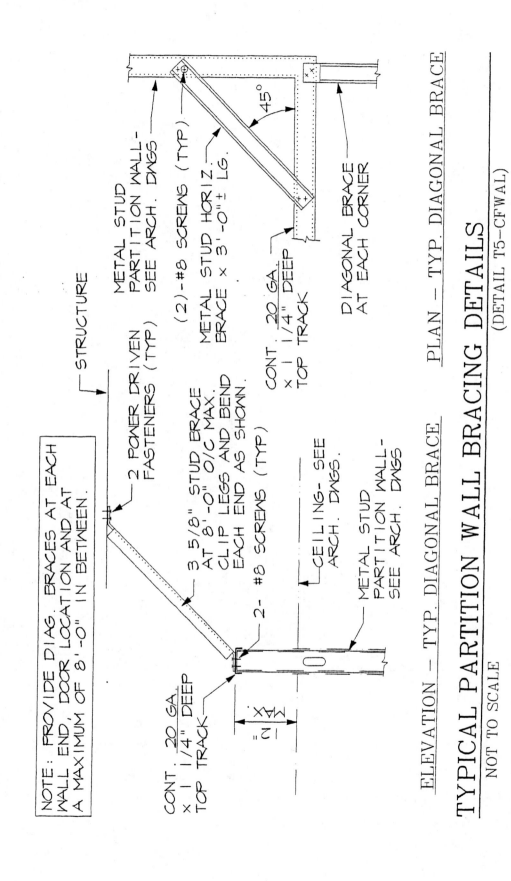

NOTE: PROVIDE DIAG. BRACES AT EACH WALL END, DOOR LOCATION AND AT A MAXIMUM OF 8'-0" IN BETWEEN.

STRUCTURE

METAL STUD PARTITION WALL - SEE ARCH. DWGS.

2 POWER DRIVEN FASTENERS (TYP)

(2)-#8 SCREWS (TYP)

METAL STUD HORIZ. BRACE × 3'-0"± LG.

45°

CONT. 20 GA. × 1 1/4" DEEP TOP TRACK

DIAGONAL BRACE AT EACH CORNER

CONT. 20 GA. × 1 1/4" DEEP TOP TRACK

3 5/8" STUD BRACE AT 8'-0" O/C MAX. CLIP LEGS AND BEND EACH END AS SHOWN.

2- #8 SCREWS (TYP)

CEILING- SEE ARCH. DWGS.

METAL STUD PARTITION WALL - SEE ARCH. DWGS.

$\frac{MAX}{1/2"}$

PLAN – TYP. DIAGONAL BRACE

ELEVATION – TYP. DIAGONAL BRACE

TYPICAL PARTITION WALL BRACING DETAILS

(DETAIL T5-CFWAL)

NOT TO SCALE

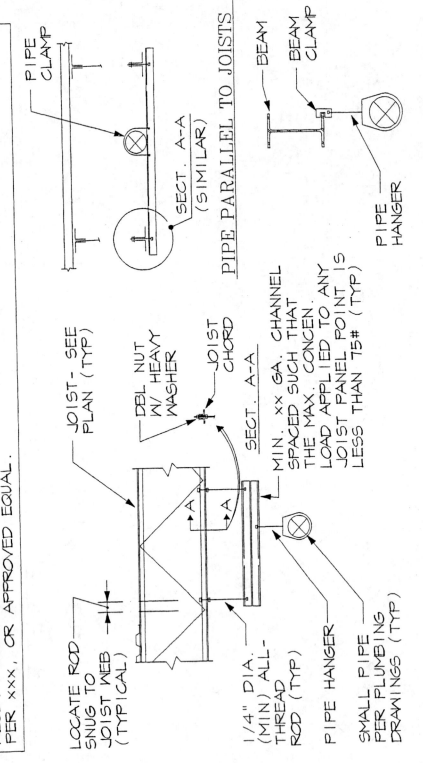

NOTE: ALL HANGER RODS, CHANNELS, PIPE HANGERS, CLAMPS AND OTHER REQD. ACCESSORIES SHALL BE DESIGNED BY THE CONTRACTOR AND SHALL BE PER xxx, OR APPROVED EQUAL.

PIPE CLAMP

SECT. A-A (SIMILAR)

PIPE PARALLEL TO JOISTS

BEAM

BEAM CLAMP

PIPE HANGER

PIPE PARALLEL TO BEAM

(DETAIL T5-HANG1)

JOIST- SEE PLAN (TYP)

DBL NUT W/ HEAVY WASHER

JOIST CHORD

SECT. A-A

MIN. xx GA. CHANNEL SPACED SUCH THAT THE MAX. CONCEN. LOAD APPLIED TO ANY JOIST PANEL POINT IS LESS THAN 75# (TYP)

LOCATE ROD SNUG TO JOIST WEB (TYPICAL)

1/4" DIA. (MIN) ALL- THREAD ROD (TYP)

PIPE HANGER

SMALL PIPE PER PLUMBING DRAWINGS (TYP)

PIPE PERPENDICULAR TO JOISTS

TYPICAL PIPE HANGER DETAILS

NOT TO SCALE

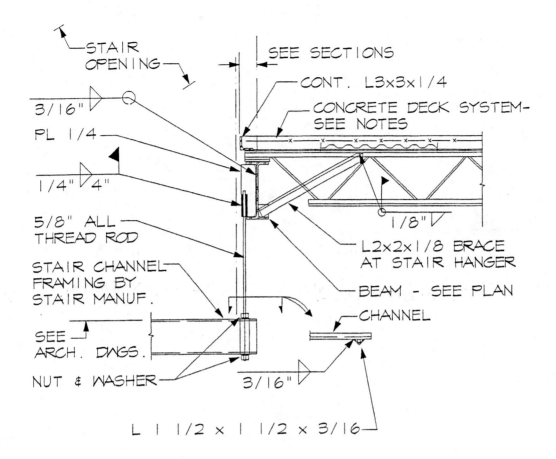

STAIR OPENING

3/16"

PL 1/4

1/4" 4"

5/8" ALL THREAD ROD

STAIR CHANNEL FRAMING BY STAIR MANUF.

SEE ARCH. DWGS.

NUT & WASHER

L 1 1/2 × 1 1/2 × 3/16

SEE SECTIONS

CONT. L3×3×1/4

CONCRETE DECK SYSTEM- SEE NOTES

1/8"

L2×2×1/8 BRACE AT STAIR HANGER

BEAM - SEE PLAN

CHANNEL

3/16"

TYP. STAIR HANGER DETAIL 1
NOT TO SCALE   (DETAIL T5-HANG2)

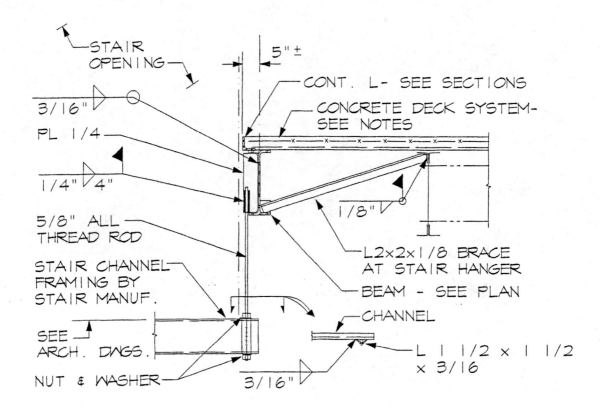

STAIR OPENING

CONT. L- SEE SECTIONS

CONCRETE DECK SYSTEM- SEE NOTES

3/16"

PL 1/4

1/4" 4"

5/8" ALL THREAD ROD

STAIR CHANNEL FRAMING BY STAIR MANUF.

SEE ARCH. DWGS.

NUT & WASHER

5" ±

1/8"

L2x2x1/8 BRACE AT STAIR HANGER

BEAM - SEE PLAN

CHANNEL

L 1 1/2 x 1 1/2 x 3/16

3/16"

TYP. STAIR HANGER DETAIL 2
NOT TO SCALE   (DETAIL T5—HANG3)

NOTE: COORD. SIZE AND LOCATION OF ROOF
OPENINGS WITH ACTUAL EQUIPMENT SELECTED.

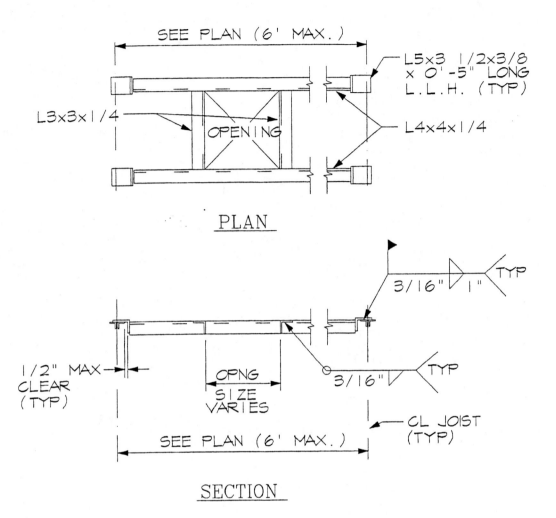

PLAN

SECTION

TYP. ROOF OPENING DETAIL 1

NOT TO SCALE   (DETAIL T5—R01)

NOTE: COORD. SIZE AND LOCATION OF ROOF OPENINGS WITH ACTUAL EQUIPMENT SELECTED.

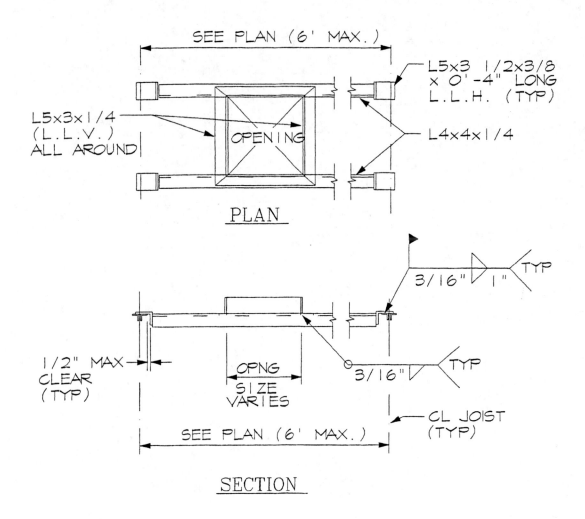

PLAN

SECTION

TYP. ROOF OPENING DETAIL 2

NOT TO SCALE   (DETAIL T5-R02)

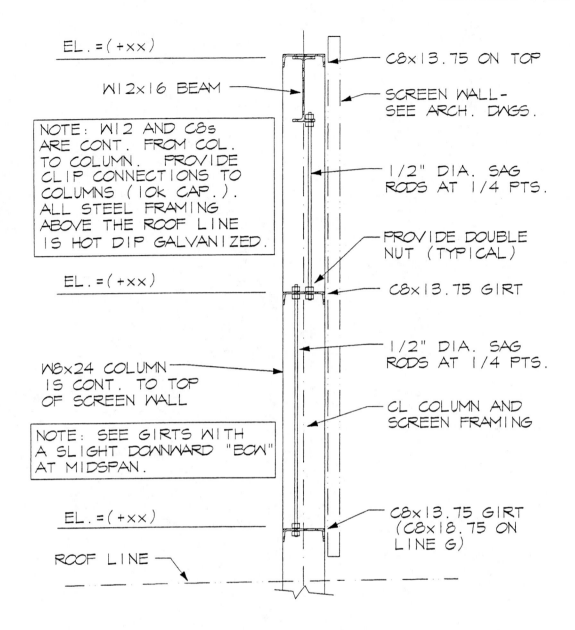

EL.=(+xx)

W12x16 BEAM

C8x13.75 ON TOP

SCREEN WALL-
SEE ARCH. DWGS.

NOTE: W12 AND C8s
ARE CONT. FROM COL.
TO COLUMN. PROVIDE
CLIP CONNECTIONS TO
COLUMNS (10k CAP.).
ALL STEEL FRAMING
ABOVE THE ROOF LINE
IS HOT DIP GALVANIZED.

1/2" DIA. SAG
RODS AT 1/4 PTS.

PROVIDE DOUBLE
NUT (TYPICAL)

EL.=(+xx)

C8x13.75 GIRT

1/2" DIA. SAG
RODS AT 1/4 PTS.

W8x24 COLUMN
IS CONT. TO TOP
OF SCREEN WALL

CL COLUMN AND
SCREEN FRAMING

NOTE: SEE GIRTS WITH
A SLIGHT DOWNWARD "BOW"
AT MIDSPAN.

EL.=(+xx)

C8x13.75 GIRT
(C8x18.75 ON
LINE G)

ROOF LINE

TYP. ROOF SCREEN DETAIL

NOT TO SCALE   (DETAIL T5-SCREN)

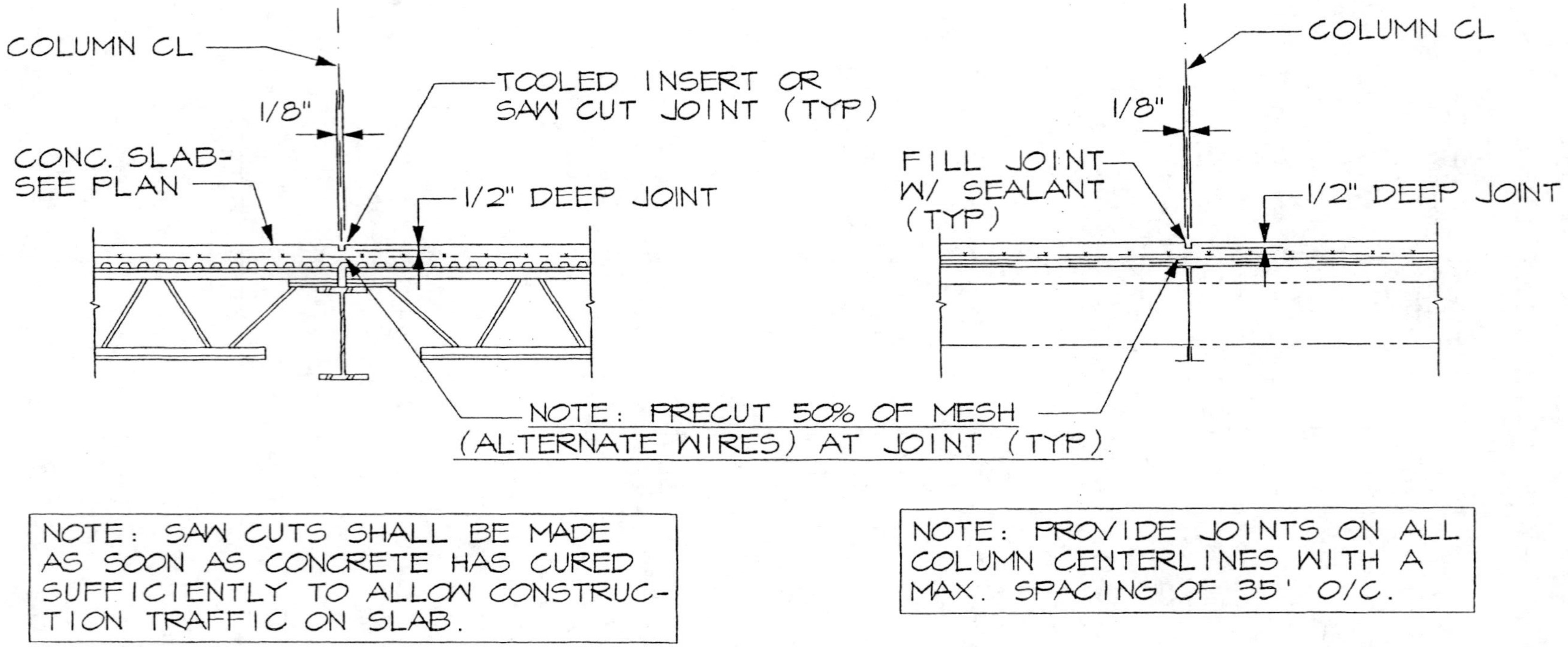

COLUMN CL

1/2" DEEP JOINT

1/8"

FILL JOINT
W/ SEALANT
(TYP)

TOOLED INSERT OR
SAW CUT JOINT (TYP)

1/2" DEEP JOINT

NOTE: PRECUT 50% OF MESH
(ALTERNATE WIRES) AT JOINT (TYP)

NOTE: PROVIDE JOINTS ON ALL
COLUMN CENTERLINES WITH A
MAX. SPACING OF 35' O/C.

PARALLEL TO JOIST

COLUMN CL

1/8"

CONC. SLAB-
SEE PLAN

NOTE: SAW CUTS SHALL BE MADE
AS SOON AS CONCRETE HAS CURED
SUFFICIENTLY TO ALLOW CONSTRUC-
TION TRAFFIC ON SLAB.

PERPENDICULAR TO JOIST

TYPICAL ELEVATED SLAB JOINT DETAILS

NOT TO SCALE

(DETAIL T5-SBJT1)

308

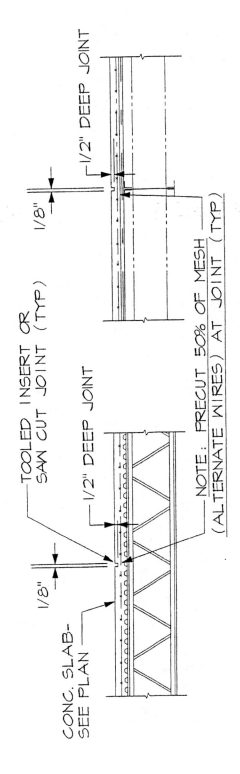

CONC. SLAB-
SEE PLAN

TOOLED INSERT OR
SAW CUT JOINT (TYP)

1/8"

1/2" DEEP JOINT

1/2" DEEP JOINT

1/8"

NOTE: PRECUT 50% OF MESH
(ALTERNATE WIRES) AT JOINT (TYP)

NOTE: SAW CUTS SHALL BE MADE
AS SOON AS CONCRETE HAS CURED
SUFFICIENTLY TO ALLOW CONSTRUC-
TION TRAFFIC ON SLAB.

NOTE: PROVIDE JOINTS ON ALL
COLUMN CENTERLINES WITH A
MAX. SPACING OF 35' O/C.

PERPENDICULAR TO JOIST

PARALLEL TO JOIST

TYPICAL ELEVATED SLAB JOINT DETAILS

NOT TO SCALE

(DETAIL T5-SBJT2)

309

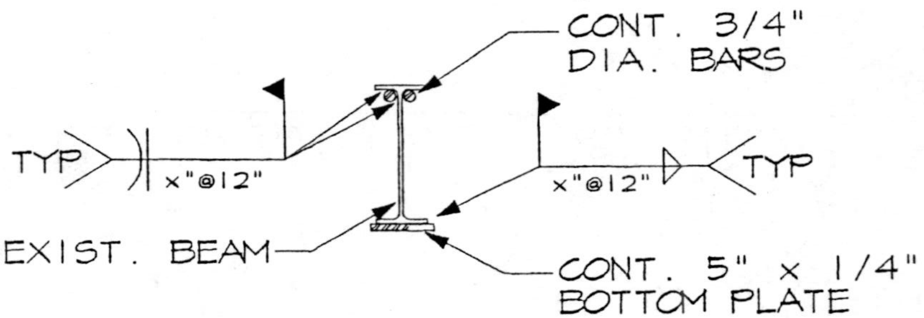

CONT. 3/4"
DIA. BARS

TYP

CONT. 5" × 1/4"
BOTTOM PLATE

TYP

x"@12"

EXIST. BEAM

x"@12"

SECTION

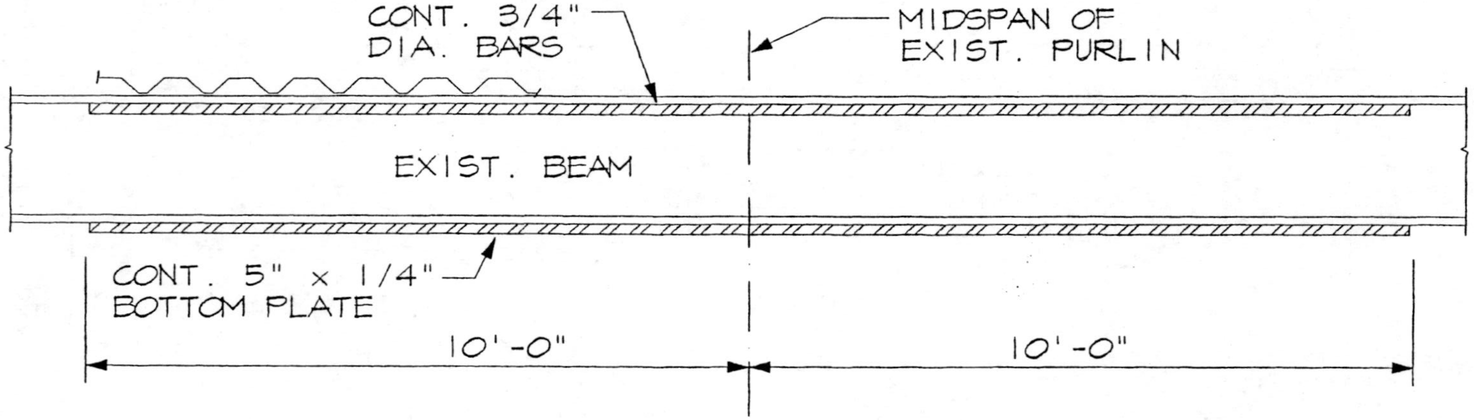

MIDSPAN OF
EXIST. PURLIN

CONT. 3/4"
DIA. BARS

EXIST. BEAM

CONT. 5" × 1/4"
BOTTOM PLATE

10'-0"

10'-0"

ELEVATION

TYPICAL BEAM REPAIR DETAIL

NOT TO SCALE

(DETAIL T5-BMREP)

310

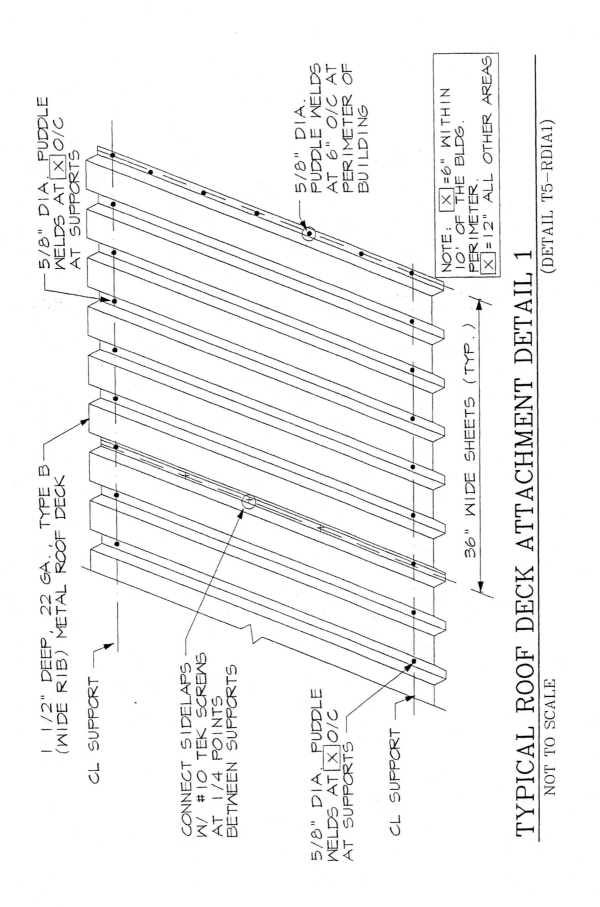

1 1/2" DEEP, 22 GA., TYPE B (WIDE RIB) METAL ROOF DECK

5/8" DIA. PUDDLE WELDS AT ⊠ O/C AT SUPPORTS

5/8" DIA. PUDDLE WELDS AT 6" O/C AT PERIMETER OF BUILDING

CL SUPPORT

CONNECT SIDELAPS W/ #10 TEK SCREWS AT 1/4 POINTS BETWEEN SUPPORTS

5/8" DIA. PUDDLE WELDS AT ⊠ O/C AT SUPPORTS

CL SUPPORT

36" WIDE SHEETS (TYP.)

NOTE: ⊠ =6" WITHIN 10' OF THE BLDG. PERIMETER. ⊠ =12" ALL OTHER AREAS

(DETAIL T5-RDIA1)

TYPICAL ROOF DECK ATTACHMENT DETAIL 1

NOT TO SCALE

311

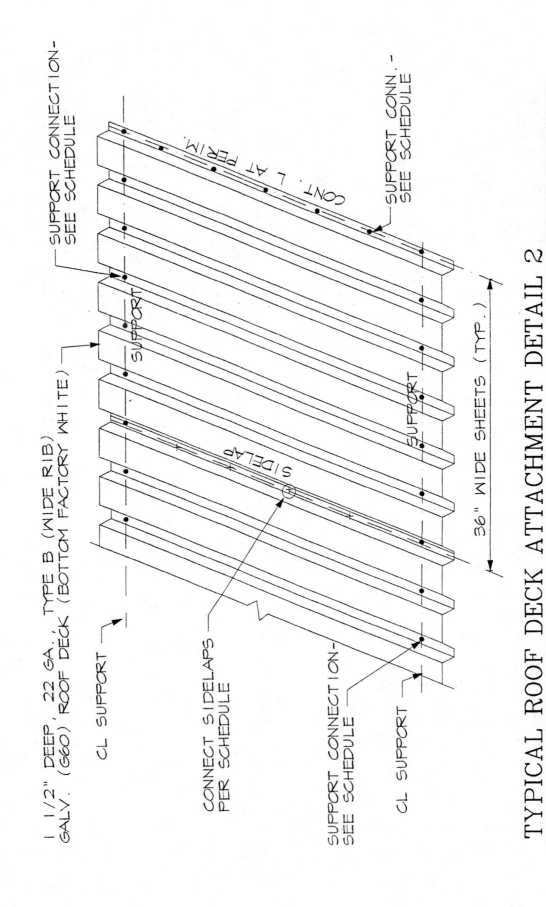

TYPICAL ROOF DECK ATTACHMENT DETAIL 2

(DETAIL T5-RDIA2)

NOT TO SCALE

1 1/2" DEEP, 22 GA., TYPE B (WIDE RIB)
GALV. (G60) ROOF DECK (BOTTOM FACTORY WHITE)

SUPPORT CONNECTION-
SEE SCHEDULE

CL SUPPORT

SUPPORT

SUPPORT CONN. -
SEE SCHEDULE

CONT. L AT PERIM.

36" WIDE SHEETS (TYP.)

CONNECT SIDELAPS
PER SCHEDULE

SIDELAP

SUPPORT

SUPPORT CONNECTION-
SEE SCHEDULE

CL SUPPORT

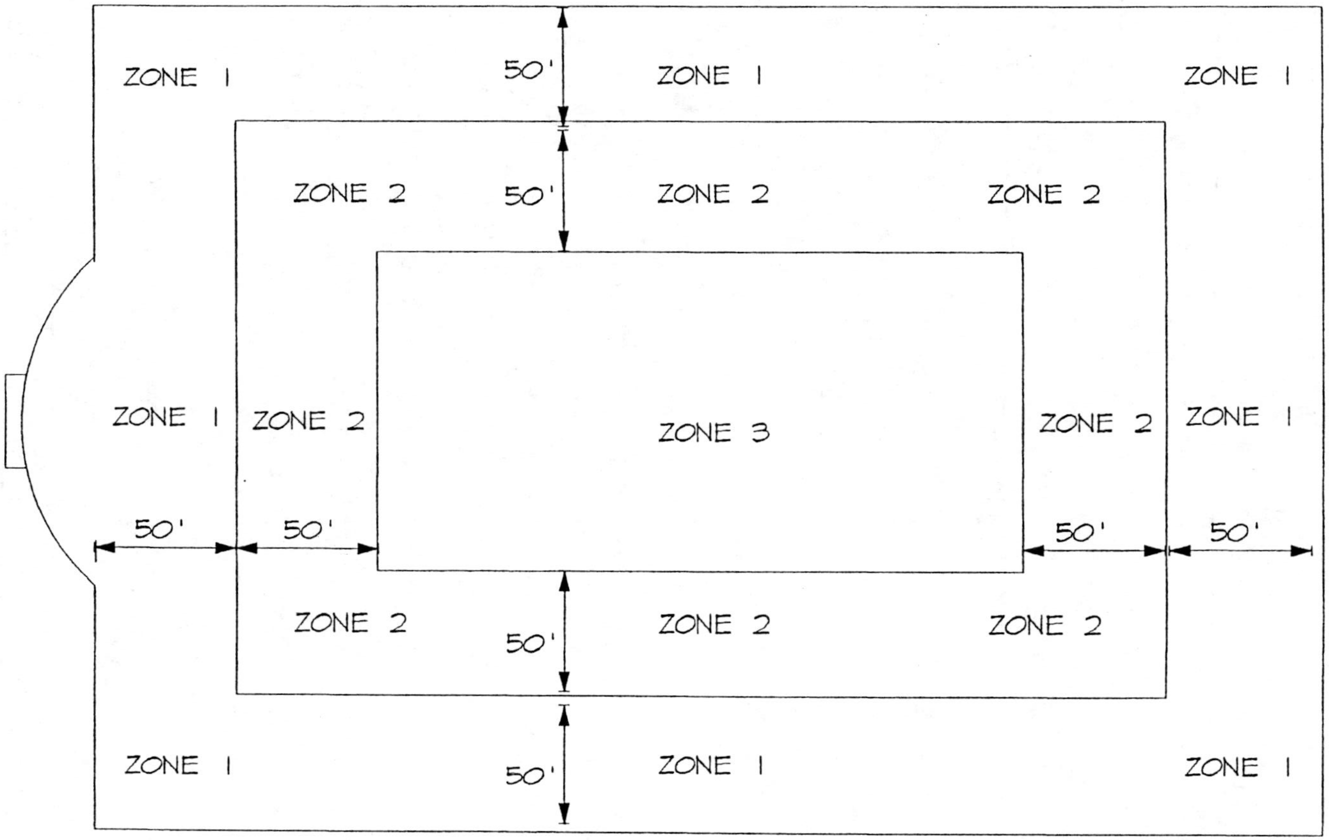

ROOF DECK ATTACHMENT ZONES

BLDG. PLAN

NOT TO SCALE

(DETAIL T5−RDIA3)

313

ROOF DECK CONNECTION SCHEDULE	
ZONE	DECK CONNECTION
ZONE 1 (DIAPH. CAPACITY= 470#/LF)	WELDS IN 36/7 PATTERN (6)-#10 TEKS @ SIDELAPS
ZONE 2 (DIAPH. CAPACITY= 290#/LF)	WELDS IN 36/7 PATTERN (3)-#10 TEKS @ SIDELAPS
ZONE 3 (DIAPH. CAPACITY= 290#/LF)	WELDS IN 36/4 PATTERN (3)-#10 TEKS @ SIDELAPS

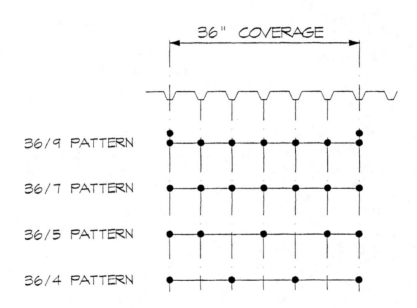

36" COVERAGE

36/9 PATTERN

36/7 PATTERN

36/5 PATTERN

36/4 PATTERN

ROOF DECK ATTACHMENT PATTERNS

NOT TO SCALE   (DETAIL T5-RDIA4)

NOTE: ROOF TOP UNIT TO BE ANCHORED
TO SUPPORT FRAMING PER MANUFACTURER'S
REQUIREMENTS TO MEET CODE REQD. WIND LOADING

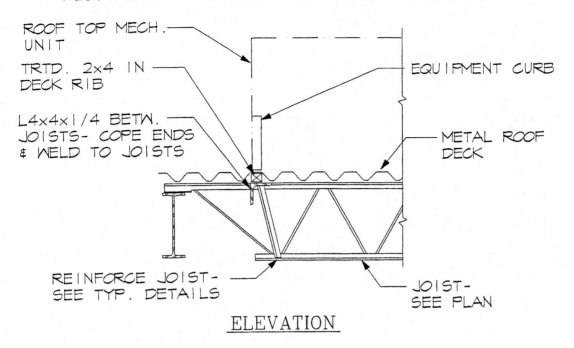

ROOF TOP MECH.
UNIT

TRTD. 2x4 IN
DECK RIB

L4x4x1/4 BETW.
JOISTS- COPE ENDS
& WELD TO JOISTS

EQUIPMENT CURB

METAL ROOF
DECK

REINFORCE JOIST-
SEE TYP. DETAILS

JOIST-
SEE PLAN

ELEVATION

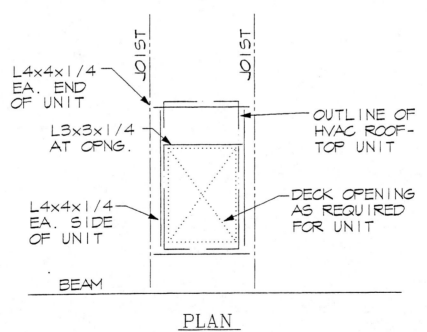

JOIST

JOIST

L4x4x1/4
EA. END
OF UNIT

L3x3x1/4
AT OPNG.

L4x4x1/4
EA. SIDE
OF UNIT

OUTLINE OF
HVAC ROOF-
TOP UNIT

DECK OPENING
AS REQUIRED
FOR UNIT

BEAM

PLAN

TYP. DETAIL AT ROOFTOP UNITS

NOT TO SCALE   (DETAIL T5—MECH1)

NOTE: ROOFTOP UNIT TO BE ANCHORED
TO SUPPORT FRAMING PER MANUFACTURER'S
REQUIREMENTS TO MEET CODE REQD. WIND LOADING

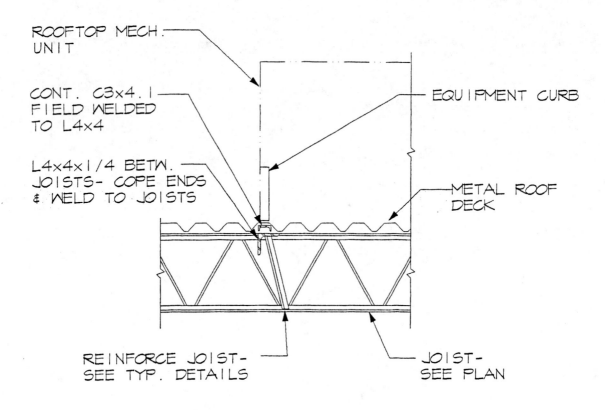

ROOFTOP MECH. UNIT

CONT. C3x4.1 FIELD WELDED TO L4x4

L4x4x1/4 BETW. JOISTS- COPE ENDS & WELD TO JOISTS

EQUIPMENT CURB

METAL ROOF DECK

REINFORCE JOIST- SEE TYP. DETAILS

JOIST- SEE PLAN

ELEVATION

TYP. DETAIL AT ROOFTOP UNITS

NOT TO SCALE   (DETAIL T5-MECH2)

Commentary—Steel Details

CONNECTIONS

BEAM SHEAR CONNECTION DETAIL #1 (DETAIL T5-CSH1)

The two shear connection details shown are both for wide-flange (WF) beams framing to tube section (TS) columns. One detail shows the beam bearing over the top of the column with a cap plate shop welded to the column. The beam has stiffener plates, which restrain the beam from rotating off the column. The stiffeners may also be needed to strengthen the beam web. The beam is field bolted to the column cap plate. The column cap plate thickness needs to be investigated, especially if there are uplift forces on the beam. The other detail shows a wide-flange beam framing into a TS column. This connection is noted as being designed by the steel fabricator. The shear force should be shown on the plan or in a schedule. The tube is noted as being slotted for the shear plate. For lightly loaded beams, a shear tab plate may be acceptable with no slotting required. The beam web is field bolted to the shear plate. A cap plate is recommended to stiffen the column and prevent rain from entering the column during construction.

BEAM SHEAR CONNECTION DETAIL #2 (DETAIL T5-CSH2)

These two connection details are similar to the previous one, but there are beams on each side of the column. One detail also shows the condition where the column runs continuous, such as at an intermediate floor level.

BEAM SHEAR CONNECTION DETAIL #3 (DETAIL T5-CSH3)

Wide-flange beams are shown framing into WF columns. One detail shows the beam framing into the column web. The other detail shows a WF beam framing into the column flange. Double angles are shop welded on the beam webs and these angles are field bolted to the column. The connections are noted as being designed by the steel fabricator.

BEAM SHEAR CONNECTION DETAIL #4 (DETAIL T5-CSH4)

Two connection details are shown. One detail shows the beam framing into the web of a column with beams on both sides of the column. The other detail shows a beam to beam shear splice. Double angles are shop welded onto the beam webs. The connections are noted as being designed by the steel fabricator.

BEAM MOMENT CONNECTION DETAIL #1 (DETAIL T5-CMO1)

The two moment connection details shown are both for WF beams framing to TS columns. One detail shows a beam framing into one side of the column. The other detail shows a beam framing into both sides. Flange continuity is provided by welding the flanges directly to the column. Shear connections are also required. These details are intended for low moments. For high moments, the tube face may be locally overstressed.

BEAM MOMENT CONNECTION DETAIL #2 (DETAIL T5-CMO2)

The two moment connection details shown are both for WF beams framing to the flanges of WF columns. One detail shows a beam framing into one side of the column. The other detail shows a beam framing into both sides. Flange continuity is provided by welding the flanges directly to the column. Shear connections are also required. Depending on the magnitude of the moment, column stiffener plates may also be required on the column webs.

BEAM MOMENT CONNECTION DETAIL #3 (DETAIL T5-CMO3)

A beam is framed into both sides of a TS column. Flange continuity is provided by welding flange plates to the beam and column. Shear connections are also required. This detail is intended for low moments. For large moments, the tube face may be locally overstressed.

BEAM MOMENT CONNECTION DETAIL #4 (DETAIL T5-CMO4)

This detail is similar to the previous one, but the flange plates allow the moment to be transferred from beam to beam, minimizing the moment transfer to the column. Flange continuity is provided by welding flange plates to the beam and column. Shear connections are also required. This detail is intended for high moment transfer from beam to beam. One alternative is to run the beam continuous and splice the column.

TYPICAL CANTILEVER ROOF BEAM DETAILS (DETAIL T5-CBM1 & 2)

The two separate details shown are essentially the same, but one detail is for TS columns and the other is for WF columns. The roof beam cantilevers past the interior column and the beams are shear spliced together. Running the beam over the column results in a beam moment reversal. Compression occurs in the bottom flange; hence, the beam bottom flange needs to be braced off at certain intervals.

TYPICAL BEAM BEARING ON MASONRY DETAILS (DETAIL T5-CMU1 & 2)

The two separate details shown are essentially the same, but one detail is for a *fixed connection* and the other is for a *slip connection*. The bearing plate distributes the beam end reaction to the wall. The bearing plate size is scheduled. No field welding is required. For roof beams, uplift force transfer needs to be carefully investigated. For the slip connection, the bearing surface should be lubricated. The nuts are also finger tightened and the exposed threads are spoiled so that the nut cannot work its way off.

TYPICAL BEAM BEARING ON EXISTING MASONRY DETAILS (DETAIL T5-CMU3)

This detail is similar to the previous two, but the beam bears on an existing masonry wall. The bearing plate is set loose onto the masonry wall. This detail is primarily intended for floor beams with light end reactions that bear on interior walls.

LINTEL BEARING ON EXISTING MASONRY (DETAIL T5-CMU4)

This detail is similar to the previous one and involves a steel lintel bearing on an existing masonry wall. The detail is intended for conditions where openings are to be provided within existing masonry walls. The wall may not need to be shored; however, this issue is normally left up to the contractor. The bearing plate is set into the new masonry that is *sawtoothed* at the opening jambs. The lintel is slipped into position and field welded to the plate. Field bolting is not practical due to limited erection clearances.

TYPICAL TUBE SECTION COLUMN SPLICE DETAILS (DETAIL T5-COL1)

The two details shown are both for splices in TS columns. The cap plate is shop welded to the lower column. The upper column is field welded to the cap plate. Fillet welding is preferred; however, penetration welds may be required for high-force transfer.

TYPICAL WIDE-FLANGE COLUMN SPLICE DETAILS (DETAIL T5-COL2)

The two details shown are both for splices in WF columns. One detail is for same-depth WF sections. The other detail shows the condition where the column depths are different. Tensile force transfer is provided by the bolted splice plates.

TYPICAL FUTURE COLUMN CONNECTION (DETAIL T5-COL3)

This detail is intended for cases where columns will be added in the future. The column cap plate is shop welded to the column. The future column would need to be field welded.

TYPICAL COLUMN BRACE DETAIL (DETAIL T5-BRCE1)

Bracing the weak axis of WF columns can significantly increase their axial load capacity. This detail is intended for cases where perimeter WF columns can be braced off to the girts.

TYPICAL KNEE BRACE DETAIL (DETAIL T5-BRCE2)

Knee braces are an alternate way to provide rigid frame action in lieu of using moment connections. No field welding is required. The forces need to be carefully investigated.

TYPICAL LIGHT-GAUGE METAL GIRT CONNEC-TION DETAIL #1 (DETAIL T5-GIRT1)

Light-gauge metal girts are typically designed by the girt supplier; however, a preliminary size and spacing should be determined. The girts are lapped at the column, providing *continuity* and lower moments. The girt to column connection should be determined by the girt supplier.

TYPICAL LIGHT-GAUGE METAL GIRT CONNEC-TION DETAIL #2 (DETAIL T5-GIRT2)

This detail is similar to the previous one, but a condition at a building *recess* is shown.

TYPICAL CHANNEL GIRT CONNECTION DETAILS (DETAIL T5-GIRT3)

The two conditions shown are both for connecting structural steel C-shaped girts to their supports. One support condition is at a TS column. The other is at a masonry wall.

TYPICAL GIRT CONNECTION DETAILS (DETAIL T5-GIRT4)

The two conditions shown are similar to the previous one. Both involve connecting structural steel C-shaped girts to WF columns. In one condition, the girt flange needs to be coped to clear the column flange.

BRACED BAYS AND WALL ELEVATIONS

TYPICAL BRACED BAY ELEVATION 1 (DETAIL T5-BAYE1)

A single-story braced bay elevation detail is shown. TS diagonals are being used. The wind axial force is shown on the detail so the fabricator can size the TS connection welds. The connections are detailed elsewhere.

TYPICAL BRACED BAY ELEVATION 2 (DETAIL T5-BAYE2)

This detail is similar to the previous one, but it is for a two-story building.

TYPICAL BRACED BAY ELEVATION 3 (DETAIL T5-BAYE3)

This detail shows a condition where a braced bay is provided, even though a column is not located at one end of the TS diagonal. The beams need to be carefully investigated for the wind loading condition. The TS members are normally neglected when sizing the beams for the gravity loading condition.

TYPICAL BRACED BAY ELEVATION 4 (DETAIL T5-BAYE4)

This detail is similar to the previous one, but a column is provided at both ends of the TS diagonal.

BRACED BAY DETAIL A (DETAIL T5-BAYD1)

The two connection details shown are both at the low end of TS diagonals, where the columns connect to the foundation. One detail shows a TS column; the other a WF column. A ¼" thick gusset plate is shown; however, the plate size needs to be investigated. The gusset plate is shop welded to the columns. Fillet welds are shown. The TS diagonals are field welded to the gusset plates. An erection bolt may be used as a temporary connection.

BRACED BAY DETAILS B AND C
(DETAIL T5-BAYD2)

These two details are similar to the previous one, but both are at the upper end of TS diagonals. One detail shows a double TS diagonal to beam condition; the other shows a single TS diagonal to beam/column condition.

BRACED BAY DETAILS D AND E
(DETAIL T5-BAYD3)

These two details are similar to the previous detail. One detail shows a single TS diagonal to beam condition; the other shows a TS diagonal to beam/column condition.

BRACED BAY DETAIL F (DETAIL T5-BAYD4)

This detail is similar to detail T5-BAYD1, but threaded rods are used in lieu of TS diagonals. A WF column is shown.

BRACED BAY DETAIL G (DETAIL T5-BAYD5)

This detail is similar to detail T5-BAYD2, but threaded rods are used in lieu of TS diagonals. A WF column is shown.

TYPICAL PERIMETER WALL ELEVATION 1
(DETAIL T5-WALL1)

Although the wall system can be shown in the wall sections, it is also advantageous to show the condition in elevation. A masonry skirt wall (8′ high) is shown supported by a C8 girt. The connection between the girt and masonry wall needs to be detailed elsewhere. Metal panels run above the masonry skirt wall. These panels are supported by light-gauge metal Z girts. An 8″ girt is indicated; however, the final design (gauge and flange width) is indicated by the girt manufacturer. Sag rods brace the girt weak axis. The sag rods can be located in a symmetrical pattern or offset as shown (were hanging from steel bar joists above). The WF columns have a weak-axis column brace near the midheight of the column.

TYPICAL PERIMETER WALL ELEVATION 2
(DETAIL T5-WALL2)

This detail is similar to the previous one, but it has no masonry skirt wall. The metal panel connects at the base to a continuous steel angle.

TYPICAL PERIMETER WALL ELEVATION 3
(DETAIL T5-WALL3)

This detail is similar to detail T5-WALL1, but there is an upper level window. An 8″ C-shaped light-gauge girt is provided above and below the window. The weight of the wall system above the window is transferred to the sag rods. The weight of the window system gets transferred to the lower girts, which span in weak-axis bending between the 6″ light-gauge struts. These struts transfer the weight of the wall system to the masonry wall.

TYPICAL PERIMETER WALL ELEVATION 4
(DETAIL T5-WALL4)

This detail is similar to the previous details, but it has a large amount of storefront system. A heavy C10 girt is provided at the storefront head.

TYPICAL PERIMETER WALL ELEVATION 5
(DETAIL T5-WALL5)

Overhead doors require special framing, and an elevation is very helpful. This detail is similar to the previous details, but it has two overhead doors framed out with structural 8″ channel sections.

JOISTS AND JOIST GIRDERS

TYPICAL K-JOIST CONNECTION DETAILS
(DETAIL T5-J1)

Two conditions are shown: one at an interior beam and the other at a perimeter beam. The K-series joists are field welded to the beams. The standard seat depth of K-series joists is 2½″. Normally, the minimum weld required by the Steel Joist Institute will suffice; however, roof joists may require more significant welding to resist wind uplift. A beam bottom flange brace is shown. The brace locations would be indicated on the plan. These braces may be required for wind uplift or for a cantilevered beam system. Simple span, interior floor beams would not normally need bottom flange braces.

TYPICAL LH-JOIST CONNECTION DETAILS
(DETAIL T5-J2)

This detail is similar to the previous one, but LH-series joists are shown. The minimum seat depth of LH-series joists is 5½″.

TYPICAL CONNECTION DETAILS AT COLUMN CENTERLINE (DETAIL T5-J3)

One detail shows the condition at a cantilevered roof beam system. The other condition is where the column runs continuous. Per the SJI specifications, the joists on the column centerlines (or nearest the column centerlines) need to be field bolted. Field welds may also be required depending on the forces involved (such as uplift).

TYPICAL CONNECTION DETAIL AT COLUMN CENTERLINE (DETAIL T5-J4)

This detail is similar to the previous one, but the joists have end moments. The joists may be part of a rigid frame. If so, the end moments need to be specified so the joist fabricator can design the joists.

TYPICAL CONCENTRATED LOAD DETAIL (DETAIL T5-J5)

Joists are intended for uniform loading conditions. Concentrated loads need to be properly accounted for. If heavy concentrated loads occur between the panel point locations, additional angles should be added to the joist.

TYPICAL JOIST TO GIRDER CONNECTION (DETAIL T5-JG1)

Two joist to girder conditions are shown. One is for K-series joists; the other is for LH-series joists. The joist girder bottom chord bracing should be determined by the girder manufacturer. This bracing is especially critical for wind uplift loading conditions.

TYPICAL DETAIL AT INTERIOR COLUMN CENTERLINE (DETAIL T5-JG2)

The joists and joist girders are field bolted over the top of the TS column. This detail shows the transverse framing condition to the following detail.

TYPICAL JOIST GIRDER CONNECTION DETAIL AT INTERIOR (DETAIL T5-JG3)

The joists and joist girders are field bolted over the top of the TS column. This detail shows the transverse framing condition to the previous detail. The bottom chord of the joist girder should not be connected to the column unless it is part of a rigid frame. In that case, the end moments would need to be given. This detail is not intended for end moments. The bottom chord angles are extended but not connected. This brace prevents the joist girder from rotating.

TYPICAL JOIST GIRDER CONNECTION DETAIL AT PERIMETER (DETAIL T5-JG4)

This detail is similar to the previous one, but a perimeter condition is shown. A WF column and WF perpendicular beam are indicated.

SPECIAL JOIST LOADING DIAGRAM 1 (DETAIL T5-JL1)

Standard joists are intended for uniform loading conditions. In this case, the joists are being provided in a shell building that may have concentrated loads added in the future. The uniform and special (potential equipment) loading conditions are provided for the joist manufacturer to design the joists. The span can be indicated on the diagrams or referred to on the plan. The loadings indicated are intended for fairly light conditions and should represent the actual loading requirements. The wind uplift loading indicates a "net" uplift, which takes the roof dead loads into account.

SPECIAL JOIST LOADING DIAGRAM 2 (DETAIL T5-JL2)

Standard joists are intended to be within a certain slope. For slopes beyond this range, the joists need to be specially designed.

SPECIAL JOIST LOADING DIAGRAM 3 (DETAIL T5-JL3)

This detail is similar to detail T5-JL1, but the concentrated loads are known in advance. The magnitude of the uniform and concentrated loads should be shown on the loading diagram or in a schedule.

COMPOSITE CONSTRUCTION

TYPICAL COMPOSITE BEAM ELEVATION 1 (DETAIL T5-COMP1)

Several different beam and deck configurations are possible. This detail shows the typical stringer elevation condition, where the beam runs perpendicular to the deck system. The headed shear studs are field welded through the metal deck to the beam. The deck is also welded. It is recommended that the composite beams be shown in a schedule.

TYPICAL COMPOSITE BEAM ELEVATION 2 (DETAIL T5-COMP2)

This detail shows the condition where the beam runs parallel to the deck system. A composite beam schedule is also shown. A ¼″ stud diameter is shown. The length of stud needs to be determined and shown on the detail.

TYPICAL COMPOSITE DECK CLOSURE DETAILS (DETAIL T5-COMP3)

A variety of deck closure details are shown: two at an edge condition and two at an interior condition. Deck closure provides a tight metal form condition preventing concrete loss.

TYPICAL COMPOSITE DECK REINFORCING DETAILS (DETAIL T5-COMP4)

Two different reinforcing conditions are shown. One indicates welded wire fabric; the other shows reinforcing bars. The latter is better suited for heavy loading conditions, such as required for forklifts. All reinforcing needs to be properly sized for the specific application.

COMPOSITE BEAM SECTION DETAILS (DETAIL T5-COMP5)

Several different conditions are shown. These details are the transverse conditions for details T5-COMP1 and T5-COMP2.

COMPOSITE BEAM SECTION DETAILS (DETAIL T5-COMP6)

The two different conditions both show reinforcing bars in the deck system. Perimeter support is provided by steel beams. These details were taken from a project that involved a pre-engineered metal building.

COMPOSITE BEAM SECTION DETAILS (DETAIL T5-COMP7 & 8)

These details are similar to the previous one, but the perimeter support condition consists of a masonry load-bearing wall.

OVERHEAD CRANE FRAMING

TYPICAL RUNWAY GIRDER DETAIL
(DETAIL T5-CRAN1)

Two different details show overhead crane framing. One shows a typical runway girder section. A channel *cap* is frequently provided on longer span girders. This provides a composite section. The welding between the girder and channel should be investigated based on the actual loading condition. The other detail shows a typical crane rail attachment using hooked bolts that are provided by the crane rail supplier.

TYPICAL GIRDER CONNECTION DETAILS
(DETAIL T5-CRAN2)

Two different conditions are shown. One is at the interior, and the girder webs should not be rigidly connected together across the splice point. The other detail is at the end framing condition. The beam and top of the column need to be braced laterally, as shown in the other details.

TYPICAL GIRDER CONNECTION DETAILS
(DETAIL T5-CRAN3)

This detail is similar to the previous one, but it is intended for a heavier capacity crane system.

TYPICAL CRANE STOP DETAILS
(DETAIL T5-CRAN4)

For light capacity cranes, a relatively simple and inexpensive crane stop can be provided. This type of crane stop is provided by the crane rail supplier and is connected directly to the crane rail. For heavy capacity cranes, the crane stop is connected directly to the runway girder and is fabricated from WF sections.

TYPICAL COLUMN TIE TO EXISTING COLUMN
(DETAIL T5-CRAN5)

This detail was taken from a project where a light capacity overhead crane was added within an existing building. A large base plate was added to the base of the column to distribute the column force to the existing slab without having to provide new foundations. The top girder and weak axis of the new column are braced back to the existing column.

TYPICAL COLUMN END BRACE DETAIL
(DETAIL T5-CRAN6)

This detail was also taken from a project where a light capacity overhead crane was added within an existing building. As with all overhead crane systems, longitudinal forces are developed in the direction of the runway girder. These forces are developed during the crane braking process. The diagonal angle bracing transfers these forces to the existing slab.

TYPICAL COLUMN BASE PLATE DETAILS
(DETAIL T5-CRAN7)

Two conditions are shown. One is for a relatively small base plate, which is intended for columns to be supported on new foundations. The other condition is for a relatively large base plate, which is intended for columns to be supported on existing slabs. The latter detail should only be considered for light capacity cranes.

ASSUMED FUTURE CRANE CONDITION
(DETAIL T5-CRAN8)

This detail was taken from a project where the TS columns were designed for a potential future overhead crane system. The assumed crane girder height, support, and loading conditions are shown. In this case, the existing columns were 50′ on center, and the detail notes that additional columns will be required in the future to reduce the girder span to 25′.

MISCELLANEOUS CRANE FRAMING DETAILS
(DETAIL T5-CRAN9)

Two details are shown. One involves a knee brace that provides rigid frame action. It may be necessary if the diagonal brace shown in detail T5-CRAN6 would be in the way. The other detail shows a typical girder bracing detail at an existing column.

ASSUMED CRANE WHEEL LOADING DIAGRAMS
(DETAIL T5-CRA10)

This detail was taken from a project where two different crane systems were involved, with 10-ton and 30-ton capacity cranes. A spacer is shown between the two cranes.

TYPICAL CRANE LOADINGS TO MAIN BUILDING
COLUMNS (DETAIL T5-CRA11)

This detail was taken from a project that involved a pre-engineered metal building. The metal building columns provided support to the crane framing system. The metal building manufacturer needs to know the magnitude of these forces to properly design the main building columns.

COLD-FORMED METAL FRAMING

TYPICAL FLOOR JOIST BEARING DETAIL (DETAIL T5-CF1)

Cold-formed metal can be used in a variety of framing applications. In a typical perimeter bearing condition, the floor joists are spaced at 16″ on center. The joist size can be scheduled, called out on the plan, or shown on the detail. The joists are supported by 6″ metal studs spaced to match the joists. The stud spacing at the upper level switches to 24″, assuming that roof framing occurs above. A continuous 16-gauge track is at the outer edge to help transfer the loads from the upper studs to the lower studs. It also prevents the joists from rolling over. The wall is framed with a single top and bottom track. The wall sill track is usually connected with #10 Tek screws. The wall head track is normally connected to each stud with a #8 Tek at each flange. The joists are field welded to the continuous top track; however, a clip connection could also be considered. The floor sheathing is ¾″ tongue and groove Sturd-I-Floor plywood.

TYPICAL FLOOR JOIST NONBEARING DETAIL (DETAIL T5-CF2)

In a typical perimeter nonbearing condition, the floor joists run in the same direction as the wall. A suspended ceiling is indicated, so metal joists out-lookers are required to brace the top of the wall. All studs are shown spaced at 24″ on center.

TYPICAL ROOF JOIST BEARING DETAIL (DETAIL T5-CF3)

Since the wall continues to a parapet, the studs are continuous. The joists are lapped and connected to the studs. This minimizes the amount of cutting; however, another option (especially for short parapets) is to bear the joists directly onto the stud wall. The parapet wall studs can then be lapped with the roof joists and cantilevered to the top of the wall.

TYPICAL ROOF JOIST NONBEARING DETAIL (DETAIL T5-CF4)

In a typical perimeter roof joist nonbearing condition, the roof joists run in the same direction as the wall. A suspended ceiling is indicated, so metal joists outlookers are required to brace the top of the wall. The studs are shown spaced at 24″ on center.

MISCELLANEOUS FRAMING DETAILS
(DETAIL T5-CF5)

Two miscellaneous framing details are shown. One shows the condition where the joists run parallel to an existing masonry wall. The floor joists are not connected to the wall, although they could be, especially if the wall is required for shearwall action. The other condition is a typical interior roof bearing condition where the rafters bear onto a steel beam. The rafters are lapped over the beam and field welded.

TYPICAL ROOF JOIST BEARING DETAIL 1
(DETAIL T5-CF6)

This detail shows a bar joist bearing condition. The stud directly below the bar joist stops below the stud and bears on a short piece of structural angle. The joist is field welded to the angle. The other studs run continuous to the parapet.

TYPICAL ROOF JOIST BEARING DETAIL 2
(DETAIL T5-CF7)

This detail is similar to the previous one, but the wall stops at the joist bearing. Four-in metal studs are shown; however, larger studs could be required depending on the span and loading conditions. The top track is noted as being field welded to the metal studs. Two #8 screws would normally be used; however, they provide minimal uplift resistance.

TYPICAL INTERIOR JOIST BEARING DETAIL
(DETAIL T5-CF8)

A typical interior roof bar joist bearing condition is shown. The bar joists bear on an interior load-bearing metal studwall. A continuous structural steel angle is provided at the top of the wall. The top track is noted as being field welded to the metal studs similar to the previous detail.

MISCELLANEOUS INTERIOR FRAMING DETAIL 1
(DETAIL T5-CF9)

In this detail, the bar joists run in the same direction as the wall. The roof diaphragm is connected to the wall, which serves as a shearwall.

MISCELLANEOUS INTERIOR FRAMING DETAIL 2
(DETAIL T5-CF10)

This detail is similar to the previous one, but the roof framing on each side of the wall is at a different elevation.

TYPICAL PERIMETER NONBEARING DETAIL (DETAIL T5-CF11)

This detail is similar to the previous two, but it is shown at a perimeter wall.

MISCELLANEOUS FRAMING

TYPICAL ROOF/FLOOR DIAPHRAGM DETAIL (DETAIL T5-CFDIA)

Horizontal roof and floor diaphragms typically consist of plywood sheathing. The plywood is laid in a staggered joint pattern. Blocking is required if noted on the plan. A ⅛″ gap is typically provided between the plywood sheets. Separate notes address the connection requirements, which typically consist of screws. An option in lieu of plywood is to use oriented strand board, which provides better weather resistance during construction.

TYPICAL PLYWOOD SHEARWALL ELEVATION DETAIL (DETAIL T5-CFSWE)

Vertical wall diaphragms typically consist of ½″ plywood sheathing. The plywood is laid in a staggered joint pattern. Blocking is required if noted in the shearwall schedule. A ⅛″ gap is typically provided between the plywood sheets. The screw connection requirements are shown in a shearwall schedule. In lieu of plywood, other options are to use oriented strand board, which provides better weather resistance during construction, or to use exterior grade gypsum panels. A hold-down anchor is shown at the end of the shearwall.

TYPICAL HOLD-DOWN ANCHOR DETAILS (DETAIL T5-CFHDA)

Hold-down anchors are typically used at the ends of shearwalls, building corners, and so on. This detail shows a typical foundation level condition, where the foundation is a continuous turned down slab edge. A clip angle serves as the hold-down anchor and is field welded to the metal studs. The hold-down anchor is expansion bolted to the concrete foundation.

TYPICAL FRAMED OPENING DETAIL (DETAIL T5-CFHDR)

Wall openings need to be carefully investigated. Headers should be sized for the specific loading conditions. The jamb conditions need to have full-height studs and jack studs, as required for the transverse wind loading requirements. The header schedule contains a description of the jamb stud and header requirements.

TYPICAL PARTITION WALL BRACING DETAILS
(DETAIL T5-CFWAL)

Partition wall framing and other nonstructural framing are normally shown on the architectural drawings; however, the partition walls need to be properly braced off. Interior studwalls are typically designed for a nominal wind loading. Therefore, all interior studwalls need to be properly braced off.

TYPICAL PIPE HANGER DETAILS
(DETAIL T5-HANG1)

Several different conditions are shown, including pipes perpendicular to bar joists, pipes parallel to bar joists, and pipes parallel to WF beams.

TYPICAL STAIR HANGER DETAIL 1
(DETAIL T5-HANG2)

Stair systems are normally shown on the architectural drawings. The stair stringers and other framing should be noted; however, in some regions, the stair manufacturer designs these systems. Regardless, the stair systems need to be properly supported by the main framing. In this detail, the intermediate stair landing level is being suspended by the $\frac{5}{8}''$ diameter all-thread rod. In general, details involving suspended framing should be very carefully investigated.

TYPICAL STAIR HANGER DETAIL 2
(DETAIL T5-HANG3)

This detail is similar to the previous one, but the beam supporting the all-thread rod is running parallel to the floor joists.

TYPICAL ROOF OPENING DETAIL 1
(DETAIL T5-RO1)

Metal roof decks can accommodate small openings without additional framing; however, many roof openings such as at skylights, roof hatches, and mechanical rooftop units are too large. Therefore, these larger openings need to be framed out with steel angles.

TYPICAL ROOF OPENING DETAIL 2
(DETAIL T5-RO2)

This detail is similar to the previous one, but the angle around the perimeter of the opening is turned up.

TYPICAL ROOF SCREEN DETAIL
(DETAIL T5-SCREN)

Many buildings require mechanical rooftop equipment. Sometimes this equipment is hidden behind roof screens. In this case, the columns at the upper level are cantilevered up beyond the roof level, and the steel channels are framed between the cantilevered columns.

TYPICAL ELEVATED SLAB JOINT DETAILS
(DETAIL T5-SBJT1)

Noncomposite floor slab systems consist of relatively thin concrete slabs. The light-gauge metal deck serves merely as a temporary form. After the concrete slab sets, the concrete serves as the structural element; hence, it is reinforced with wire fabric. Elevated slabs are frequently poured in large areas and concrete shrinkage will occur, resulting in cracks. Saw-cutting joints into the elevated slab will create artificially weakened planes so that the cracks form in these locations instead of at random.

TYPICAL ELEVATED SLAB JOINT DETAILS
(DETAIL T5-SBJT2)

This detail is essentially the same as the previous one, but an interior beam condition is shown.

TYPICAL BEAM REPAIR DETAIL
(DETAIL T5-BMREP)

Existing steel beams can be strengthened and reinforced in place. This detail shows a flexural reinforcement that involves adding round bars below the top flange and a flat plate below the bottom flange.

TYPICAL ROOF DECK ATTACHMENT DETAIL 1
(DETAIL T5-RDIA1)

Metal roof decks are typically used as a structural diaphragm. The shear capacity of the diaphragm depends on the type of attachment. Several options for connecting the metal deck to supports are available; however, field welding is preferred. The welds are spaced more closely within the building perimeter to account for increased wind uplift. The sidelap fastening of the metal deck also has an effect on the diaphragm capacity. Sidelaps are typically connected with #10 Tek screws.

TYPICAL ROOF DECK ATTACHMENT DETAIL 2 (DETAIL T5-RDIA2)

This detail is similar to the previous one, but the connection requirements are shown in a connection schedule. The connection schedule is shown in detail T5-RDIA4.

ROOF DECK ATTACHMENT ZONES (DETAIL T5-RDIA3)

This detail is intended to work with the previous one and with detail T5-RDIA4.

ROOF DECK ATTACHMENT PATTERNS (DETAIL T5-RDIA4)

Several different support attachment patterns are available. The two simplest patterns are 36/7 and 36/4. Pattern 36/7 has an attachment at every rib (6" on center). Pattern 36/4 has an attachment at every other rib (12" on center).

TYPICAL DETAIL AT ROOFTOP UNITS (DETAIL T5-MECH1)

This detail shows two views (a plan and elevation) of the framing condition at a mechanical rooftop unit. A treated 2×4 is placed below the curb so that the weight of the mechanical unit does not crush the metal deck.

TYPICAL DETAIL AT ROOFTOP UNITS (DETAIL T5-MECH2)

This detail is similar to the previous one, but a steel channel is used in lieu of the treated 2×4.

Wood

(CSI Division 6)

Discussion—Wood

TYPES OF DETAILS INCLUDED

WOOD FRAMING DETAILS ARE PROVIDED IN
THIS CHAPTER FOR THE FOLLOWING:

- ▶ *2 × Framing*
- ▶ *Premanufactured Wood I-Joist Framing*
- ▶ *Pre-Engineered Wood Truss Framing*
- ▶ *Diaphragms*
- ▶ *Hold-Down Anchors*
- ▶ *Miscellaneous Details*

2× FRAMING

Many small low-rise commercial buildings can be successfully designed using standard dimension lumber, referred to as 2× framing. The most commonly used dimension lumber includes 2 × 4, 2 × 6, 2 × 8, 2 × 10, and 2 × 12 members. These are nominal sizes, in inches. The actual dressed sizes are smaller. For instance, a standard 2 × 4 is actually 1.5″ × 3.5″ and a standard 2 × 12 is actually 1.5″ × 11.25″. Dimension lumber is readily available and is usually the most economical way to frame relatively short spans. However, for longer spans or heavy loadings, premanufactured wood framing systems may be required.

For walls, 2 × 4 and 2 × 6 studs can frequently be used. For most exterior walls, 2 × 6 studs are recommended due to wind or seismic lateral loading requirements. However, larger studs may be required due to any unusual span or loading requirements. For interior walls, 2 × 4 studs may suffice. Heavily loaded interior walls may need to be upgraded to 2 × 6 or 2 × 8 studs.

It is preferable to space the studs to match and align with the framing being supported. For studs supporting floor framing, a maximum spacing of 16″ is recommended. For studs supporting roof framing, a maximum spacing of 24″ is recommended. The 24″ maximum spacing easily accommodates the spanning capacity of most exterior sheathing systems.

One of the most serious considerations with using 2× wall framing is to ensure that all studs are continuous from support to support. The bottom is typically supported on a single sill plate, which is attached to the foundation or floor diaphragm. The top is typically supported by a double top plate, which is attached to the floor or roof framing. Stud walls are normally constructed using the *platform construction method*. With platform construction, the walls are field fabricated in a horizontal position and then lifted up into a vertical position. This works well for most conditions; however, the wall framing for open rooms (such as for rooms with cathedral ceilings) and stairwell walls should be carefully investigated. A structural weakness can occur if these types of walls are constructed using platform construction because the double top plates are not properly braced off by permanent construction.

Another commonly encountered problem with using 2× wall framing is at the jambs of wall openings. Additional transverse wind (or seismic) loads from the window and door openings get transferred to the jambs, which need to have additional full-height studs to accommodate the increased loads. For interior walls, the jambs of larger openings may need to have additional jack studs to accommodate the additional vertical loads.

Floor joist and roof rafter framing need to be sized for the specific loading and performance requirements. Long-term deflection, or "creep," should also be investigated. The spacing of 2× framing is limited by the spanning capacity of the sheathing diaphragm. Floor diaphragms typically use ¾″ plywood, and a maximum joist spacing of 16″ is recommended. Greater spacings can be used; however, 16″ usually results in a higher performance floor system. Roof diaphragms typically use ⅝″ plywood, and a maximum rafter spacing of 24″ is recommended. Thinner roof sheathing, such as ½″, can be used; however, roof waviness may result.

Since 2× framing can be easily cut, notched, and altered, unintended structural weaknesses can be introduced in the field. Many subcontractors (plumbers, electricians, etc.) mistakenly weaken 2× framing systems long after the wood framing subcontractor has left the site. Field quality control is necessary to ensure that these problems are discovered and corrected prior to being covered up.

Special nailing and connection requirements should be shown on the specific details; however, it is too cumbersome to note every nail on every detail. The model codes provide minimum nailing requirements. It is recommended that these requirements be summarized on a nailing schedule and included on the drawings.

PREMANUFACTURED WOOD I-JOIST FRAMING

Due to the span and loading limitations of dimension lumber, premanufactured wood framing systems were developed. Wood I-joists are very popular due to their inherent structural efficiency. The efficiency is provided by locating more material, usually high-grade laminated veneer lumber (LVL) where it is most needed, away from the neutral axis at the chords, similar to wide-flange steel beams. Wood I-joists typically use plywood or oriented strand board for the webs and LVL for the chords. I-joists are premanufactured under strict quality control procedures to ensure that the webs and chords are properly bonded.

Compared to 2× framing, I-joist products can also more easily accommodate openings for electrical and mechanical utilities. Many manufacturers provide small knockouts in the webs for electrical utilities. Larger openings for mechanical ductwork can also be provided in the webs; however, all web openings must be in accordance with the I-joist manufacturer's requirements. Wood I-joist lengths can be cut to fit in the field, similar to 2× framing; however, the chords should never be notched or cut without the manufacturer's approval. Another popular product is laminated veneer lumber, typically used for headers and beams. It provides high-grade material to handle higher load and longer span conditions.

PRE-ENGINEERED WOOD TRUSS FRAMING

Pre-engineered wood trusses are typically used for long span roof conditions. The trusses are engineered and designed by truss manufacturers. The information necessary for a manufacturer to design the trusses must be indicated, either on the drawings or in the specifications. The information required includes the overall truss geometry, governing building code, required loadings, and so on. The actual final interior configuration of the truss is normally left up to the truss manufacturer; however, an exception would be if there are any clear opening requirements, such as for mechanical ductwork.

Wood roof trusses are an economical and practical design solution for spans up to around 50'. Longer spans can be built; however, they require special handling during transportation and erection. One-piece trusses can be built to an overall height of about 12'. Taller trusses are usually built in two pieces, with the second truss being "piggybacked." Wood trusses typically use dimension lumber for the chords and webs. The truss top chord can be pitched to accommodate the roof slope. The bottom chords can be flat or pitched.

Truss chords and webs are usually connected at the panel points with light-gauge metal connector plates. All wood trusses should be plant fabricated under strict quality control procedures. The metal connector plates are installed using special equipment. Although used primarily for roof systems, parallel chord trusses can also be used for floors. Due to the large openings between members, utilities can be easily accommodated without needing to cut or notch truss members.

Trusses require permanent bracing, which is supplied after the trusses are field erected. Truss top chords are in compression under normal (gravity) loading conditions. Truss bottom chords are in compression under wind (uplift) loading conditions. Truss webs are also in compression, some during gravity loading and some during wind loading. Permanent bracing is required for all chord members under compression. Webs may be able to span between chords without additional permanent bracing. Top chord permanent bracing is normally provided by the plywood roof sheathing. However, during major wind events, the plywood may be lost in some areas of the roof; therefore, permanent continuous 2 × 4 bracing at each top chord panel point should also be considered. Truss bottom chords will be braced by the ceiling diaphragm if it is directly applied to the truss. For suspended ceilings, permanent continuous 2 × 4 bracing is recommended as required by the truss manufacturer to suit the uplift loading condition. Permanent bottom chord bracing may also be prudent even if the ceiling is directly applied to the chords. Many drywall ceilings are completely destroyed during hurricanes due to roofing system failure and the subsequent water damage.

Until all permanent bracing is in place, temporary truss bracing is required. The Truss Plate Institute provides minimum temporary and erection truss bracing guidelines. Trusses should only be erected by experienced crews. Serious construction failures can occur when trusses are improperly erected or braced.

A-type trusses with pitched top and bottom chords are called *scissor trusses*. The scissor truss bottom chord pitch should preferably not exceed one-half of the top chord pitch. Scissor trusses present several potential design problems. Due to long-term deflections, there will be a natural tendency for scissor trusses to "kick out" horizontally at the supports. This will tend to push out the supports. The trusses must be predesigned to limit this long-term potential horizontal movement to an acceptable level. Another problem with scissor trusses is that they can result in gable end wall conditions with wall hinges. Wall hinges must be prevented by running the studs continuous to the eave. Another option is to make sure that the wall double plates are properly sized and braced.

DIAPHRAGMS

The structural integrity of wood framed buildings relies on the roof and floor diaphragms to transfer the lateral loads to the lateral load resisting system, typically plywood shearwalls. Floor and roof diaphragms need to be investigated for the actual loading conditions. The diaphragm stresses should be determined. The allowable stresses of plywood diaphragms depend on several factors, including plywood thickness, nail size, nail spacing, support framing width, and so forth. One way to increase the diaphragm capacity is to add blocking at all unsupported plywood edges. Another way is to decrease the nail spacing.

All plywood sizes are nominal. For ¾, ⅝, and ½″ plywood, the actual thicknesses are, respectively, ²³⁄₃₂, ¹⁹⁄₃₂, and ¹⁵⁄₃₂″. For floor diaphragms, a minimum of ¾″ nominal tongue and groove Sturd-I-Floor plywood is recommended. Sturd-I-Floor is a registered trademark of the American Plywood Association. All floor sheathing should be glued and nailed to supports. Glue helps eliminate floor squeaks and provides several other benefits. However, glue that sets too long prior to laying the sheathing will not be effective. Screws can also be used; however, nails perform well and are less expensive. Pneumatic nail guns have become very popular and may be acceptable for most floor sheathing applications. But overdriven nails can cause problems, especially for roof and wall sheathing.

For roof diaphragms, a minimum of ⅜″ nominal plywood is recommended. The free edges are normally connected together with metal clips. All roof sheathing should be properly nailed to the supports. Glue is not normally used. It is recommended that all roof sheathing nails be hand driven. Pneumatic nail guns tend to overdrive the nails. Overdriven nails can seriously reduce diaphragm strength. The plywood can also be more easily lifted over the nails during wind uplift.

For wall diaphragms, a minimum of ½″ nominal plywood is recommended. All wall sheathing should be properly nailed to the wall studs. Glue is not normally used. It is recommended that all wall sheathing nails be hand driven. Overdriven nails can seriously reduce diaphragm strength. Gypboard shearwall diaphragms can also be used, especially for interior walls. However, their structural effectiveness during major cyclical loadings, such as earthquakes, is questionable.

HOLD-DOWN ANCHORS

The structural integrity of wood framed buildings relies on all systems being properly connected. Hurricanes, tornadoes, and earthquakes cause major lateral loading conditions. These loads are typically resisted by plywood shearwalls. Shearwalls can develop large endwall chord and overturning forces, requiring hold-down anchors at the floor framing levels and foundation. Hold-down anchors are fabricated using galvanized steel. A variety of manufacturers provide hold-down anchors, rated capacities, and recommendations for their use.

MISCELLANEOUS DETAILS

A variety of other details are also presented to address headers, built-up framing, partition wall bracing, joist loading diagrams, diagonal bracing, floor girders, and relocating existing buildings to new pile supports.

GENERAL COMMENT

Practices vary from region to region and the details that follow have generally been taken from specific applications. Any member sizes or connections shown were intended for that particular application. Accordingly, the applicability of all details should be investigated and all member sizes and connections should be determined prior to use.

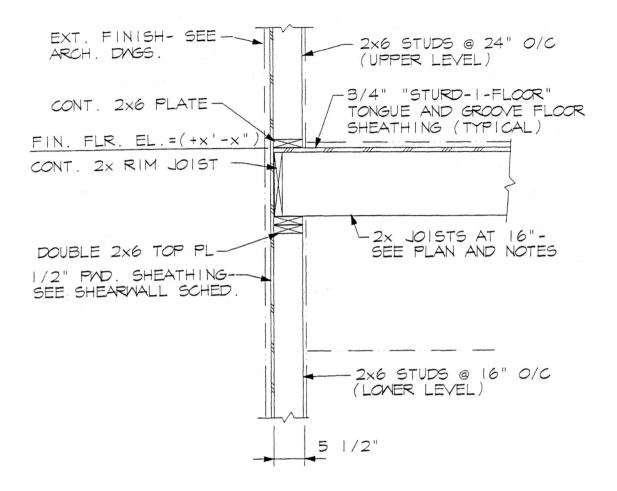

EXT. FINISH- SEE ARCH. DWGS.

2x6 STUDS @ 24" O/C (UPPER LEVEL)

3/4" "STURD-I-FLOOR" TONGUE AND GROOVE FLOOR SHEATHING (TYPICAL)

CONT. 2x6 PLATE

FIN. FLR. EL.=(+x'-x")

CONT. 2x RIM JOIST

2x JOISTS AT 16"- SEE PLAN AND NOTES

DOUBLE 2x6 TOP PL

1/2" PWD. SHEATHING- SEE SHEARWALL SCHED.

2x6 STUDS @ 16" O/C (LOWER LEVEL)

5 1/2"

TYP. 2x JOIST PERIMETER BEARING

NOT TO SCALE   (DETAIL T6-2x1)

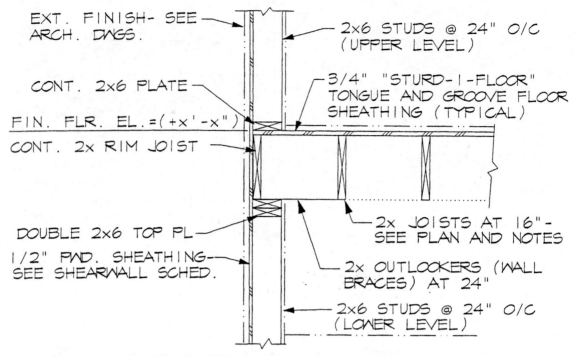

EXT. FINISH- SEE ARCH. DWGS.

2x6 STUDS @ 24" O/C (UPPER LEVEL)

CONT. 2x6 PLATE

3/4" "STURD-I-FLOOR" TONGUE AND GROOVE FLOOR SHEATHING (TYPICAL)

FIN. FLR. EL.=(+x'-x")

CONT. 2x RIM JOIST

2x JOISTS AT 16"- SEE PLAN AND NOTES

DOUBLE 2x6 TOP PL

1/2" PWD. SHEATHING- SEE SHEARWALL SCHED.

2x OUTLOOKERS (WALL BRACES) AT 24"

2x6 STUDS @ 24" O/C (LOWER LEVEL)

TYP. 2x JOIST PERIMETER NONBEARING
NOT TO SCALE   (DETAIL T6-2x2)

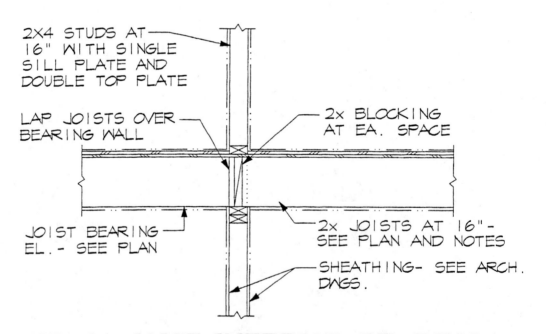

2X4 STUDS AT 16" WITH SINGLE SILL PLATE AND DOUBLE TOP PLATE

LAP JOISTS OVER BEARING WALL

2x BLOCKING AT EA. SPACE

JOIST BEARING EL.- SEE PLAN

2x JOISTS AT 16"- SEE PLAN AND NOTES

SHEATHING- SEE ARCH. DWGS.

TYP. 2x JOIST INTERIOR BEARING
NOT TO SCALE   (DETAIL T6-2x3)

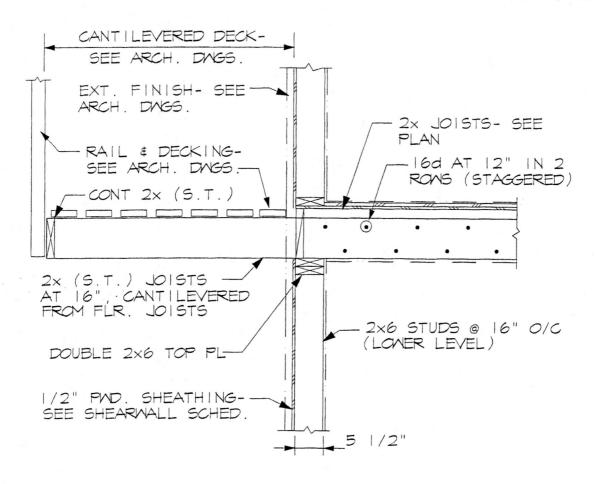

CANTILEVERED DECK-
SEE ARCH. DWGS.

EXT. FINISH- SEE
ARCH. DWGS.

RAIL & DECKING-
SEE ARCH. DWGS.

CONT 2x (S.T.)

2x JOISTS- SEE
PLAN

16d AT 12" IN 2
ROWS (STAGGERED)

2x (S.T.) JOISTS
AT 16", CANTILEVERED
FROM FLR. JOISTS

DOUBLE 2x6 TOP PL

1/2" PWD. SHEATHING-
SEE SHEARWALL SCHED.

2x6 STUDS @ 16" O/C
(LOWER LEVEL)

5 1/2"

TYP. 2x CANTILEVERED DECK JOISTS
NOT TO SCALE   (DETAIL T6-2x4)

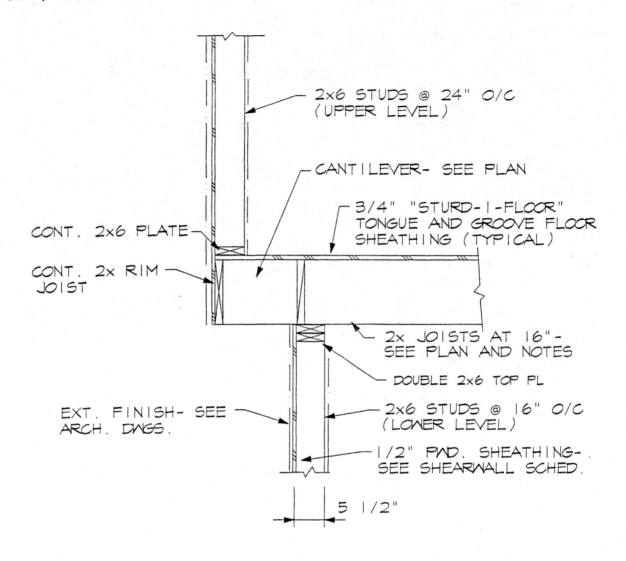

2x6 STUDS @ 24" O/C
(UPPER LEVEL)

CANTILEVER- SEE PLAN

3/4" "STURD-I-FLOOR"
TONGUE AND GROOVE FLOOR
SHEATHING (TYPICAL)

CONT. 2x6 PLATE

CONT. 2x RIM
JOIST

2x JOISTS AT 16"-
SEE PLAN AND NOTES

DOUBLE 2x6 TOP PL

EXT. FINISH- SEE
ARCH. DWGS.

2x6 STUDS @ 16" O/C
(LOWER LEVEL)

1/2" PWD. SHEATHING-
SEE SHEARWALL SCHED.

5 1/2"

TYP. 2x CANTILEVERED JOIST DETAIL
NOT TO SCALE   (DETAIL T6-2x5)

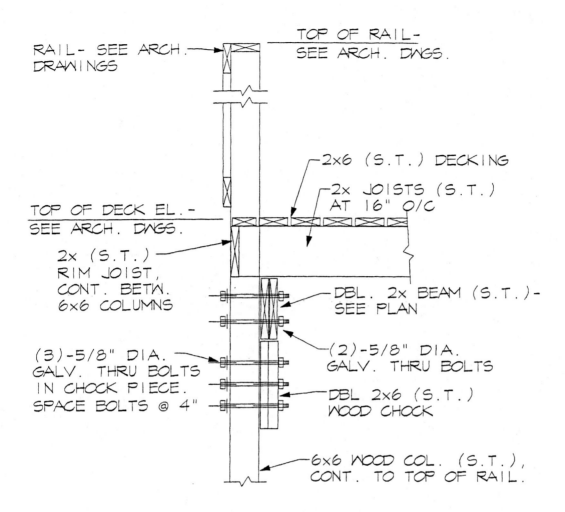

RAIL - SEE ARCH. DRAWINGS

TOP OF RAIL - SEE ARCH. DWGS.

2×6 (S.T.) DECKING

2× JOISTS (S.T.) AT 16" O/C

TOP OF DECK EL. - SEE ARCH. DWGS.

2× (S.T.) RIM JOIST, CONT. BETW. 6×6 COLUMNS

DBL. 2× BEAM (S.T.) - SEE PLAN

(3)-5/8" DIA. GALV. THRU BOLTS IN CHOCK PIECE. SPACE BOLTS @ 4"

(2)-5/8" DIA. GALV. THRU BOLTS

DBL 2×6 (S.T.) WOOD CHOCK

6×6 WOOD COL. (S.T.), CONT. TO TOP OF RAIL.

TYP. DECK PERIMETER FRAMING DETAIL
NOT TO SCALE   (DETAIL T6-2x6)

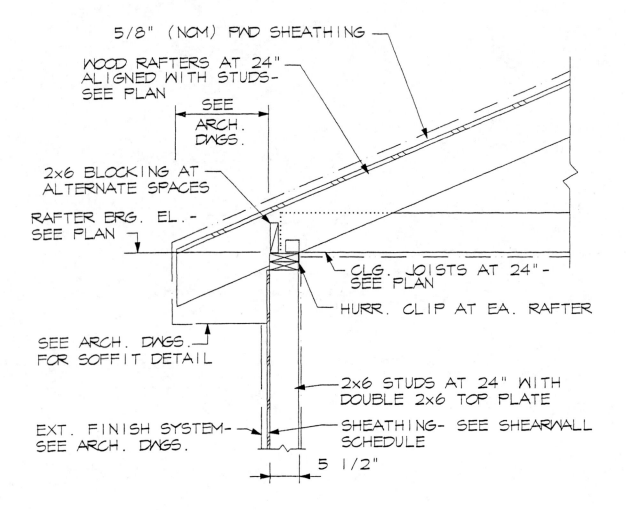

5/8" (NOM) PWD SHEATHING

WOOD RAFTERS AT 24"
ALIGNED WITH STUDS-
SEE PLAN

SEE
ARCH.
DWGS.

2x6 BLOCKING AT
ALTERNATE SPACES

RAFTER BRG. EL.-
SEE PLAN

CLG. JOISTS AT 24"-
SEE PLAN

HURR. CLIP AT EA. RAFTER

SEE ARCH. DWGS.
FOR SOFFIT DETAIL

2x6 STUDS AT 24" WITH
DOUBLE 2x6 TOP PLATE

SHEATHING- SEE SHEARWALL
SCHEDULE

EXT. FINISH SYSTEM-
SEE ARCH. DWGS.

5 1/2"

TYP. 2x RAFTER BRG. ON STUDWALL

NOT TO SCALE   (DETAIL T6-R2x1)

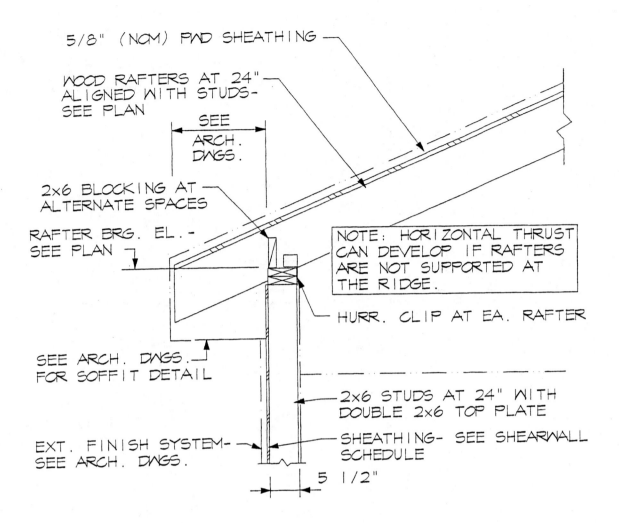

5/8" (NOM) PWD SHEATHING

WOOD RAFTERS AT 24"
ALIGNED WITH STUDS-
SEE PLAN

SEE
ARCH.
DWGS.

2x6 BLOCKING AT
ALTERNATE SPACES

RAFTER BRG. EL.-
SEE PLAN

NOTE: HORIZONTAL THRUST
CAN DEVELOP IF RAFTERS
ARE NOT SUPPORTED AT
THE RIDGE.

HURR. CLIP AT EA. RAFTER

SEE ARCH. DWGS.
FOR SOFFIT DETAIL

2x6 STUDS AT 24" WITH
DOUBLE 2x6 TOP PLATE

SHEATHING- SEE SHEARWALL
SCHEDULE

EXT. FINISH SYSTEM-
SEE ARCH. DWGS.

5 1/2"

TYP. 2x RAFTER BRG. ON STUDWALL

NOT TO SCALE   (DETAIL T6-R2x2)

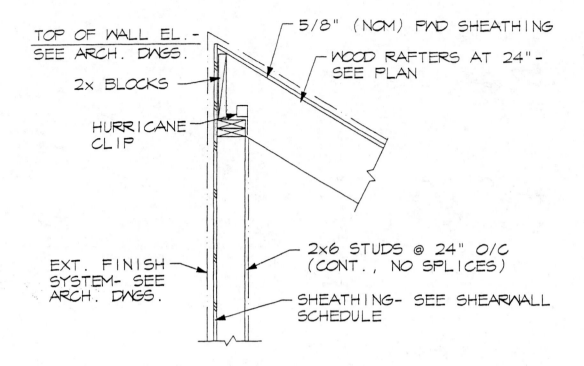

TOP OF WALL EL.-
SEE ARCH. DWGS.

2x BLOCKS

HURRICANE
CLIP

5/8" (NOM) PWD SHEATHING

WOOD RAFTERS AT 24"-
SEE PLAN

EXT. FINISH
SYSTEM- SEE
ARCH. DWGS.

2x6 STUDS @ 24" O/C
(CONT., NO SPLICES)

SHEATHING- SEE SHEARWALL
SCHEDULE

TYP. 2x RAFTER BRG. ON STUDWALL

NOT TO SCALE   (DETAIL T6-R2x3)

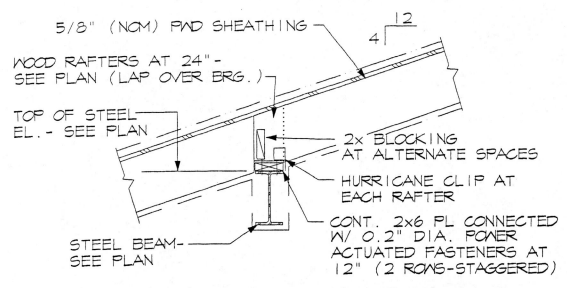

5/8" (NOM) PWD SHEATHING

WOOD RAFTERS AT 24" - SEE PLAN (LAP OVER BRG.)

TOP OF STEEL EL. - SEE PLAN

$\frac{12}{4}$

2x BLOCKING AT ALTERNATE SPACES

HURRICANE CLIP AT EACH RAFTER

STEEL BEAM- SEE PLAN

CONT. 2x6 PL CONNECTED W/ 0.2" DIA. POWER ACTUATED FASTENERS AT 12" (2 ROWS-STAGGERED)

DETAIL AT INTERIOR STEEL BEAMS

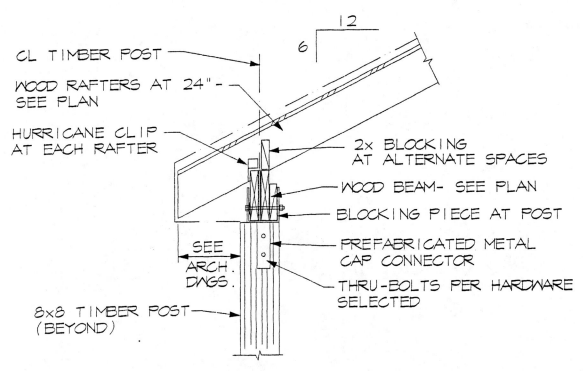

CL TIMBER POST

WOOD RAFTERS AT 24" - SEE PLAN

HURRICANE CLIP AT EACH RAFTER

$\frac{12}{6}$

2x BLOCKING AT ALTERNATE SPACES

WOOD BEAM- SEE PLAN

BLOCKING PIECE AT POST

PREFABRICATED METAL CAP CONNECTOR

THRU-BOLTS PER HARDWARE SELECTED

SEE ARCH. DWGS.

8x8 TIMBER POST (BEYOND)

DETAIL AT PERIMETER WOOD BEAMS

MISC. 2x RAFTER BEARING DETAILS

NOT TO SCALE   (DETAIL T6-R2x4)

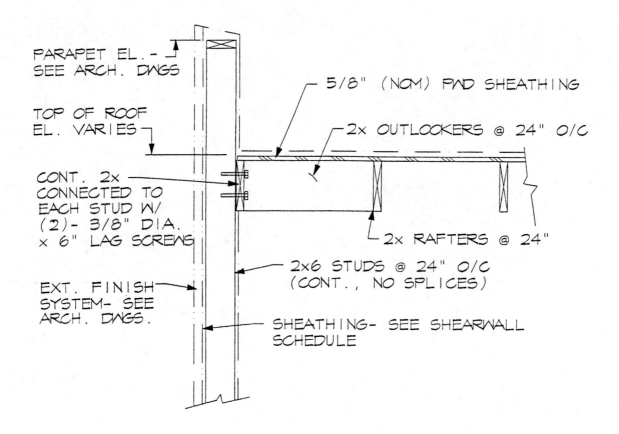

PARAPET EL. -
SEE ARCH. DWGS

TOP OF ROOF
EL. VARIES

CONT. 2x
CONNECTED TO
EACH STUD W/
(2)- 3/8" DIA.
x 6" LAG SCREWS

EXT. FINISH
SYSTEM- SEE
ARCH. DWGS.

5/8" (NOM) PWD SHEATHING

2x OUTLOOKERS @ 24" O/C

2x RAFTERS @ 24"

2x6 STUDS @ 24" O/C
(CONT., NO SPLICES)

SHEATHING- SEE SHEARWALL
SCHEDULE

TYP. 2x RAFTER NONBEARING DETAIL
NOT TO SCALE   (DETAIL T6-R2x5)

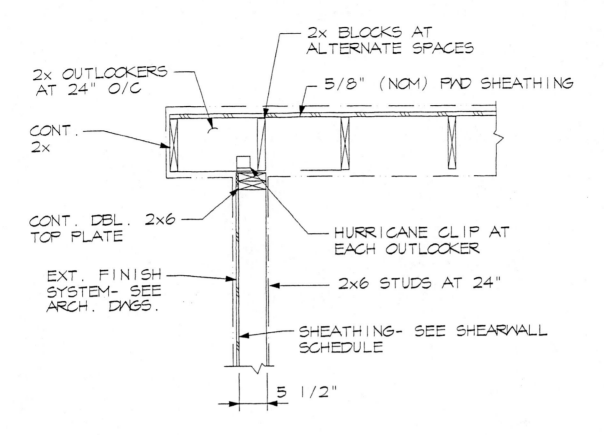

2x BLOCKS AT ALTERNATE SPACES

2x OUTLOOKERS AT 24" O/C

5/8" (NOM) PWD SHEATHING

CONT. 2x

CONT. DBL. 2x6 TOP PLATE

HURRICANE CLIP AT EACH OUTLOOKER

EXT. FINISH SYSTEM- SEE ARCH. DWGS.

2x6 STUDS AT 24"

SHEATHING- SEE SHEARWALL SCHEDULE

5 1/2"

TYP. 2x RAFTER NONBEARING DETAIL

NOT TO SCALE   (DETAIL T6-R2x6)

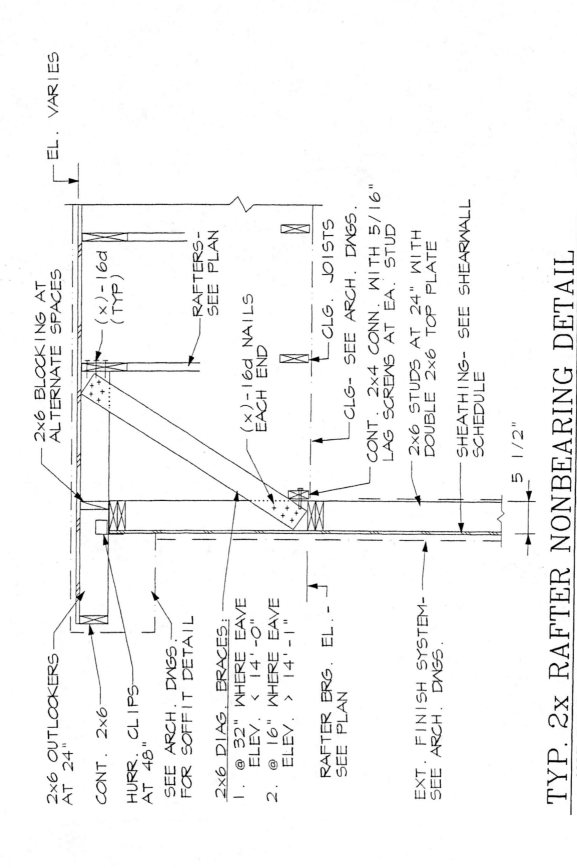

2x6 OUTLOOKERS
AT 24"

2x6 BLOCKING AT
ALTERNATE SPACES

CONT. 2x6

(x)-16d
(TYP)

HURR. CLIPS
AT 48"

RAFTERS-
SEE PLAN

SEE ARCH. DWGS.
FOR SOFFIT DETAIL

(x)-16d NAILS
EACH END

2x6 DIAG. BRACES:

1. @ 32" WHERE EAVE
   ELEV. < 14'-0"

2. @ 16" WHERE EAVE
   ELEV. > 14'-1"

CLG. JOISTS

CLG- SEE ARCH. DWGS.

RAFTER BRG. EL. -
SEE PLAN

CONT. 2x4 CONN. WITH 5/16"
LAG SCREWS AT EA. STUD

2x6 STUDS AT 24" WITH
DOUBLE 2x6 TOP PLATE

EXT. FINISH SYSTEM-
SEE ARCH. DWGS.

SHEATHING- SEE SHEARWALL
SCHEDULE

5 1/2"

EL. VARIES

TYP. 2x RAFTER NONBEARING DETAIL

NOT TO SCALE

(DETAIL T6-R2x7)

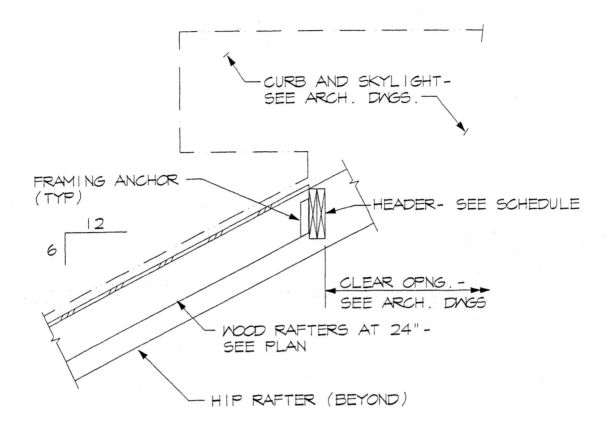

CURB AND SKYLIGHT-
SEE ARCH. DWGS.

FRAMING ANCHOR
(TYP)

12
6

HEADER- SEE SCHEDULE

CLEAR OPNG. -
SEE ARCH. DWGS

WOOD RAFTERS AT 24"-
SEE PLAN

HIP RAFTER (BEYOND)

2x RAFTER FRMG. DETAIL @ SKYLIGHT

NOT TO SCALE   (DETAIL T6—R2x8)

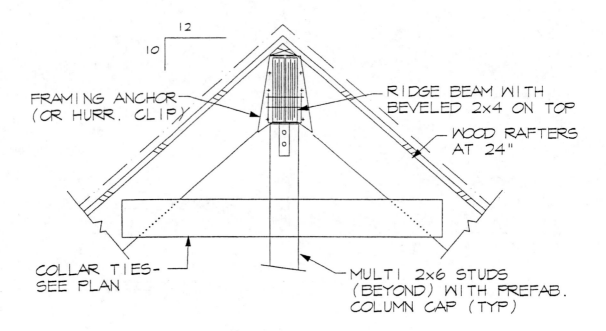

12
10

FRAMING ANCHOR
(OR HURR. CLIP)

RIDGE BEAM WITH
BEVELED 2x4 ON TOP

WOOD RAFTERS
AT 24"

COLLAR TIES-
SEE PLAN

MULTI 2x6 STUDS
(BEYOND) WITH PREFAB.
COLUMN CAP (TYP)

DETAIL AT BEAM RIDGE

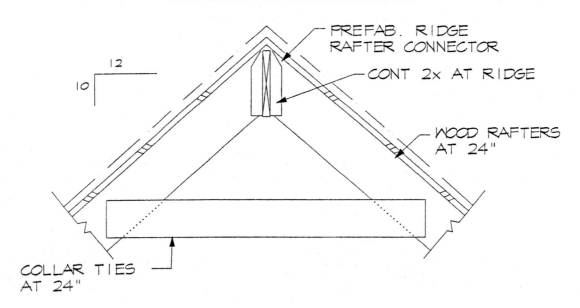

12
10

PREFAB. RIDGE
RAFTER CONNECTOR

CONT 2x AT RIDGE

WOOD RAFTERS
AT 24"

COLLAR TIES
AT 24"

DETAIL AT 2x RIDGE

TYP. 2x FRAMING DETAILS AT RIDGE
NOT TO SCALE   (DETAIL T6-R2x9)

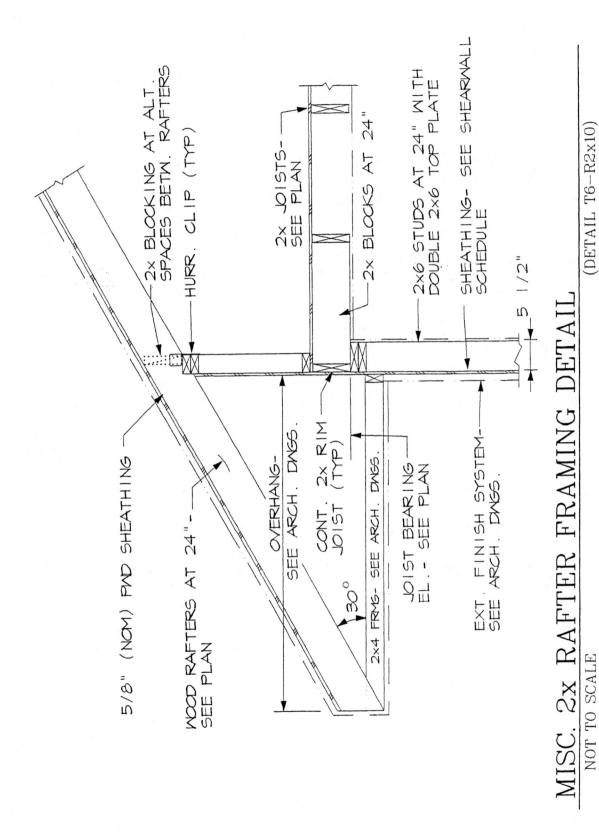

5/8" (NOM) PWD SHEATHING

WOOD RAFTERS AT 24" - SEE PLAN

2x BLOCKING AT ALT. SPACES BETW. RAFTERS

HURR. CLIP (TYP)

2x JOISTS - SEE PLAN

2x BLOCKS AT 24"

2x6 STUDS AT 24" WITH DOUBLE 2x6 TOP PLATE

SHEATHING - SEE SHEARWALL SCHEDULE

5 1/2"

OVERHANG - SEE ARCH. DWGS.

CONT. 2x RIM JOIST (TYP)

30°

2x4 FRMG - SEE ARCH. DWGS.

JOIST BEARING EL. - SEE PLAN

EXT. FINISH SYSTEM - SEE ARCH. DWGS.

MISC. 2x RAFTER FRAMING DETAIL

NOT TO SCALE

(DETAIL T6-R2x10)

355

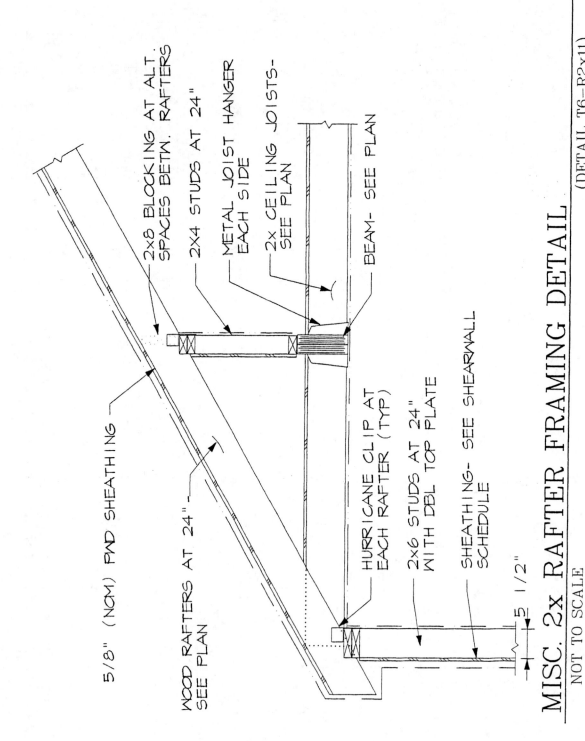

5/8" (NOM) PWD SHEATHING

WOOD RAFTERS AT 24"-
SEE PLAN

2x8 BLOCKING AT ALT.
SPACES BETW. RAFTERS

2X4 STUDS AT 24"

METAL JOIST HANGER
EACH SIDE

2x CEILING JOISTS-
SEE PLAN

BEAM- SEE PLAN

HURRICANE CLIP AT
EACH RAFTER (TYP)

2x6 STUDS AT 24"
WITH DBL TOP PLATE

SHEATHING- SEE SHEARWALL
SCHEDULE

5 1/2"

MISC. 2x RAFTER FRAMING DETAIL

NOT TO SCALE

(DETAIL T6-R2x11)

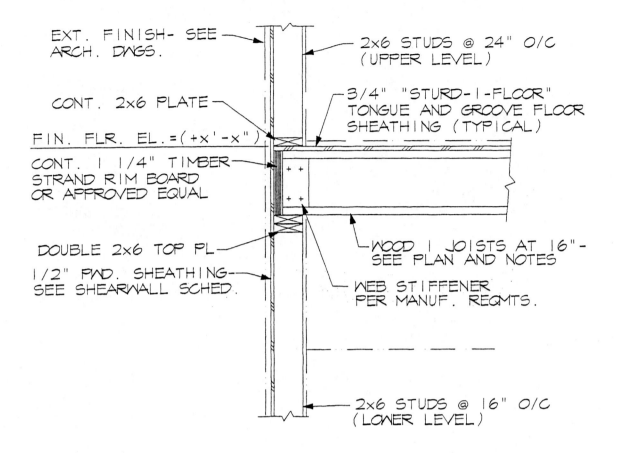

EXT. FINISH- SEE ARCH. DWGS.

2x6 STUDS @ 24" O/C (UPPER LEVEL)

CONT. 2x6 PLATE

3/4" "STURD-I-FLOOR" TONGUE AND GROOVE FLOOR SHEATHING (TYPICAL)

FIN. FLR. EL.=(+x'-x")

CONT. 1 1/4" TIMBER STRAND RIM BOARD OR APPROVED EQUAL

DOUBLE 2x6 TOP PL.

WOOD I JOISTS AT 16"- SEE PLAN AND NOTES

1/2" PWD. SHEATHING- SEE SHEARWALL SCHED.

WEB STIFFENER PER MANUF. REQMTS.

2x6 STUDS @ 16" O/C (LOWER LEVEL)

TYP. I-JOIST BEARING ON STUDWALL

NOT TO SCALE   (DETAIL T6-IJ1)

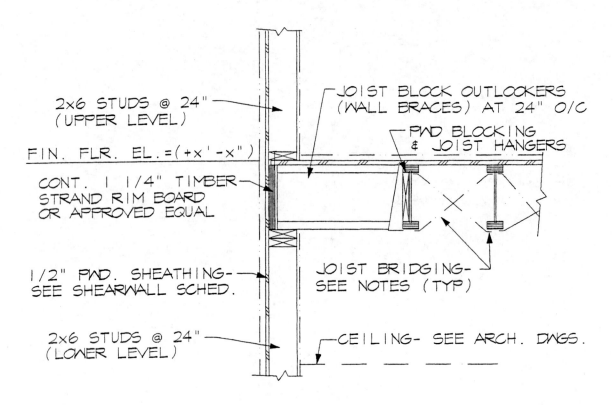

2x6 STUDS @ 24"
(UPPER LEVEL)

FIN. FLR. EL.=(+x'-x")

CONT. 1 1/4" TIMBER
STRAND RIM BOARD
OR APPROVED EQUAL

1/2" PWD. SHEATHING-
SEE SHEARWALL SCHED.

2x6 STUDS @ 24"
(LOWER LEVEL)

JOIST BLOCK OUTLOOKERS
(WALL BRACES) AT 24" O/C

PWD BLOCKING
& JOIST HANGERS

JOIST BRIDGING-
SEE NOTES (TYP)

CEILING- SEE ARCH. DWGS.

TYP. I-JOIST PERIMETER FRMG. DETAIL

NOT TO SCALE   (DETAIL T6-IJ2)

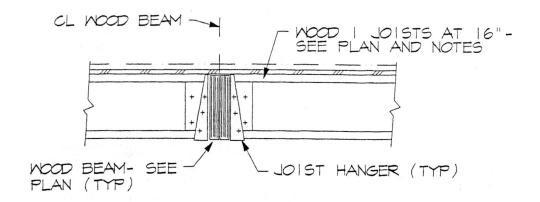

DETAIL AT INTERIOR WOOD BEAMS

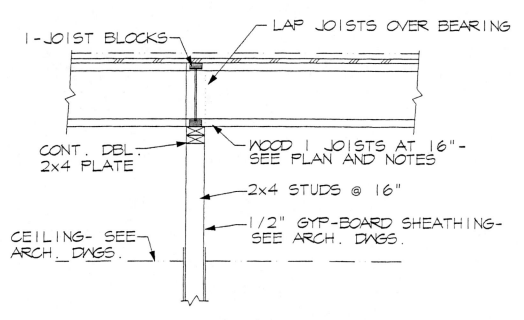

DETAIL AT INTERIOR STUDWALL

TYP. I-JOIST INTERIOR BRG. DETAILS

NOT TO SCALE   (DETAIL T6-IJ3)

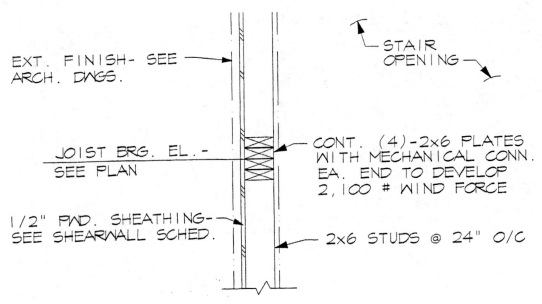

EXT. FINISH- SEE ARCH. DWGS.

JOIST BRG. EL.- SEE PLAN

1/2" PWD. SHEATHING- SEE SHEARWALL SCHED.

STAIR OPENING

CONT. (4)-2x6 PLATES WITH MECHANICAL CONN. EA. END TO DEVELOP 2,100 # WIND FORCE

2x6 STUDS @ 24" O/C

DETAIL AT EXTERIOR STUDWALL

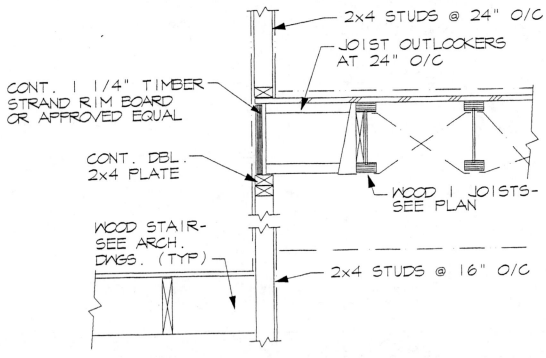

CONT. 1 1/4" TIMBER STRAND RIM BOARD OR APPROVED EQUAL

CONT. DBL. 2x4 PLATE

WOOD STAIR- SEE ARCH. DWGS. (TYP)

2x4 STUDS @ 24" O/C

JOIST OUTLOOKERS AT 24" O/C

WOOD I JOISTS- SEE PLAN

2x4 STUDS @ 16" O/C

DETAIL AT INTERIOR STUDWALL

TYPICAL FRAMING DETAILS AT STAIRS
NOT TO SCALE   (DETAIL T6-IJ4)

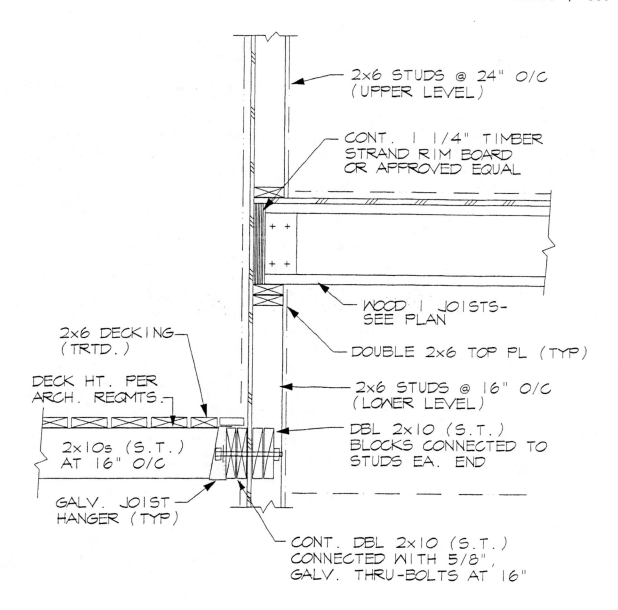

2x6 STUDS @ 24" O/C
(UPPER LEVEL)

CONT. 1 1/4" TIMBER
STRAND RIM BOARD
OR APPROVED EQUAL

WOOD I JOISTS-
SEE PLAN

DOUBLE 2x6 TOP PL (TYP)

2x6 STUDS @ 16" O/C
(LOWER LEVEL)

DBL 2x10 (S.T.)
BLOCKS CONNECTED TO
STUDS EA. END

CONT. DBL 2x10 (S.T.)
CONNECTED WITH 5/8",
GALV. THRU-BOLTS AT 16"

2x6 DECKING
(TRTD.)

DECK HT. PER
ARCH. REQMTS.

2x10s (S.T.)
AT 16" O/C

GALV. JOIST
HANGER (TYP)

TYP. I—JOIST & EXTERIOR DECK DETAIL
NOT TO SCALE   (DETAIL T6-IJ5)

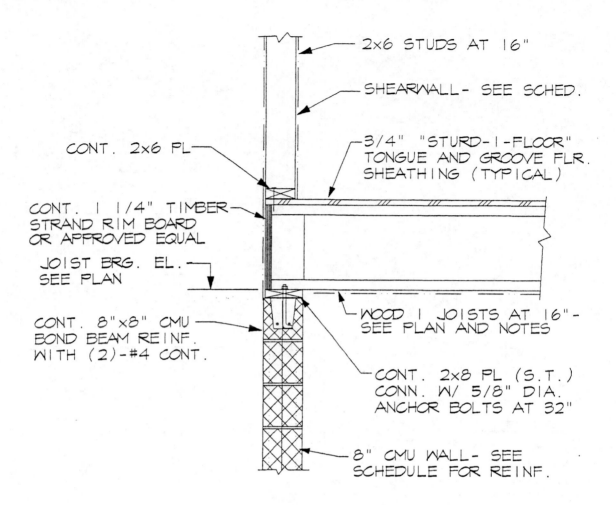

CONT. 2x6 PL

CONT. 1 1/4" TIMBER
STRAND RIM BOARD
OR APPROVED EQUAL

JOIST BRG. EL.
SEE PLAN

CONT. 8"x8" CMU
BOND BEAM REINF.
WITH (2)-#4 CONT.

2x6 STUDS AT 16"

SHEARWALL- SEE SCHED.

3/4" "STURD-1-FLOOR"
TONGUE AND GROOVE FLR.
SHEATHING (TYPICAL)

WOOD 1 JOISTS AT 16"-
SEE PLAN AND NOTES

CONT. 2x8 PL (S.T.)
CONN. W/ 5/8" DIA.
ANCHOR BOLTS AT 32"

8" CMU WALL- SEE
SCHEDULE FOR REINF.

TYP. I–JOIST BEARING ON CMU WALL
NOT TO SCALE   (DETAIL T6–IJ6)

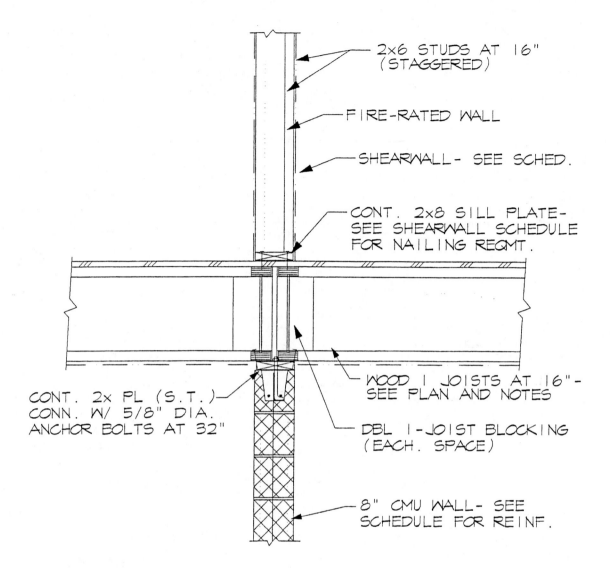

2x6 STUDS AT 16"
(STAGGERED)

FIRE-RATED WALL

SHEARWALL- SEE SCHED.

CONT. 2x8 SILL PLATE-
SEE SHEARWALL SCHEDULE
FOR NAILING REQMT.

WOOD I JOISTS AT 16"-
SEE PLAN AND NOTES

DBL I-JOIST BLOCKING
(EACH. SPACE)

8" CMU WALL- SEE
SCHEDULE FOR REINF.

CONT. 2x PL (S.T.)
CONN. W/ 5/8" DIA.
ANCHOR BOLTS AT 32"

TYP. I—JOIST BEARING ON INT. CMU

NOT TO SCALE   (DETAIL T6—IJ7)

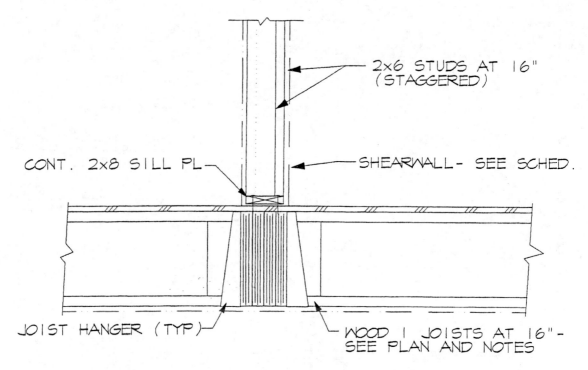

2x6 STUDS AT 16"
(STAGGERED)

SHEARWALL- SEE SCHED.

CONT. 2x8 SILL PL

JOIST HANGER (TYP)

WOOD I JOISTS AT 16"-
SEE PLAN AND NOTES

DETAIL AT INTERIOR WOOD BEAMS

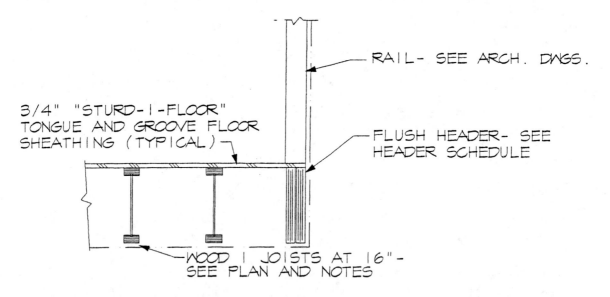

RAIL- SEE ARCH. DWGS.

3/4" "STURD-I-FLOOR"
TONGUE AND GROOVE FLOOR
SHEATHING (TYPICAL)

FLUSH HEADER- SEE
HEADER SCHEDULE

WOOD I JOISTS AT 16"-
SEE PLAN AND NOTES

DETAIL AT RAILING

TYP. I-JOIST INTERIOR FRMG. DETAILS
NOT TO SCALE   (DETAIL T6-IJ8)

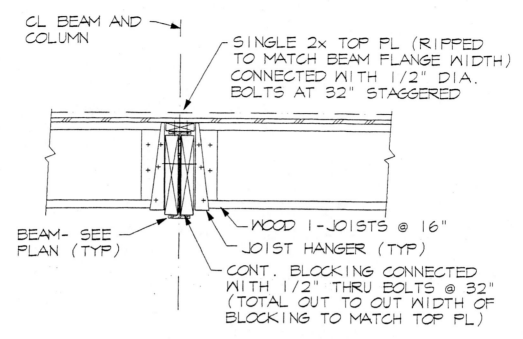

CL BEAM AND COLUMN

SINGLE 2x TOP PL (RIPPED TO MATCH BEAM FLANGE WIDTH) CONNECTED WITH 1/2" DIA. BOLTS AT 32" STAGGERED

BEAM- SEE PLAN (TYP)

WOOD I-JOISTS @ 16"

JOIST HANGER (TYP)

CONT. BLOCKING CONNECTED WITH 1/2" THRU BOLTS @ 32" (TOTAL OUT TO OUT WIDTH OF BLOCKING TO MATCH TOP PL)

DETAIL AT FLOOR LEVELS

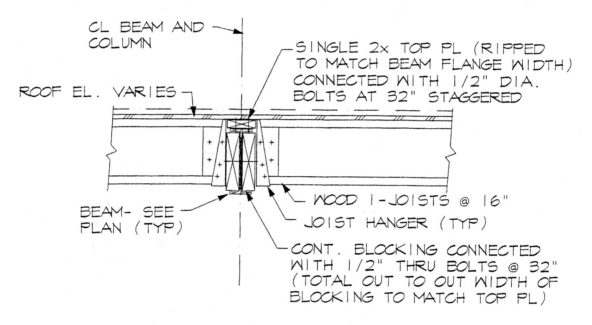

CL BEAM AND COLUMN

ROOF EL. VARIES

SINGLE 2x TOP PL (RIPPED TO MATCH BEAM FLANGE WIDTH) CONNECTED WITH 1/2" DIA. BOLTS AT 32" STAGGERED

BEAM- SEE PLAN (TYP)

WOOD I-JOISTS @ 16"

JOIST HANGER (TYP)

CONT. BLOCKING CONNECTED WITH 1/2" THRU BOLTS @ 32" (TOTAL OUT TO OUT WIDTH OF BLOCKING TO MATCH TOP PL)

DETAIL AT ROOF LEVEL

TYP. I-JOIST CONNECTIONS TO STEEL

NOT TO SCALE   (DETAIL T6-IJ9)

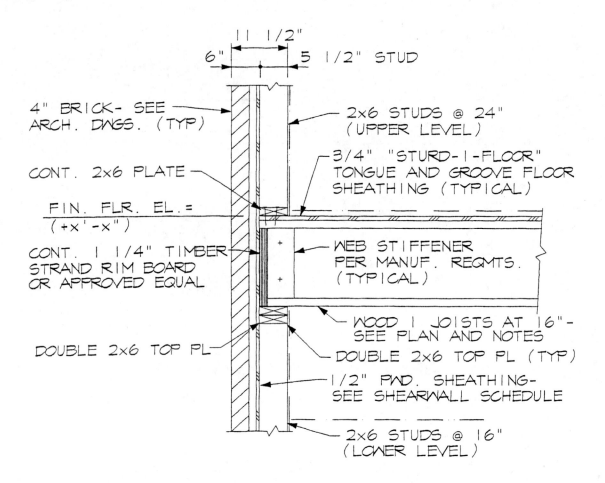

11 1/2"

6"   5 1/2" STUD

4" BRICK- SEE ARCH. DWGS. (TYP)

2x6 STUDS @ 24" (UPPER LEVEL)

3/4" "STURD-I-FLOOR" TONGUE AND GROOVE FLOOR SHEATHING (TYPICAL)

CONT. 2x6 PLATE

FIN. FLR. EL. = (+x'-x")

WEB STIFFENER PER MANUF. REQMTS. (TYPICAL)

CONT. 1 1/4" TIMBER STRAND RIM BOARD OR APPROVED EQUAL

WOOD I JOISTS AT 16"- SEE PLAN AND NOTES

DOUBLE 2x6 TOP PL

DOUBLE 2x6 TOP PL (TYP)

1/2" PWD. SHEATHING- SEE SHEARWALL SCHEDULE

2x6 STUDS @ 16" (LOWER LEVEL)

TYP. I—JOIST BEARING ON STUDWALL

NOT TO SCALE   (DETAIL T6—IJ10)

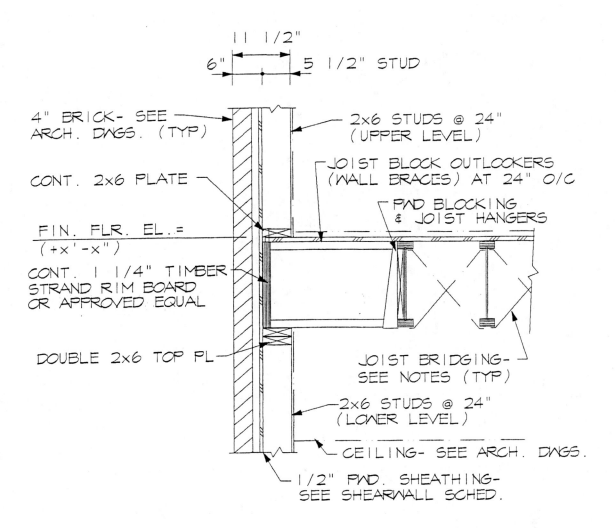

11 1/2"

6" | 5 1/2" STUD

4" BRICK- SEE ARCH. DWGS. (TYP)

2x6 STUDS @ 24" (UPPER LEVEL)

CONT. 2x6 PLATE

JOIST BLOCK OUTLOOKERS (WALL BRACES) AT 24" O/C

PWD BLOCKING & JOIST HANGERS

FIN. FLR. EL.= (+x'-x")

CONT. 1 1/4" TIMBER STRAND RIM BOARD OR APPROVED EQUAL

JOIST BRIDGING- SEE NOTES (TYP)

DOUBLE 2x6 TOP PL

2x6 STUDS @ 24" (LOWER LEVEL)

CEILING- SEE ARCH. DWGS.

1/2" PWD. SHEATHING- SEE SHEARWALL SCHED.

TYP. I–JOIST PERIMETER FRMG. DETAIL

NOT TO SCALE
(DETAIL T6–IJ11)

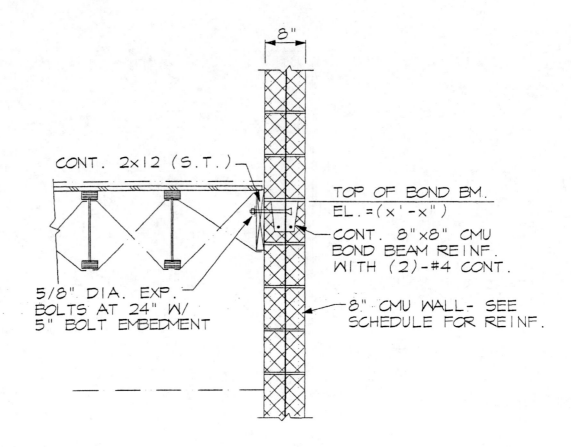

8"

CONT. 2x12 (S.T.)

TOP OF BOND BM.
EL.=(x'-x")

CONT. 8"x8" CMU
BOND BEAM REINF.
WITH (2)-#4 CONT.

5/8" DIA. EXP.
BOLTS AT 24" W/
5" BOLT EMBEDMENT

8" CMU WALL- SEE
SCHEDULE FOR REINF.

TYP. I-JOIST PARALLEL TO CMU WALL
NOT TO SCALE   (DETAIL T6-IJ12)

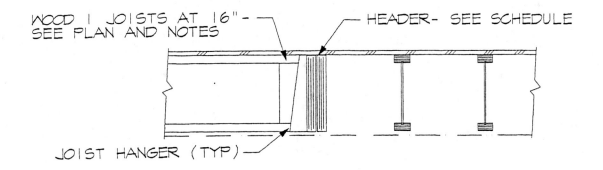

WOOD I JOISTS AT 16"- SEE PLAN AND NOTES

HEADER- SEE SCHEDULE

JOIST HANGER (TYP)

DETAIL AT JOIST DIRECTION CHANGE

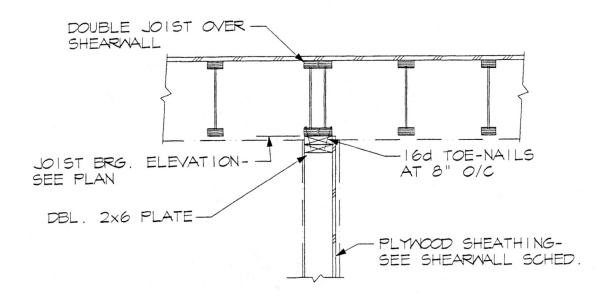

DOUBLE JOIST OVER SHEARWALL

JOIST BRG. ELEVATION- SEE PLAN

DBL. 2x6 PLATE

16d TOE-NAILS AT 8" O/C

PLYWOOD SHEATHING- SEE SHEARWALL SCHED.

DETAIL AT JOISTS PARALLEL TO SHEARWALL

TYP. I-JOIST INTERIOR FRMG. DETAILS

NOT TO SCALE   (DETAIL T6-IJ13)

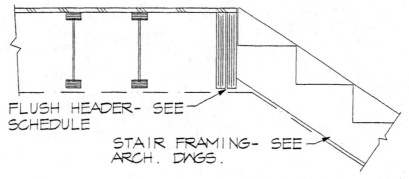

FLUSH HEADER- SEE
SCHEDULE

STAIR FRAMING- SEE
ARCH. DWGS.

DETAIL AT MAIN FLOOR LEVEL

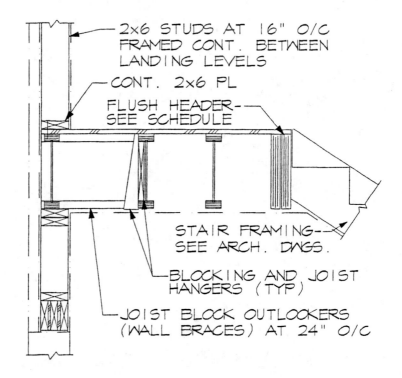

2x6 STUDS AT 16" O/C
FRAMED CONT. BETWEEN
LANDING LEVELS

CONT. 2x6 PL

FLUSH HEADER-
SEE SCHEDULE

STAIR FRAMING-
SEE ARCH. DWGS.

BLOCKING AND JOIST
HANGERS (TYP)

JOIST BLOCK OUTLOOKERS
(WALL BRACES) AT 24" O/C

DETAIL AT INTERMEDIATE LANDING LEVEL

TYP. I-JOIST FRAMING AT STAIRS

NOT TO SCALE   (DETAIL T6-IJ14)

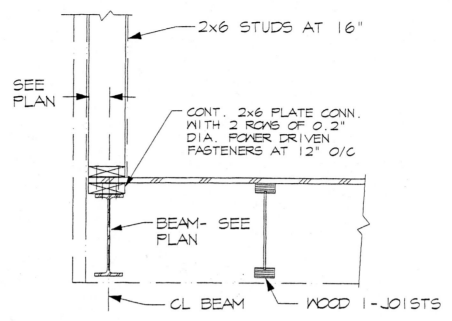

2x6 STUDS AT 16"

SEE PLAN

CONT. 2x6 PLATE CONN. WITH 2 ROWS OF 0.2" DIA. POWER DRIVEN FASTENERS AT 12" O/C

BEAM- SEE PLAN

CL BEAM

WOOD I-JOISTS

DETAIL AT JOISTS PARALLEL TO STEEL BEAM

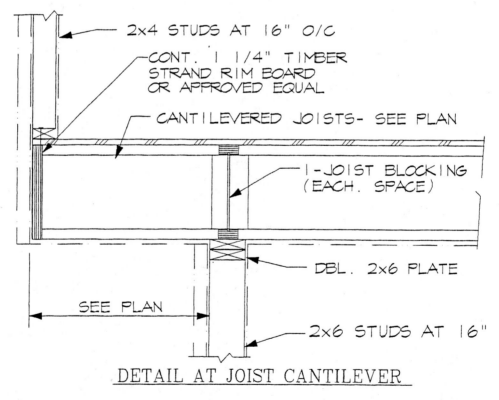

2x4 STUDS AT 16" O/C

CONT. 1 1/4" TIMBER STRAND RIM BOARD OR APPROVED EQUAL

CANTILEVERED JOISTS- SEE PLAN

I-JOIST BLOCKING (EACH. SPACE)

DBL. 2x6 PLATE

SEE PLAN

2x6 STUDS AT 16"

DETAIL AT JOIST CANTILEVER

TYP. I–JOIST PERIMETER FRMG. DETAILS

NOT TO SCALE   (DETAIL T6–IJ15)

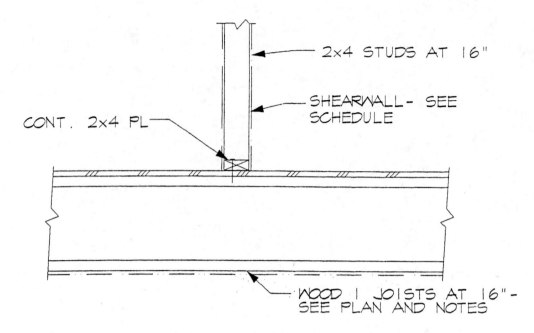

CONT. 2x4 PL

2x4 STUDS AT 16"

SHEARWALL- SEE SCHEDULE

WOOD I JOISTS AT 16"- SEE PLAN AND NOTES

DETAIL AT INTERIOR SHEARWALL

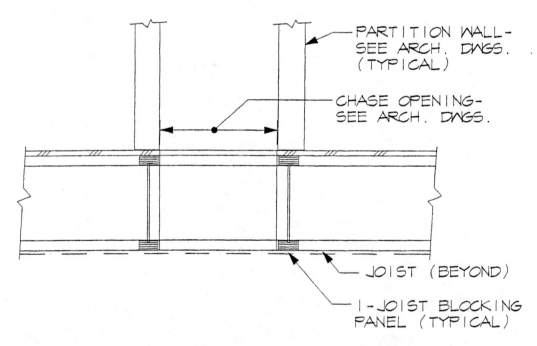

PARTITION WALL- SEE ARCH. DWGS. (TYPICAL)

CHASE OPENING- SEE ARCH. DWGS.

JOIST (BEYOND)

I-JOIST BLOCKING PANEL (TYPICAL)

DETAIL AT MECHANICAL DUCT CHASE

MISC. I-JOIST FRAMING DETAILS

NOT TO SCALE   (DETAIL T6-IJ16)

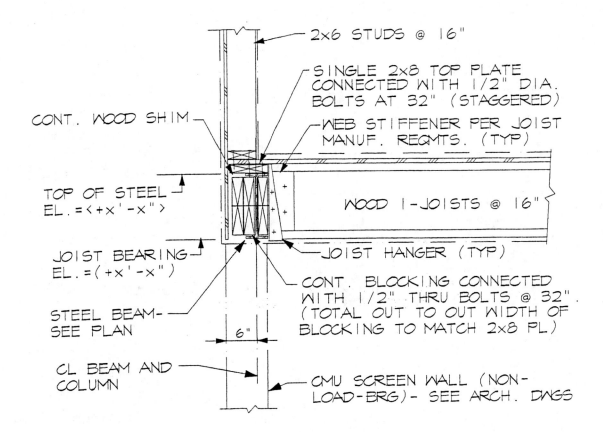

2×6 STUDS @ 16"

SINGLE 2×8 TOP PLATE CONNECTED WITH 1/2" DIA. BOLTS AT 32" (STAGGERED)

WEB STIFFENER PER JOIST MANUF. REQMTS. (TYP)

CONT. WOOD SHIM

TOP OF STEEL EL.=<+x'-x">

WOOD I-JOISTS @ 16"

JOIST BEARING EL.=(+x'-x")

JOIST HANGER (TYP)

STEEL BEAM-SEE PLAN

CONT. BLOCKING CONNECTED WITH 1/2" THRU BOLTS @ 32". (TOTAL OUT TO OUT WIDTH OF BLOCKING TO MATCH 2×8 PL)

6"

CL BEAM AND COLUMN

CMU SCREEN WALL (NON-LOAD-BRG)- SEE ARCH. DWGS

DETAIL AT PERIMETER, LOADBEARING JOISTS

MISC. I-JOIST CONNECTIONS TO STEEL

NOT TO SCALE   (DETAIL T6-IJ17)

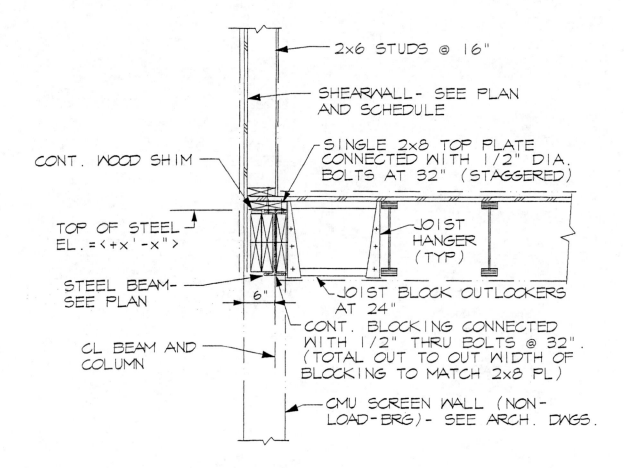

2x6 STUDS @ 16"

SHEARWALL- SEE PLAN AND SCHEDULE

SINGLE 2x8 TOP PLATE CONNECTED WITH 1/2" DIA. BOLTS AT 32" (STAGGERED)

CONT. WOOD SHIM

TOP OF STEEL EL.=<+x'-x">

STEEL BEAM- SEE PLAN

6"

CL BEAM AND COLUMN

JOIST HANGER (TYP)

JOIST BLOCK OUTLOOKERS AT 24"

CONT. BLOCKING CONNECTED WITH 1/2" THRU BOLTS @ 32". (TOTAL OUT TO OUT WIDTH OF BLOCKING TO MATCH 2x8 PL)

CMU SCREEN WALL (NON-LOAD-BRG)- SEE ARCH. DWGS.

DETAIL AT PERIMETER, NONLOADBEARING JOISTS

MISC. I-JOIST CONNECTIONS TO STEEL

NOT TO SCALE   (DETAIL T6-IJ18)

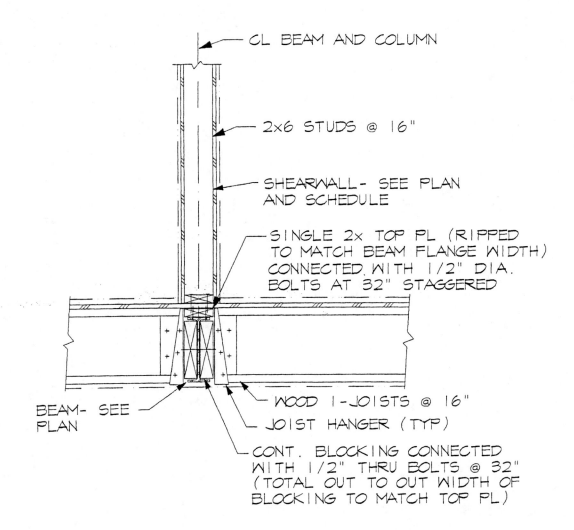

CL BEAM AND COLUMN

2×6 STUDS @ 16"

SHEARWALL- SEE PLAN AND SCHEDULE

SINGLE 2× TOP PL (RIPPED TO MATCH BEAM FLANGE WIDTH) CONNECTED WITH 1/2" DIA. BOLTS AT 32" STAGGERED

BEAM- SEE PLAN

WOOD I-JOISTS @ 16"

JOIST HANGER (TYP)

CONT. BLOCKING CONNECTED WITH 1/2" THRU BOLTS @ 32" (TOTAL OUT TO OUT WIDTH OF BLOCKING TO MATCH TOP PL)

DETAIL AT INTERIOR, LOAD-BEARING WALL ABOVE

MISC. I-JOIST CONNECTIONS TO STEEL

NOT TO SCALE   (DETAIL T6-IJ19)

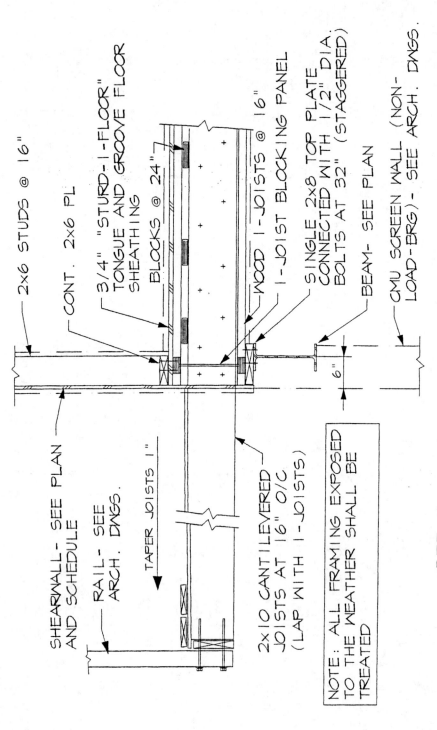

2x6 STUDS @ 16"

CONT. 2x6 PL.

3/4" "STURD-I-FLOOR" TONGUE AND GROOVE FLOOR SHEATHING

BLOCKS @ 24"

WOOD I-JOISTS @ 16"

I-JOIST BLOCKING PANEL

SINGLE 2x8 TOP PLATE CONNECTED WITH 1/2" DIA. BOLTS AT 32" (STAGGERED)

BEAM- SEE PLAN

CMU SCREEN WALL (NON-LOAD-BRG)- SEE ARCH. DWGS.

6"

SHEARWALL- SEE PLAN AND SCHEDULE

RAIL- SEE ARCH. DWGS.

TAPER JOISTS 1"

2x10 CANTILEVERED JOISTS AT 16" O/C (LAP WITH I-JOISTS)

NOTE: ALL FRAMING EXPOSED TO THE WEATHER SHALL BE TREATED

DETAIL AT STEEL BEAM SUPPORT

TYPICAL CANTILEVERED DECK JOIST DETAIL

(DETAIL T6-IJ20)

NOT TO SCALE

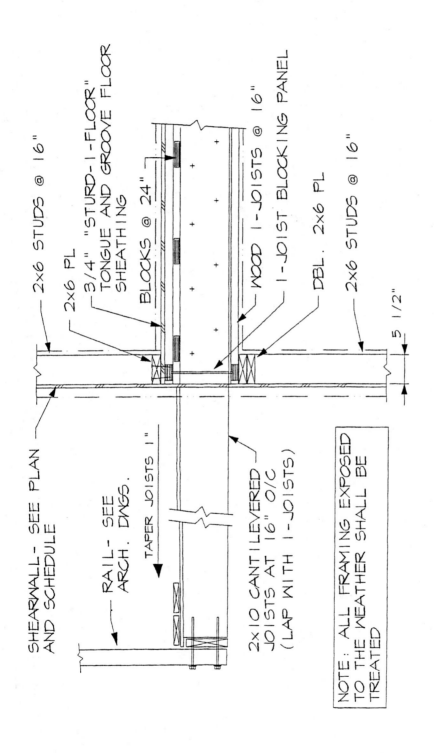

2x6 STUDS @ 16"

2x6 PL

3/4" "STURD-I-FLOOR"
TONGUE AND GROOVE FLOOR
SHEATHING

BLOCKS @ 24"

WOOD I-JOISTS @ 16"

I-JOIST BLOCKING PANEL

DBL. 2x6 PL

2x6 STUDS @ 16"

5 1/2"

SHEARWALL- SEE PLAN
AND SCHEDULE

RAIL- SEE
ARCH. DWGS.

TAPER JOISTS 1"

2x10 CANTILEVERED
JOISTS AT 16" O/C
(LAP WITH I-JOISTS)

NOTE: ALL FRAMING EXPOSED
TO THE WEATHER SHALL BE
TREATED

DETAIL AT STUDWALL SUPPORT

TYPICAL CANTILEVERED DECK JOIST DETAIL

(DETAIL T6-IJ21)

NOT TO SCALE

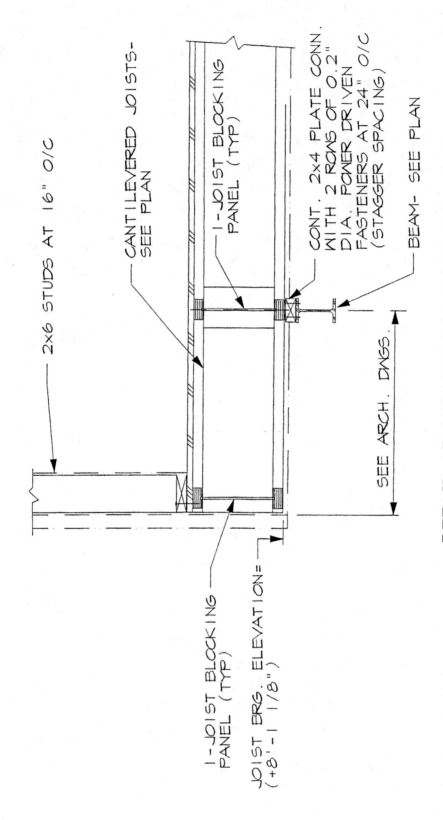

2x6 STUDS AT 16" O/C

CANTILEVERED JOISTS-
SEE PLAN

I-JOIST BLOCKING
PANEL (TYP)

CONT. 2x4 PLATE CONN.
WITH 2 ROWS OF 0.2"
DIA. POWER DRIVEN
FASTENERS AT 24" O/C
(STAGGER SPACING)

BEAM- SEE PLAN

SEE ARCH. DWGS.

I-JOIST BLOCKING
PANEL (TYP)

JOIST BRG. ELEVATION=
(+8'-1 1/8")

DETAIL AT STEEL BEAM SUPPORT

TYPICAL CANTILEVERED WOOD I-JOIST DETAIL

(DETAIL T6-IJ22)

NOT TO SCALE

378

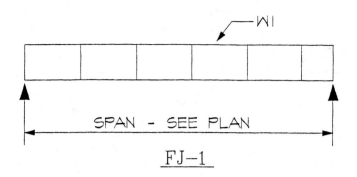

FJ-1

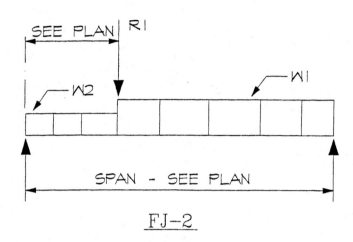

FJ-2

NOTATIONS:

WI = 10#/LF + 100#/LF = 110 #/LF
    (FLR. DL) (FLR. LL)

W2 = 10#/LF + 30#/LF = 40 #/LF
    (FLR. DL) (FLR. LL)

RI = 125# + 200# = 325#
    (ROOF DL) (ROOF LL)

NOTE: ALL LOADS ARE FOR A 1'
      WIDE STRIP.

LIMIT LL DEFLECTION TO L/360.

WOOD I-JOIST LOADING DIAGRAMS
NOT TO SCALE   (DETAIL T6-FJOIS)

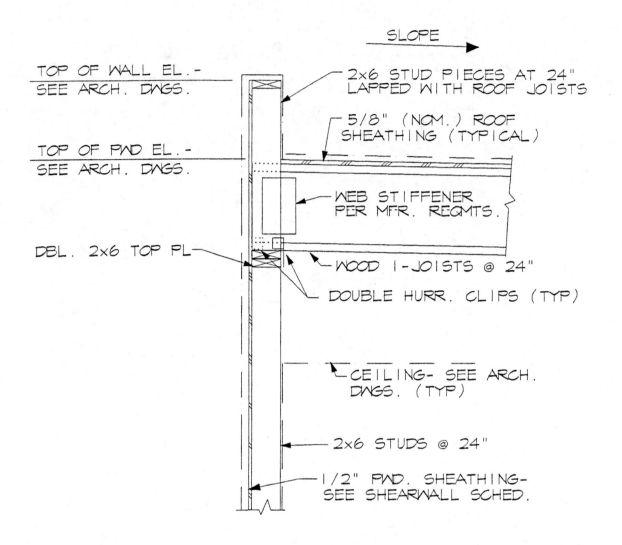

SLOPE

TOP OF WALL EL. -
SEE ARCH. DWGS.

2x6 STUD PIECES AT 24"
LAPPED WITH ROOF JOISTS

5/8" (NOM.) ROOF
SHEATHING (TYPICAL)

TOP OF PWD EL. -
SEE ARCH. DWGS.

WEB STIFFENER
PER MFR. REQMTS.

DBL. 2x6 TOP PL

WOOD I-JOISTS @ 24"

DOUBLE HURR. CLIPS (TYP)

CEILING- SEE ARCH.
DWGS. (TYP)

2x6 STUDS @ 24"

1/2" PWD. SHEATHING-
SEE SHEARWALL SCHED.

TYP. ROOF I–JOIST BEARING DETAIL
NOT TO SCALE   (DETAIL T6–RJ1)

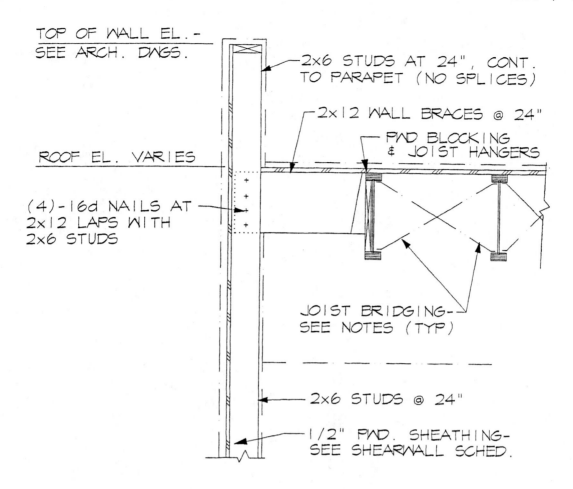

TOP OF WALL EL.-
SEE ARCH. DWGS.

2x6 STUDS AT 24", CONT.
TO PARAPET (NO SPLICES)

2x12 WALL BRACES @ 24"

PWD BLOCKING
& JOIST HANGERS

ROOF EL. VARIES

(4)-16d NAILS AT
2x12 LAPS WITH
2x6 STUDS

JOIST BRIDGING-
SEE NOTES (TYP)

2x6 STUDS @ 24"

1/2" PWD. SHEATHING-
SEE SHEARWALL SCHED.

TYP. ROOF I-JOIST PARALLEL TO WALL

NOT TO SCALE (DETAIL T6-RJ2)

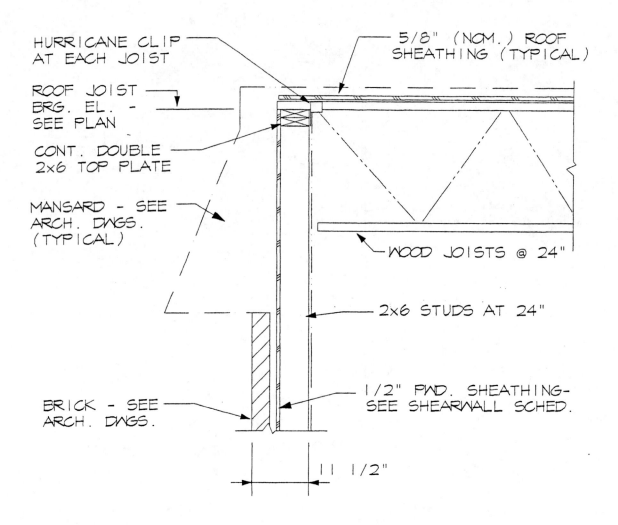

HURRICANE CLIP
AT EACH JOIST

ROOF JOIST
BRG. EL. -
SEE PLAN

CONT. DOUBLE
2x6 TOP PLATE

MANSARD - SEE
ARCH. DWGS.
(TYPICAL)

BRICK - SEE
ARCH. DWGS.

5/8" (NOM.) ROOF
SHEATHING (TYPICAL)

WOOD JOISTS @ 24"

2x6 STUDS AT 24"

1/2" PWD. SHEATHING-
SEE SHEARWALL SCHED.

11 1/2"

TYP. ROOF JOIST BEARING DETAIL
NOT TO SCALE   (DETAIL T6-RJ3)

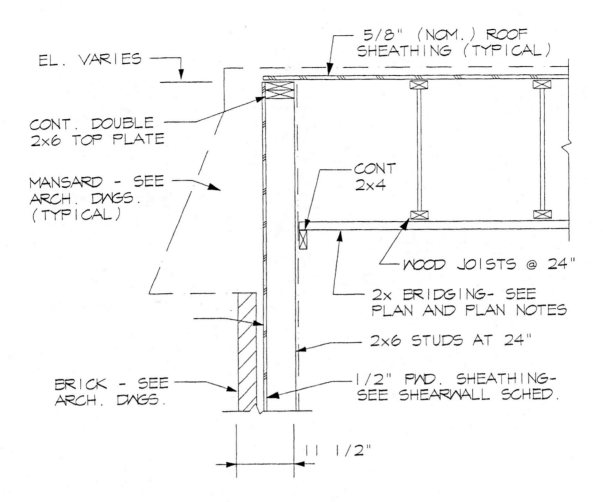

EL. VARIES

5/8" (NOM.) ROOF SHEATHING (TYPICAL)

CONT. DOUBLE 2x6 TOP PLATE

MANSARD - SEE ARCH. DWGS. (TYPICAL)

CONT 2x4

WOOD JOISTS @ 24"

2x BRIDGING- SEE PLAN AND PLAN NOTES

2x6 STUDS AT 24"

BRICK - SEE ARCH. DWGS.

1/2" PWD. SHEATHING- SEE SHEARWALL SCHED.

11 1/2"

TYP. ROOF JOIST PARALLEL TO WALL

NOT TO SCALE   (DETAIL T6-RJ4)

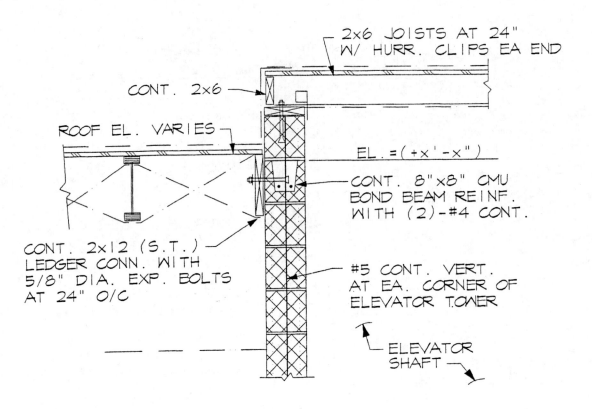

CONT. 2x6

ROOF EL. VARIES

2x6 JOISTS AT 24"
W/ HURR. CLIPS EA END

EL. = (+x' -x")

CONT. 8"x8" CMU
BOND BEAM REINF.
WITH (2)-#4 CONT.

CONT. 2x12 (S.T.)
LEDGER CONN. WITH
5/8" DIA. EXP. BOLTS
AT 24" O/C

#5 CONT. VERT.
AT EA. CORNER OF
ELEVATOR TOWER

ELEVATOR
SHAFT

ROOF JOIST DETAIL @ ELEVATOR WALL

NOT TO SCALE

(DETAIL T6-RJ5)

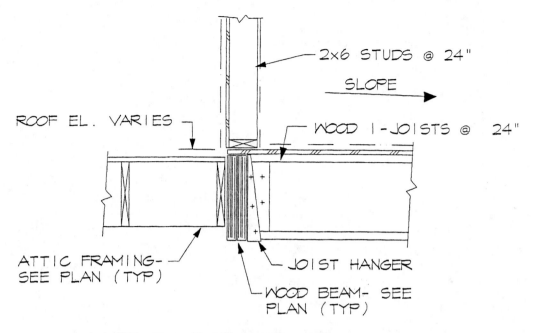

ROOF EL. VARIES

2x6 STUDS @ 24"

SLOPE

WOOD I-JOISTS @ 24"

ATTIC FRAMING-
SEE PLAN (TYP)

JOIST HANGER

WOOD BEAM- SEE
PLAN (TYP)

DETAIL AT FRAMING LEVEL ABOVE

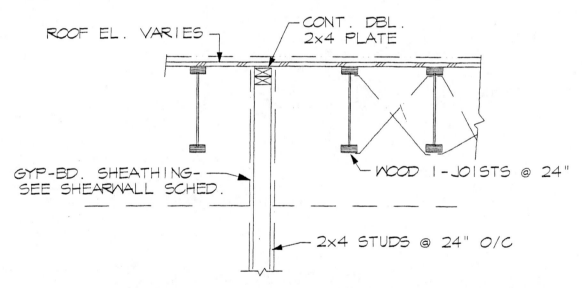

ROOF EL. VARIES

CONT. DBL.
2x4 PLATE

GYP-BD. SHEATHING-
SEE SHEARWALL SCHED.

WOOD I-JOISTS @ 24"

2x4 STUDS @ 24" O/C

DETAIL AT INTERIOR SHEARWALL

MISC. ROOF I-JOIST DETAILS
NOT TO SCALE   (DETAIL T6-RJ6)

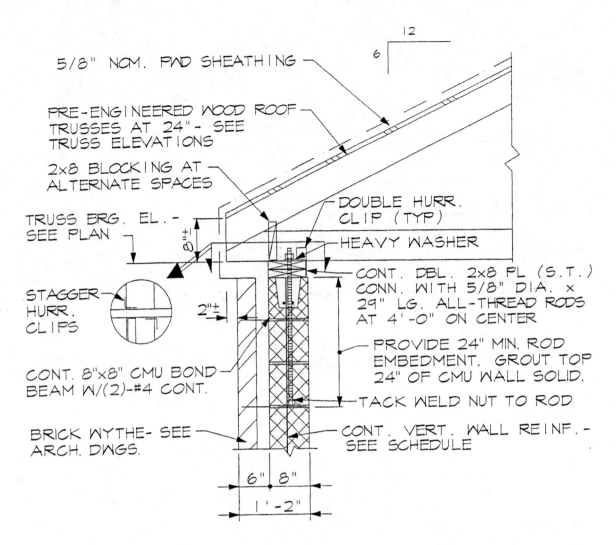

5/8" NOM. PWD SHEATHING

PRE-ENGINEERED WOOD ROOF TRUSSES AT 24"- SEE TRUSS ELEVATIONS

2×8 BLOCKING AT ALTERNATE SPACES

TRUSS BRG. EL.- SEE PLAN

8"±

DOUBLE HURR. CLIP (TYP)

HEAVY WASHER

CONT. DBL. 2×8 PL (S.T.) CONN. WITH 5/8" DIA. × 29" LG. ALL-THREAD RODS AT 4'-0" ON CENTER

STAGGER HURR. CLIPS

2"±

PROVIDE 24" MIN. ROD EMBEDMENT. GROUT TOP 24" OF CMU WALL SOLID.

TACK WELD NUT TO ROD

CONT. 8"×8" CMU BOND BEAM W/(2)-#4 CONT.

BRICK WYTHE- SEE ARCH. DWGS.

CONT. VERT. WALL REINF.- SEE SCHEDULE

6" 8"

1'-2"

12

6

TYPICAL ROOF TRUSS BEARING DETAIL
NOT TO SCALE   (DETAIL T6-TR1)

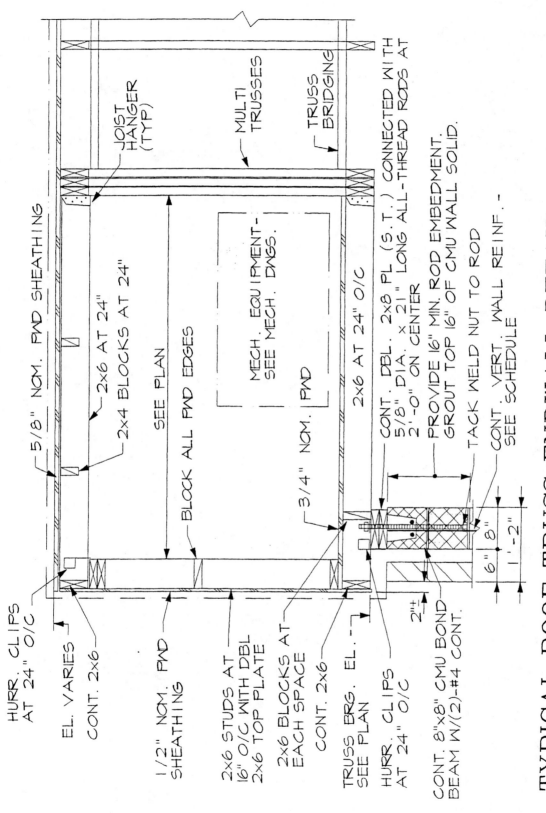

HURR. CLIPS AT 24" O/C

EL. VARIES

CONT. 2x6

5/8" NOM. PWD SHEATHING

2x6 AT 24"

2x4 BLOCKS AT 24"

SEE PLAN

BLOCK ALL PWD EDGES

MECH. EQUIPMENT - SEE MECH. DWGS.

JOIST HANGER (TYP)

MULTI TRUSSES

TRUSS BRIDGING

3/4" NOM. PWD

2x6 AT 24" O/C

CONT. DBL. 2x8 PL (S.T.) CONNECTED WITH 5/8" DIA. x 21" LONG ALL-THREAD RODS AT 2'-0" ON CENTER

PROVIDE 16" MIN. ROD EMBEDMENT. GROUT TOP 16" OF CMU WALL SOLID.

TACK WELD NUT TO ROD

CONT. VERT. WALL REINF. - SEE SCHEDULE

6"

8"

1'-2"

1/2" NOM. PWD SHEATHING

2x6 STUDS AT 16" O/C WITH DBL 2x6 TOP PLATE

2x6 BLOCKS AT EACH SPACE

CONT. 2x6

TRUSS BRG. EL. - SEE PLAN

HURR. CLIPS AT 24" O/C

2"±

CONT. 8"x8" CMU BOND BEAM W/(2)-#4 CONT.

TYPICAL ROOF TRUSS ENDWALL DETAIL

(DETAIL T6-TR2)

NOT TO SCALE

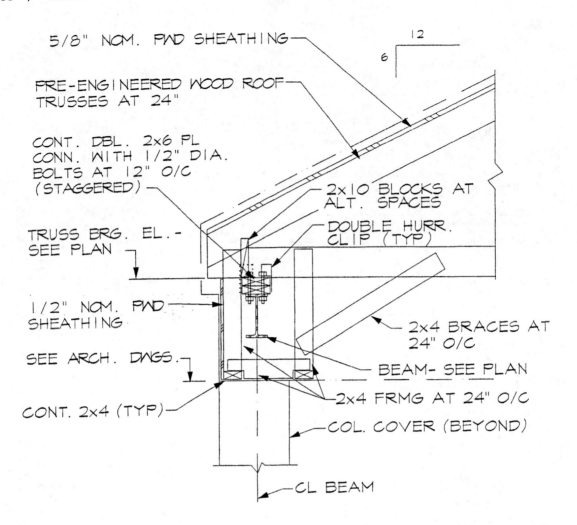

5/8" NOM. PWD SHEATHING

PRE-ENGINEERED WOOD ROOF TRUSSES AT 24"

CONT. DBL. 2x6 PL CONN. WITH 1/2" DIA. BOLTS AT 12" O/C (STAGGERED)

12
6

2x10 BLOCKS AT ALT. SPACES

DOUBLE HURR. CLIP (TYP)

TRUSS BRG. EL.- SEE PLAN

1/2" NOM. PWD SHEATHING

SEE ARCH. DWGS.

2x4 BRACES AT 24" O/C

BEAM- SEE PLAN

2x4 FRMG AT 24" O/C

CONT. 2x4 (TYP)

COL. COVER (BEYOND)

CL BEAM

CANOPY ROOF TRUSS BEARING DETAIL
NOT TO SCALE   (DETAIL T6–TR3)

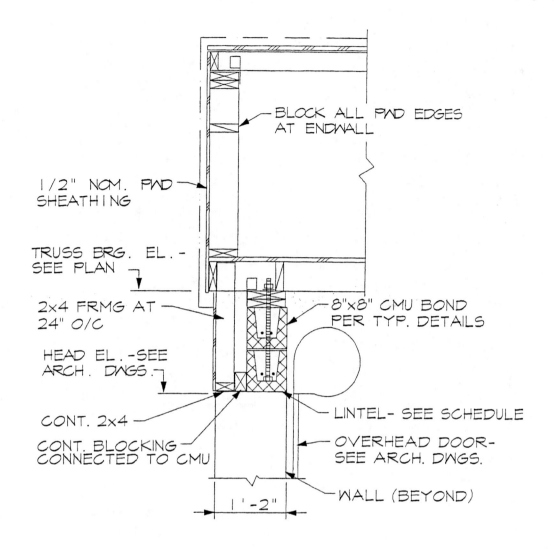

BLOCK ALL PWD EDGES AT ENDWALL

1/2" NOM. PWD SHEATHING

TRUSS BRG. EL.- SEE PLAN

2x4 FRMG AT 24" O/C

8"x8" CMU BOND PER TYP. DETAILS

HEAD EL.-SEE ARCH. DWGS.

CONT. 2x4

CONT. BLOCKING CONNECTED TO CMU

LINTEL- SEE SCHEDULE

OVERHEAD DOOR- SEE ARCH. DWGS.

WALL (BEYOND)

1'-2"

ROOF TRUSS ENDWALL DETAIL AT DOOR
NOT TO SCALE   (DETAIL T6-TR4)

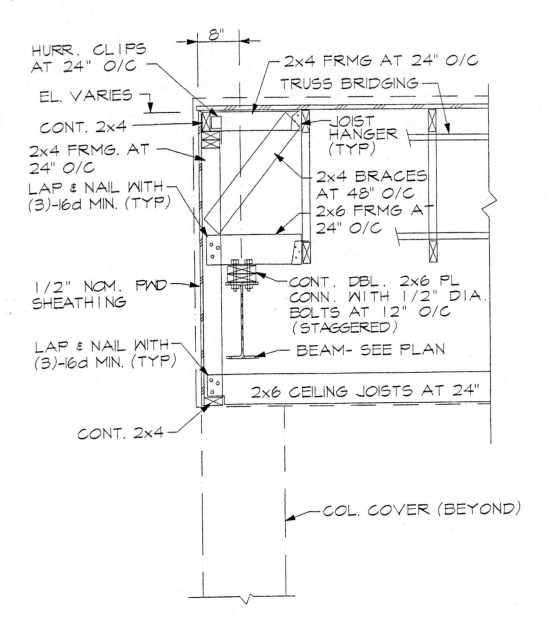

8"

HURR. CLIPS AT 24" O/C

EL. VARIES

CONT. 2x4

2x4 FRMG. AT 24" O/C

LAP & NAIL WITH (3)-16d MIN. (TYP)

1/2" NOM. PWD SHEATHING

LAP & NAIL WITH (3)-16d MIN. (TYP)

CONT. 2x4

2x4 FRMG AT 24" O/C

TRUSS BRIDGING

JOIST HANGER (TYP)

2x4 BRACES AT 48" O/C

2x6 FRMG AT 24" O/C

CONT. DBL. 2x6 PL CONN. WITH 1/2" DIA. BOLTS AT 12" O/C (STAGGERED)

BEAM- SEE PLAN

2x6 CEILING JOISTS AT 24"

COL. COVER (BEYOND)

TYPICAL CANOPY ENDWALL DETAIL
NOT TO SCALE   (DETAIL T6-TR5)

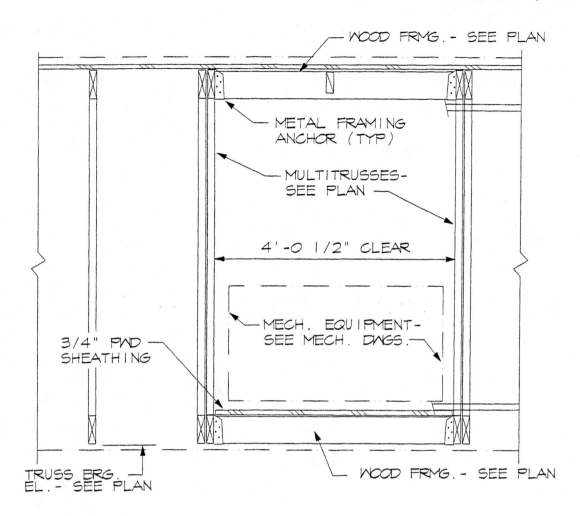

WOOD FRMG. - SEE PLAN

METAL FRAMING ANCHOR (TYP)

MULTITRUSSES- SEE PLAN —

4'-0 1/2" CLEAR

MECH. EQUIPMENT- SEE MECH. DWGS.

3/4" PWD SHEATHING

TRUSS BRG. EL. - SEE PLAN

WOOD FRMG. - SEE PLAN

TYPICAL MECHANICAL CHASE DETAIL

NOT TO SCALE   (DETAIL T6-TR6)

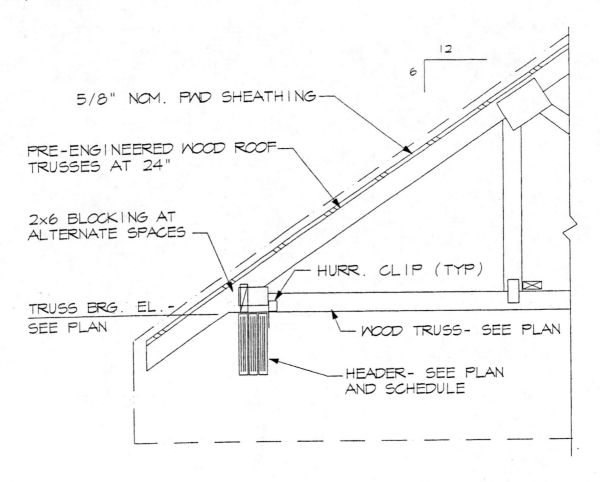

5/8" NOM. PWD SHEATHING

PRE-ENGINEERED WOOD ROOF-
TRUSSES AT 24"

2x6 BLOCKING AT
ALTERNATE SPACES

TRUSS BRG. EL.-
SEE PLAN

HURR. CLIP (TYP)

WOOD TRUSS- SEE PLAN

HEADER- SEE PLAN
AND SCHEDULE

12
6

TYPICAL TRUSS BEARING ON HEADER
NOT TO SCALE (DETAIL T6-TR7)

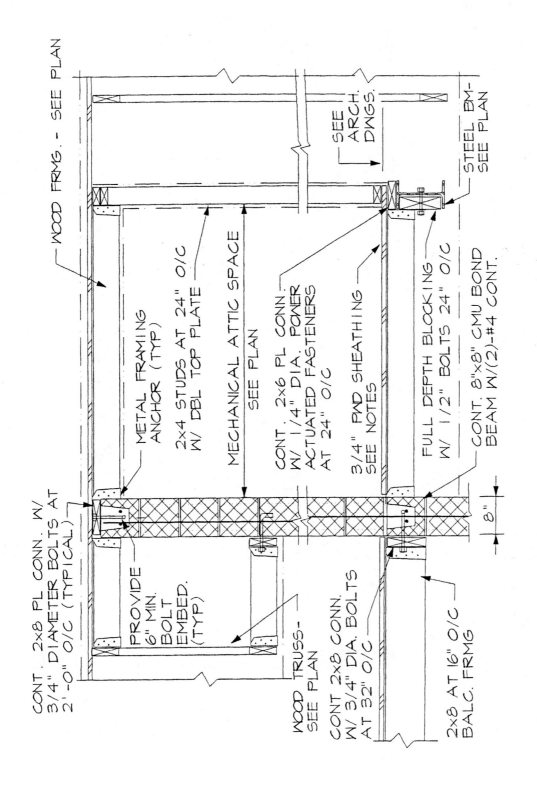

CONT. 2x8 PL CONN. W/
3/4" DIAMETER BOLTS AT
2'-0" O/C (TYPICAL)

WOOD FRMG. - SEE PLAN

METAL FRAMING
ANCHOR (TYP)

2x4 STUDS AT 24" O/C
W/ DBL TOP PLATE

MECHANICAL ATTIC SPACE
SEE PLAN

CONT. 2x6 PL CONN.
W/ 1/4" DIA. POWER
ACTUATED FASTENERS
AT 24" O/C

3/4" PWD SHEATHING
SEE NOTES

SEE
ARCH.
DWGS.

STEEL BM-
SEE PLAN

FULL DEPTH BLOCKING
W/ 1/2" BOLTS 24" O/C

CONT. 8"x8" CMU BOND
BEAM W/(2)-#4 CONT.

PROVIDE
6" MIN.
BOLT
EMBED.
(TYP)

WOOD TRUSS-
SEE PLAN

CONT 2x8 CONN.
W/ 3/4" DIA. BOLTS
AT 32" O/C

2x8 AT 16" O/C
BALC. FRMG

8"

TYPICAL ATTIC/BALCONY DETAIL

NOT TO SCALE

(DETAIL T6-TR8)

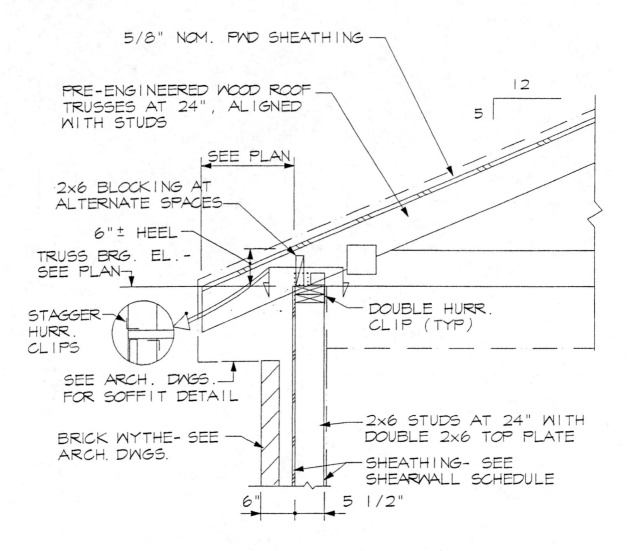

5/8" NOM. PWD SHEATHING

PRE-ENGINEERED WOOD ROOF TRUSSES AT 24", ALIGNED WITH STUDS

SEE PLAN

12

5

2x6 BLOCKING AT ALTERNATE SPACES

6"± HEEL

TRUSS BRG. EL.- SEE PLAN

STAGGER HURR. CLIPS

DOUBLE HURR. CLIP (TYP)

SEE ARCH. DWGS. FOR SOFFIT DETAIL

2x6 STUDS AT 24" WITH DOUBLE 2x6 TOP PLATE

SHEATHING- SEE SHEARWALL SCHEDULE

BRICK WYTHE- SEE ARCH. DWGS.

6"   5 1/2"

TYPICAL TRUSS BEARING ON STUDWALL

NOT TO SCALE   (DETAIL T6-TR9)

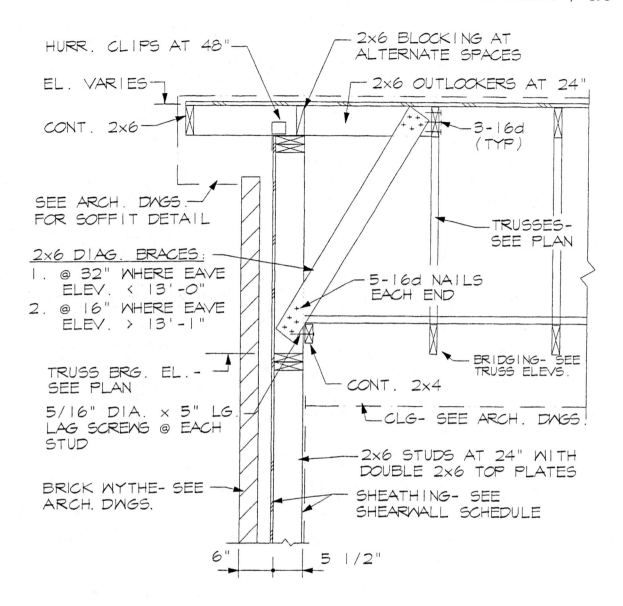

HURR. CLIPS AT 48"

EL. VARIES

CONT. 2×6

2×6 BLOCKING AT ALTERNATE SPACES

2×6 OUTLOCKERS AT 24"

3-16d (TYP)

SEE ARCH. DWGS. FOR SOFFIT DETAIL

TRUSSES- SEE PLAN

2×6 DIAG. BRACES:
1. @ 32" WHERE EAVE ELEV. < 13'-0"
2. @ 16" WHERE EAVE ELEV. > 13'-1"

5-16d NAILS EACH END

TRUSS BRG. EL.- SEE PLAN

5/16" DIA. × 5" LG. LAG SCREWS @ EACH STUD

CONT. 2×4

BRIDGING- SEE TRUSS ELEVS.

CLG- SEE ARCH. DWGS.

BRICK WYTHE- SEE ARCH. DWGS.

2×6 STUDS AT 24" WITH DOUBLE 2×6 TOP PLATES

SHEATHING- SEE SHEARWALL SCHEDULE

6"   5 1/2"

TYPICAL ENDWALL FRAMING DETAIL
NOT TO SCALE   (DETAIL T6–TR10)

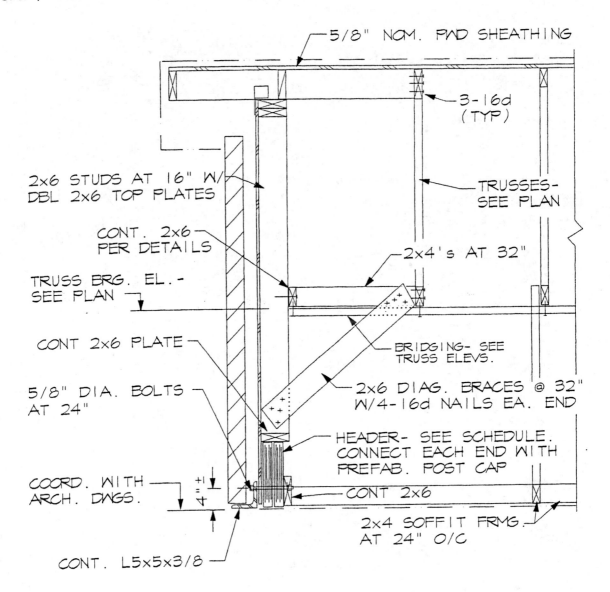

5/8" NOM. PWD SHEATHING

3-16d (TYP)

TRUSSES- SEE PLAN

2x6 STUDS AT 16" W/ DBL 2x6 TOP PLATES

CONT. 2x6 PER DETAILS

2x4's AT 32"

TRUSS BRG. EL.- SEE PLAN

CONT 2x6 PLATE

BRIDGING- SEE TRUSS ELEVS.

5/8" DIA. BOLTS AT 24"

2x6 DIAG. BRACES @ 32" W/4-16d NAILS EA. END

HEADER- SEE SCHEDULE. CONNECT EACH END WITH PREFAB. POST CAP

COORD. WITH ARCH. DWGS.

CONT 2x6

2x4 SOFFIT FRMG. AT 24" O/C

CONT. L5x5x3/8

ENDWALL FRAMING DETAIL AT ENTRY
NOT TO SCALE   (DETAIL T6-TR11)

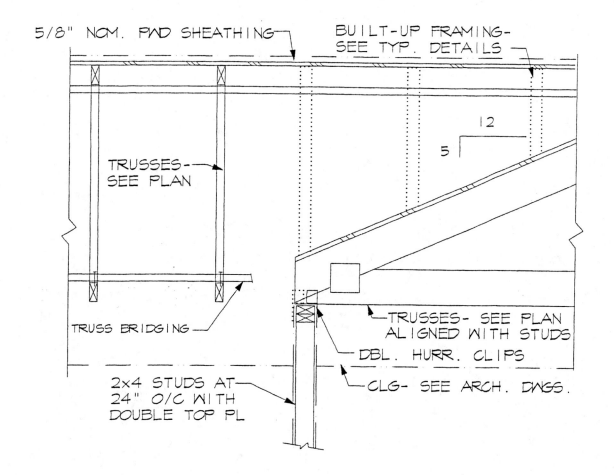

5/8" NOM. PWD SHEATHING

BUILT-UP FRAMING-
SEE TYP. DETAILS

TRUSSES-
SEE PLAN

12
5

TRUSS BRIDGING

TRUSSES- SEE PLAN
ALIGNED WITH STUDS

DBL. HURR. CLIPS

CLG- SEE ARCH. DWGS.

2×4 STUDS AT
24" O/C WITH
DOUBLE TOP PL

DETAIL AT TRUSS DIRECTION CHANGE
NOT TO SCALE   (DETAIL T6—TR12)

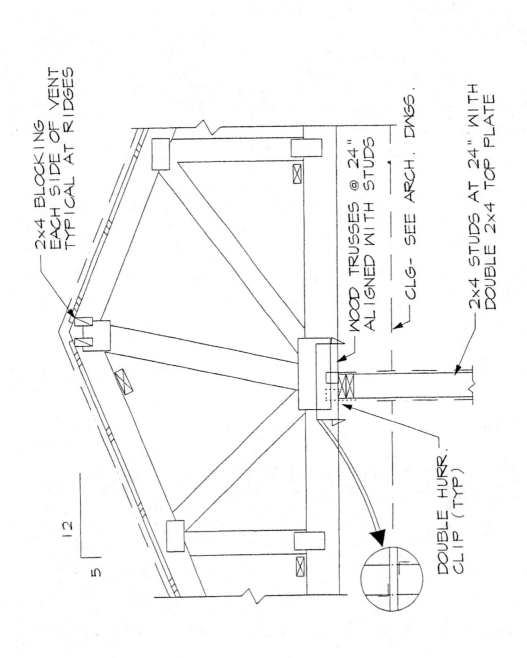

2x4 BLOCKING EACH SIDE OF VENT TYPICAL AT RIDGES

WOOD TRUSSES @ 24" ALIGNED WITH STUDS

CLG- SEE ARCH. DWGS.

2x4 STUDS AT 24" WITH DOUBLE 2x4 TOP PLATE

DOUBLE HURR. CLIP (TYP)

12
5

TYPICAL INTERIOR TRUSS BEARING DETAIL
(DETAIL T6—TR13)

NOT TO SCALE

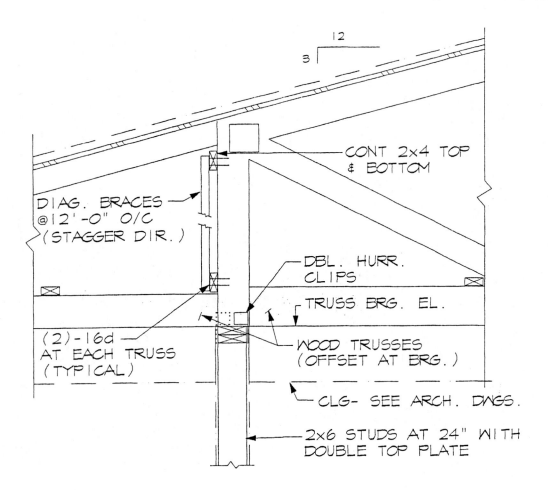

12
3

CONT 2x4 TOP & BOTTOM

DIAG. BRACES
@12'-0" O/C
(STAGGER DIR.)

DBL. HURR.
CLIPS

TRUSS BRG. EL.

(2)-16d
AT EACH TRUSS
(TYPICAL)

WOOD TRUSSES
(OFFSET AT BRG.)

CLG- SEE ARCH. DWGS.

2x6 STUDS AT 24" WITH
DOUBLE TOP PLATE

TYP. INTERIOR TRUSS BEARING DETAIL
NOT TO SCALE   (DETAIL T6–TR14)

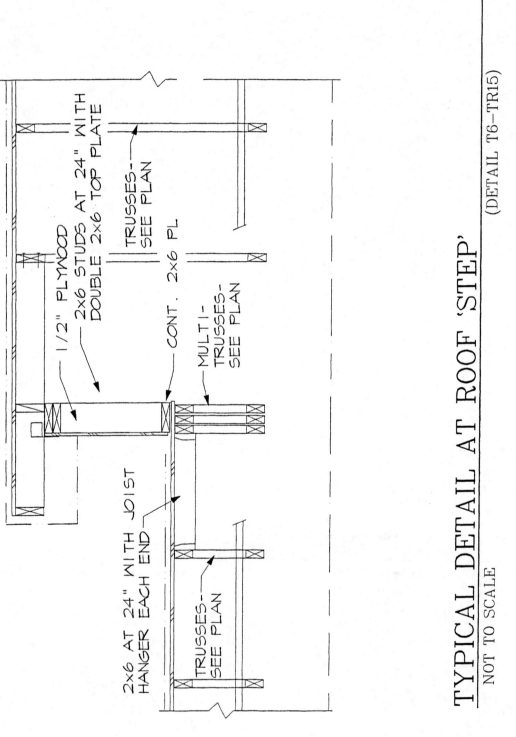

1/2" PLYWOOD

2x6 STUDS AT 24" WITH
DOUBLE 2x6 TOP PLATE

TRUSSES-
SEE PLAN

CONT. 2x6 PL.

MULTI-
TRUSSES-
SEE PLAN

2x6 AT 24" WITH JOIST
HANGER EACH END

TRUSSES-
SEE PLAN

TYPICAL DETAIL AT ROOF 'STEP'

NOT TO SCALE

(DETAIL T6-TR15)

400

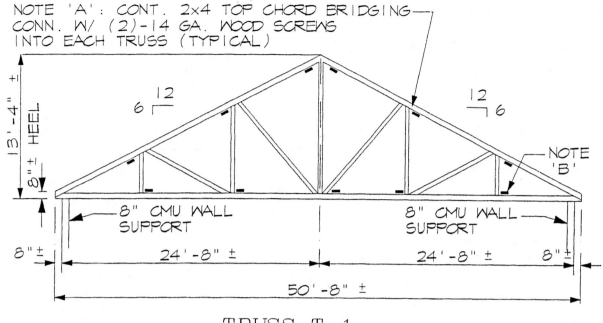

NOTE 'A': CONT. 2x4 TOP CHORD BRIDGING
CONN. W/ (2)-14 GA. WOOD SCREWS
INTO EACH TRUSS (TYPICAL)

TRUSS T-1

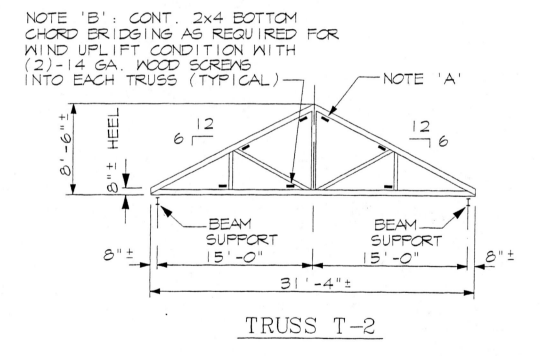

NOTE 'B': CONT. 2x4 BOTTOM
CHORD BRIDGING AS REQUIRED FOR
WIND UPLIFT CONDITION WITH
(2)-14 GA. WOOD SCREWS
INTO EACH TRUSS (TYPICAL)

TRUSS T-2

ENGINEERED WOOD TRUSS ELEVATIONS
NOT TO SCALE   (DETAIL T6-TEL1)

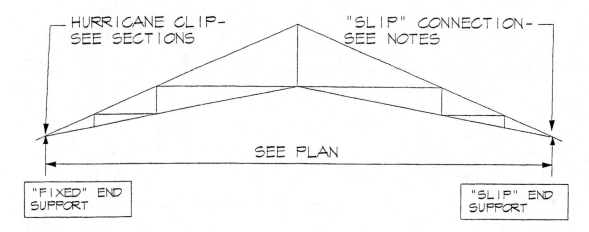

SCISSOR TRUSS NOTES:

1. SEE THE ARCHITECTURAL DRAWINGS FOR THE CHORD PITCHES. THE BOTTOM CHORD PITCH IS RECOMMENDED TO BE NO MORE THAN 1/2 OF THE TOP CHORD PITCH.

2. TRUSS INTERIOR CONFIGURATION IS SCHEMATIC AND MAY BE MODIFIED BY THE MANUF. TO SUIT THE ACTUAL DESIGN.

3. TRUSS MANUFACTURER SHALL DESIGN THE SCISSOR TRUSSES WITH ONE END FIXED AND THE OTHER END NOT RESTRAINED HORIZONTALLY. THE SUPPORT STRUCTURES HAVE NOT BEEN DESIGNED TO PROVIDE HORIZONTAL THRUST RESISTANCE.

4. CONNECT TRUSS AT FIXED END WITH HURRICANE CLIP. CONNECT TRUSS AT SLIP END WITH SLIP CONNECTOR, INSTALLED SUCH THAT HORIZONTAL MOVEMENT CAN OCCUR DURING CONSTRUCTION.

5. DURING CONSTRUCTION, BRACE BOTH SUPPORTING WALLS AND APPLY ALL DEAD LOADS (OMIT STRIP OF CEILING AT SLIP).

6. AFTER ALL DEAD LOADS ARE IN PLACE, PROVIDE NAILS AT SLIP CONNECTION SO THAT IT IS NOW FIXED.

7. SOME FUTURE LONG-TERM HORIZONTAL DISPLACEMENT WILL OCCUR DUE TO ROOF LIVE LOADS AND LONG TERM CREEP. THE TRUSS MANUFACTURER SHALL DESIGN THE SCISSOR TRUSSES TO LIMIT THE LONG-TERM HORIZONTAL DEFLECTION TO BE LESS THAN 1.00 INCH TOTAL. LONG-TERM DEFLECTION SHALL BE CALCULATED USING A MIN. OF (1.5 x DEAD LOAD DEFL.) + (LIVE LOAD DEFL.).

WOOD SCISSOR TRUSS REQUIREMENTS
NOT TO SCALE   (DETAIL T6-SCISS)

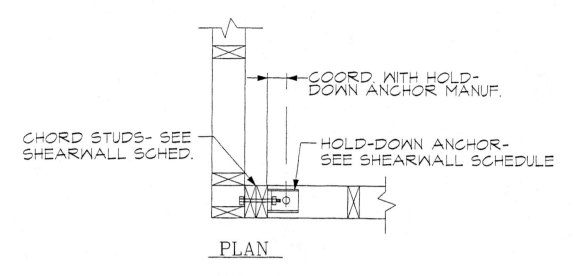

COORD. WITH HOLD-
DOWN ANCHOR MANUF.

CHORD STUDS- SEE
SHEARWALL SCHED.

HOLD-DOWN ANCHOR-
SEE SHEARWALL SCHEDULE

PLAN

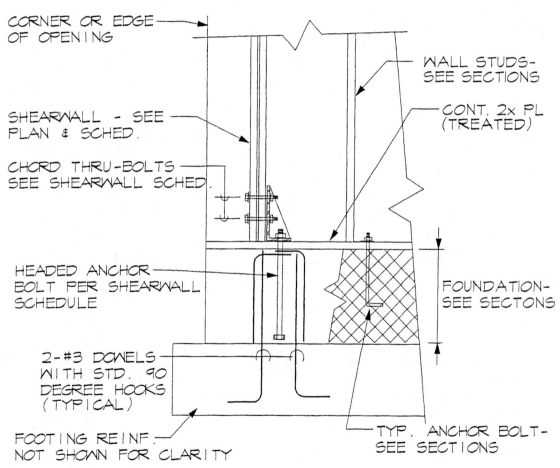

CORNER OR EDGE
OF OPENING

WALL STUDS-
SEE SECTIONS

SHEARWALL - SEE
PLAN & SCHED.

CONT. 2x PL
(TREATED)

CHORD THRU-BOLTS
SEE SHEARWALL SCHED.

HEADED ANCHOR
BOLT PER SHEARWALL
SCHEDULE

FOUNDATION-
SEE SECTONS

2-#3 DOWELS
WITH STD. 90
DEGREE HOOKS
(TYPICAL)

FOOTING REINF.
NOT SHOWN FOR CLARITY

TYP. ANCHOR BOLT-
SEE SECTIONS

ELEVATION AT FOUNDATION LEVEL

TYP. HOLD-DOWN ANCHOR DETAILS
NOT TO SCALE   (DETAIL T6-HDA)

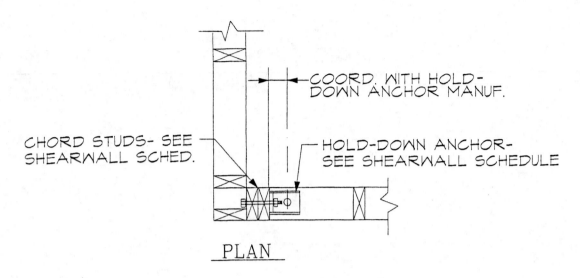

COORD. WITH HOLD-
DOWN ANCHOR MANUF.

CHORD STUDS- SEE
SHEARWALL SCHED.

HOLD-DOWN ANCHOR-
SEE SHEARWALL SCHEDULE

PLAN

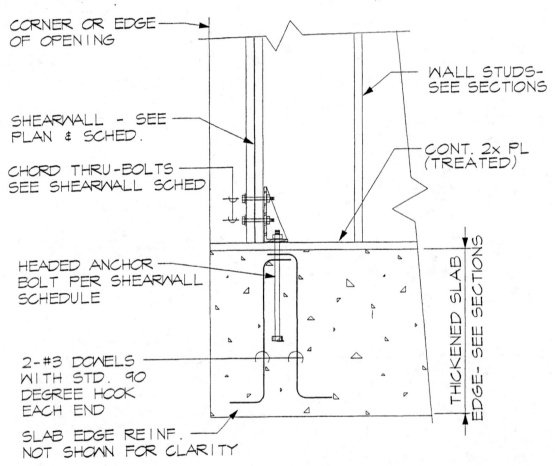

CORNER OR EDGE
OF OPENING

WALL STUDS-
SEE SECTIONS

SHEARWALL - SEE
PLAN & SCHED.

CHORD THRU-BOLTS
SEE SHEARWALL SCHED.

CONT. 2x PL
(TREATED)

HEADED ANCHOR
BOLT PER SHEARWALL
SCHEDULE

2-#3 DOWELS
WITH STD. 90
DEGREE HOOK
EACH END

SLAB EDGE REINF.
NOT SHOWN FOR CLARITY

THICKENED SLAB
EDGE- SEE SECTIONS

ELEVATION AT FOUNDATION LEVEL

TYP. HOLD-DOWN ANCHOR DETAILS

NOT TO SCALE   (DETAIL T6-HDA2)

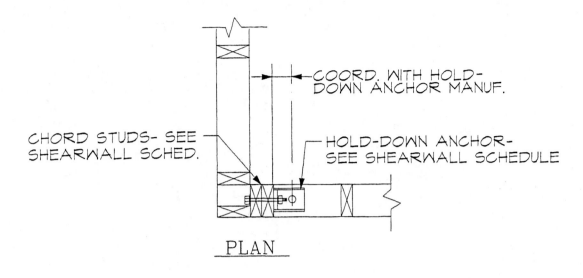

COORD. WITH HOLD-DOWN ANCHOR MANUF.

CHORD STUDS- SEE SHEARWALL SCHED.

HOLD-DOWN ANCHOR- SEE SHEARWALL SCHEDULE

PLAN

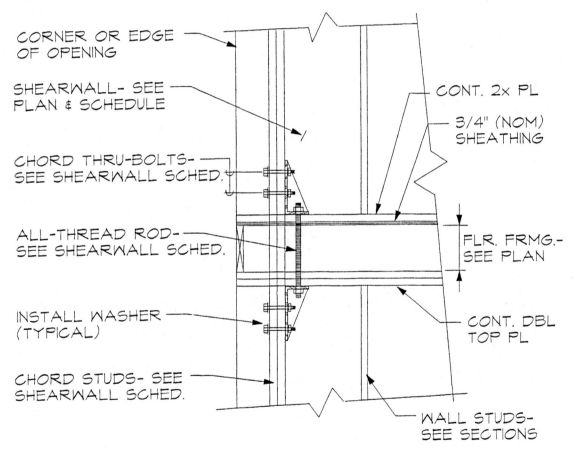

CORNER OR EDGE OF OPENING

SHEARWALL- SEE PLAN & SCHEDULE

CHORD THRU-BOLTS- SEE SHEARWALL SCHED.

ALL-THREAD ROD- SEE SHEARWALL SCHED.

INSTALL WASHER (TYPICAL)

CHORD STUDS- SEE SHEARWALL SCHED.

CONT. 2x PL

3/4" (NOM) SHEATHING

FLR. FRMG.- SEE PLAN

CONT. DBL TOP PL

WALL STUDS- SEE SECTIONS

ELEVATION AT ELEVATED FLOORS

TYP. HOLD—DOWN ANCHOR DETAILS
NOT TO SCALE   (DETAIL T6—EHD)

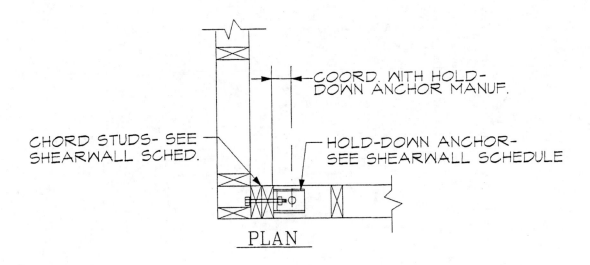

COORD. WITH HOLD-DOWN ANCHOR MANUF.

CHORD STUDS- SEE SHEARWALL SCHED.

HOLD-DOWN ANCHOR- SEE SHEARWALL SCHEDULE

PLAN

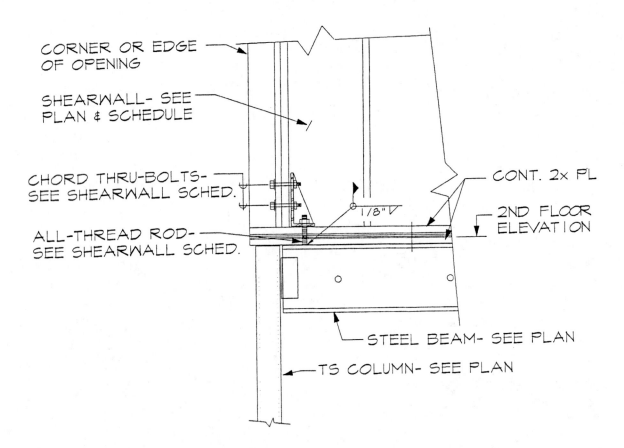

CORNER OR EDGE OF OPENING

SHEARWALL- SEE PLAN & SCHEDULE

CHORD THRU-BOLTS- SEE SHEARWALL SCHED.

ALL-THREAD ROD- SEE SHEARWALL SCHED.

CONT. 2x PL

2ND FLOOR ELEVATION

1/8"∨

STEEL BEAM- SEE PLAN

TS COLUMN- SEE PLAN

ELEVATION AT 2ND FLOOR LEVEL

TYP. HOLD–DOWN ANCHOR DETAILS
NOT TO SCALE   (DETAIL T6–EHD2)

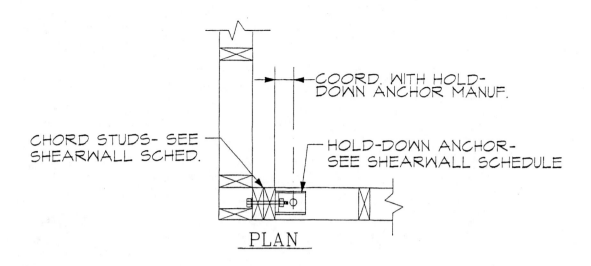

PLAN

COORD. WITH HOLD-DOWN ANCHOR MANUF.

CHORD STUDS- SEE SHEARWALL SCHED.

HOLD-DOWN ANCHOR- SEE SHEARWALL SCHEDULE

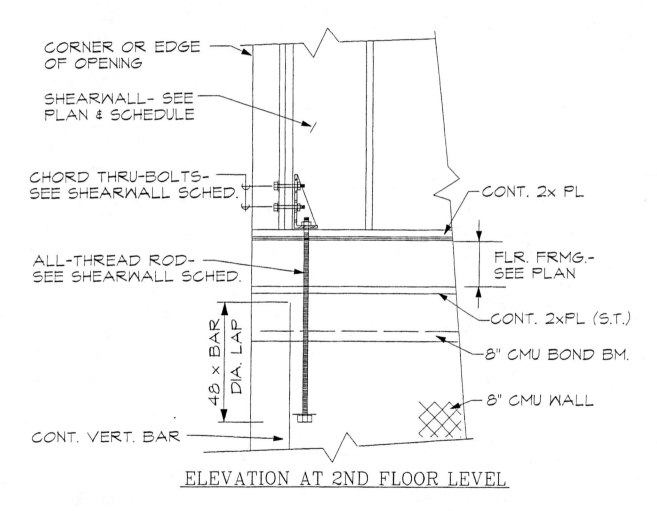

CORNER OR EDGE OF OPENING

SHEARWALL- SEE PLAN & SCHEDULE

CHORD THRU-BOLTS- SEE SHEARWALL SCHED.

ALL-THREAD ROD- SEE SHEARWALL SCHED.

48 x BAR DIA. LAP

CONT. VERT. BAR

CONT. 2x PL

FLR. FRMG.- SEE PLAN

CONT. 2xPL (S.T.)

8" CMU BOND BM.

8" CMU WALL

ELEVATION AT 2ND FLOOR LEVEL

TYP. HOLD-DOWN ANCHOR DETAILS
NOT TO SCALE   (DETAIL T6-EHD3)

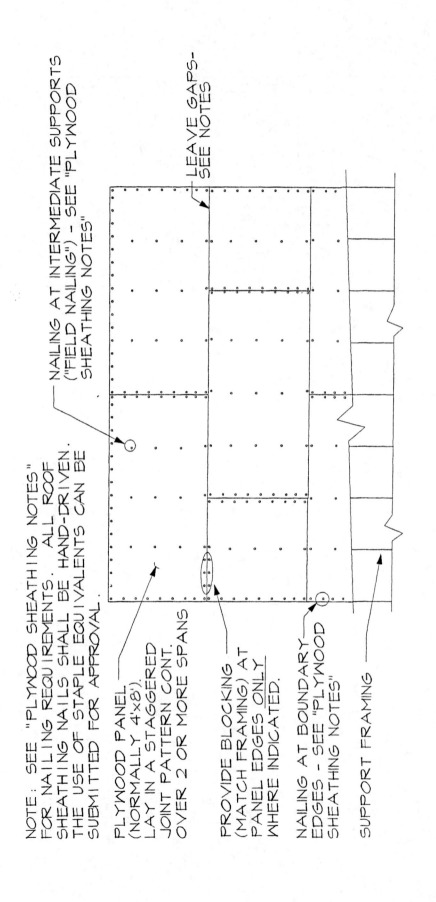

NOTE: SEE "PLYWOOD SHEATHING NOTES" FOR NAILING REQUIREMENTS. ALL ROOF SHEATHING NAILS SHALL BE HAND-DRIVEN. THE USE OF STAPLE EQUIVALENTS CAN BE SUBMITTED FOR APPROVAL.

PLYWOOD PANEL (NORMALLY 4'×8'). LAY IN A STAGGERED JOINT PATTERN CONT. OVER 2 OR MORE SPANS

PROVIDE BLOCKING (MATCH FRAMING) AT PANEL EDGES ONLY WHERE INDICATED.

NAILING AT BOUNDARY EDGES - SEE "PLYWOOD SHEATHING NOTES"

SUPPORT FRAMING

NAILING AT INTERMEDIATE SUPPORTS ("FIELD NAILING") - SEE "PLYWOOD SHEATHING NOTES"

LEAVE GAPS - SEE NOTES

TYPICAL ROOF/FLOOR DIAPHRAGM DETAIL
(DETAIL T6-RDIA)

NOT TO SCALE

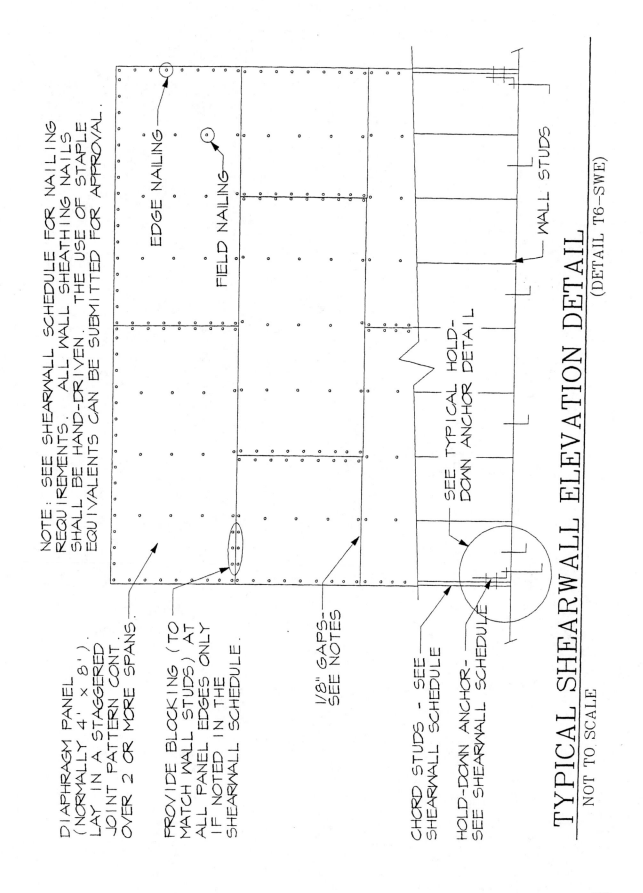

NOTE: SEE SHEARWALL SCHEDULE FOR NAILING REQUIREMENTS. ALL WALL SHEATHING NAILS SHALL BE HAND-DRIVEN. THE USE OF STAPLE EQUIVALENTS CAN BE SUBMITTED FOR APPROVAL.

EDGE NAILING

FIELD NAILING

DIAPHRAGM PANEL (NORMALLY 4' x 8'). LAY IN A STAGGERED JOINT PATTERN CONT. OVER 2 OR MORE SPANS.

PROVIDE BLOCKING (TO MATCH WALL STUDS) AT ALL PANEL EDGES ONLY IF NOTED IN THE SHEARWALL SCHEDULE.

1/8" GAPS - SEE NOTES

WALL STUDS

SEE TYPICAL HOLD-DOWN ANCHOR DETAIL

CHORD STUDS - SEE SHEARWALL SCHEDULE

HOLD-DOWN ANCHOR - SEE SHEARWALL SCHEDULE

TYPICAL SHEARWALL ELEVATION DETAIL

NOT TO SCALE

(DETAIL T6-SWE)

HEADER SCHEDULE

MARK	SECTION	SIZE	JAMB STUDS	
			JACK	FULL-HT.
H-1		(3)- 2x6 WITH 1/2" PWD PLATES AND LOOSE SHELF L	SINGLE	SINGLE
H-2		(3)- 2x10 WITH 1/2" PWD PLATES AND LOOSE SHELF L	SINGLE	DOUBLE
H-3		(2)- 2x10 WITH 1/2" PWD PLATE	SINGLE	SINGLE
H-4		(2)- 1 3/4" x 11 7/8" MICRO LAMS AND BOLTED SHELF L	DOUBLE	DOUBLE

NOTE: SEE OTHER DETAILS FOR BRICK SHELF ANGLES.

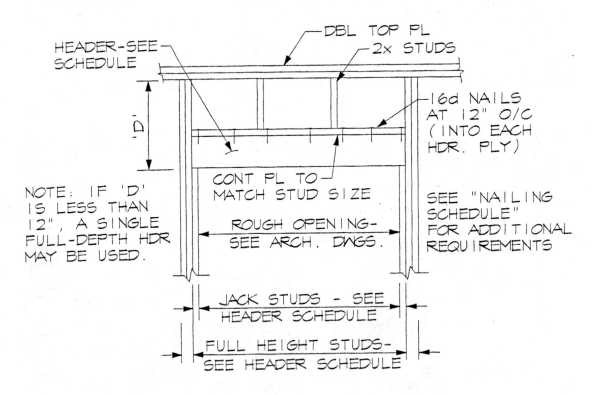

HEADER-SEE SCHEDULE

DBL TOP PL

2x STUDS

'D'

16d NAILS AT 12" O/C (INTO EACH HDR. PLY)

CONT PL TO MATCH STUD SIZE

NOTE: IF 'D' IS LESS THAN 12", A SINGLE FULL-DEPTH HDR MAY BE USED.

ROUGH OPENING- SEE ARCH. DWGS.

SEE "NAILING SCHEDULE" FOR ADDITIONAL REQUIREMENTS

JACK STUDS - SEE HEADER SCHEDULE

FULL HEIGHT STUDS- SEE HEADER SCHEDULE

TYP. FRAMED OPENING DETAIL

NOT TO SCALE   (DETAIL T6-HDR)

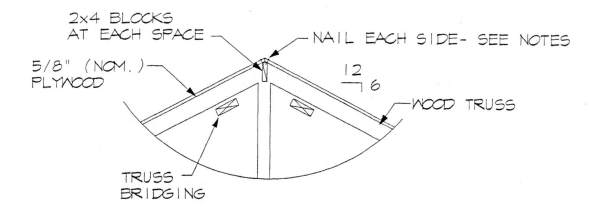

2x4 BLOCKS
AT EACH SPACE

NAIL EACH SIDE- SEE NOTES

5/8" (NOM.)
PLYWOOD

12
6

WOOD TRUSS

TRUSS
BRIDGING

TYP. DETAIL AT RIDGE
NOT TO SCALE

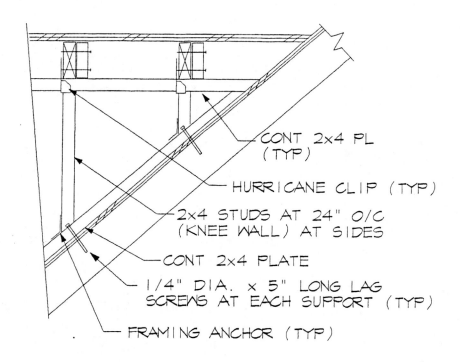

CONT 2x4 PL
(TYP)

HURRICANE CLIP (TYP)

2x4 STUDS AT 24" O/C
(KNEE WALL) AT SIDES

CONT 2x4 PLATE

1/4" DIA. x 5" LONG LAG
SCREWS AT EACH SUPPORT (TYP)

FRAMING ANCHOR (TYP)

TYP. DORMER ATTACHMENT DETAIL
NOT TO SCALE   (DETAIL T6-DORMR)

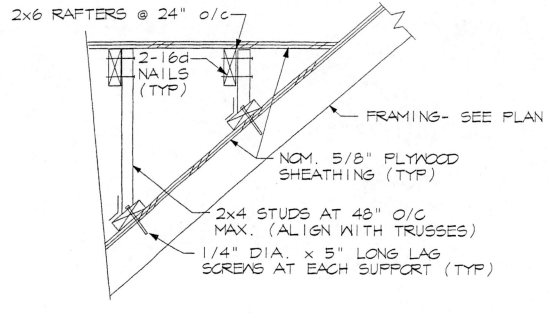

2x6 RAFTERS @ 24" O/C

2-16d NAILS (TYP)

FRAMING- SEE PLAN

NOM. 5/8" PLYWOOD SHEATHING (TYP)

2x4 STUDS AT 48" O/C MAX. (ALIGN WITH TRUSSES)

1/4" DIA. x 5" LONG LAG SCREWS AT EACH SUPPORT (TYP)

BETW. RIDGE OF BUILT-UP FRAMING

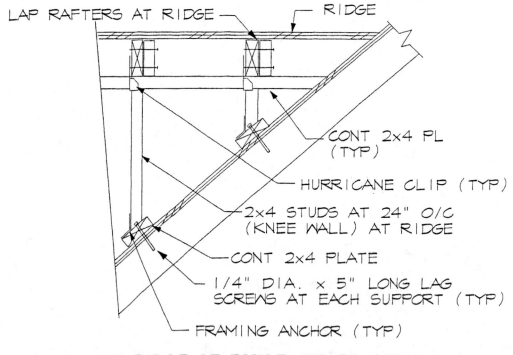

LAP RAFTERS AT RIDGE

RIDGE

CONT 2x4 PL (TYP)

HURRICANE CLIP (TYP)

2x4 STUDS AT 24" O/C (KNEE WALL) AT RIDGE

CONT 2x4 PLATE

1/4" DIA. x 5" LONG LAG SCREWS AT EACH SUPPORT (TYP)

FRAMING ANCHOR (TYP)

AT RIDGE OF BUILT-UP FRAMING

TYP. BUILT-UP ROOF FRAMING DETAILS
NOT TO SCALE   (DETAIL T6-BURF)

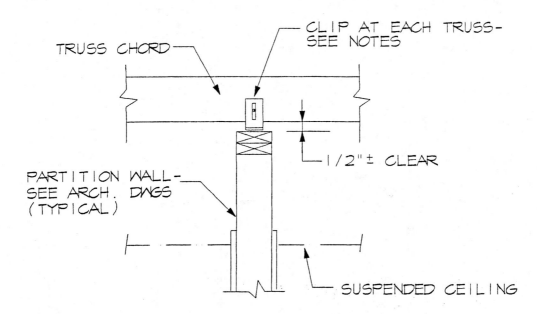

TRUSS CHORD

CLIP AT EACH TRUSS-
SEE NOTES

1/2"± CLEAR

PARTITION WALL-
SEE ARCH. DWGS
(TYPICAL)

SUSPENDED CEILING

WALL PERPENDICULAR TO ROOF TRUSSES

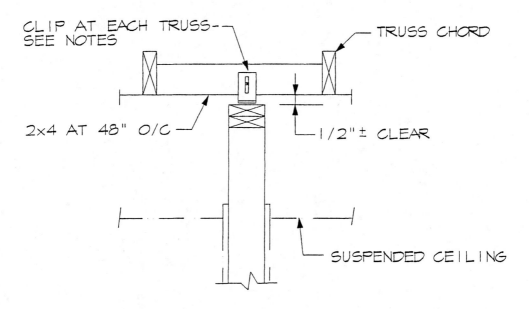

CLIP AT EACH TRUSS-
SEE NOTES

TRUSS CHORD

2x4 AT 48" O/C

1/2"± CLEAR

SUSPENDED CEILING

WALL PERPENDICULAR TO ROOF TRUSSES

TYP. PARTITION WALL BRACING DETAILS

NOT TO SCALE   (DETAIL T6-PWALL)

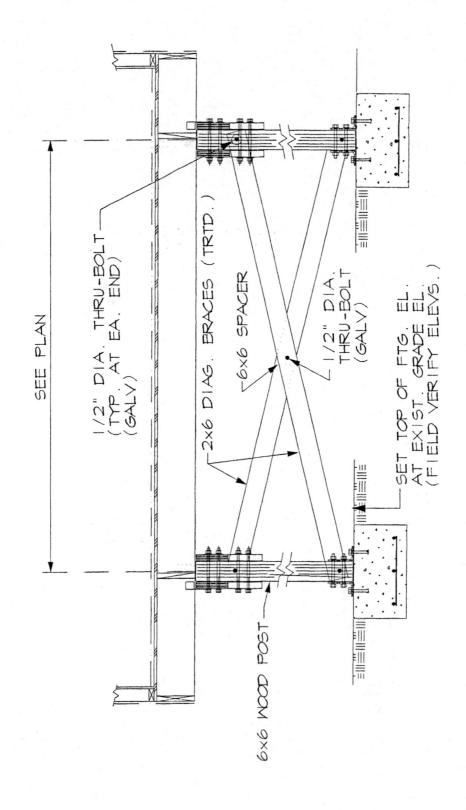

SEE PLAN

1/2" DIA. THRU-BOLT (TYP. AT EA. END) (GALV)

2x6 DIAG. BRACES (TRTD.)

6x6 SPACER

1/2" DIA. THRU-BOLT (GALV)

SET TOP OF FTG. EL. AT EXIST. GRADE EL. (FIELD VERIFY ELEVS.)

6x6 WOOD POST

TYPICAL DIAGONAL BRACE DETAIL

NOT TO SCALE

(DETAIL T6-DBR)

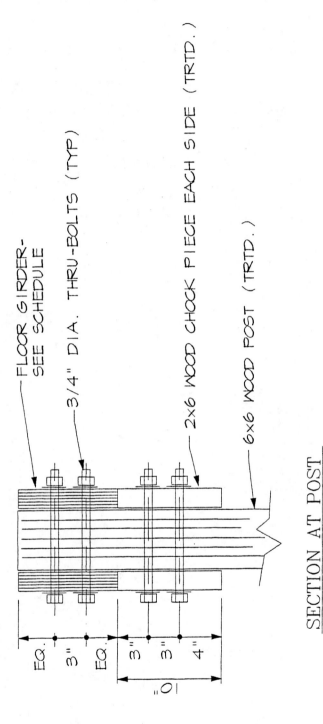

FLOOR GIRDER-
SEE SCHEDULE

3/4" DIA. THRU-BOLTS (TYP)

2x6 WOOD CHOCK PIECE EACH SIDE (TRTD.)

6x6 WOOD POST (TRTD.)

EQ.
3"
EQ.
3"
3"
4"

1'-0"

SECTION AT POST

TYPICAL FLOOR GIRDER CONNECTION DETAILS

(DETAIL T6-GRD1)

NOT TO SCALE

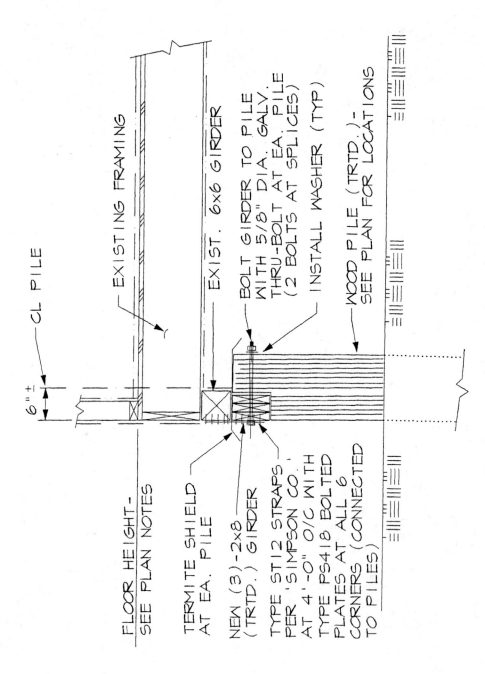

CL PILE

FLOOR HEIGHT -
SEE PLAN NOTES

6"±

EXISTING FRAMING

TERMITE SHIELD
AT EA. PILE

EXIST. 6x6 GIRDER

NEW (3)-2x8
(TRTD.) GIRDER

BOLT GIRDER TO PILE
WITH 5/8" DIA. GALV.
THRU-BOLT AT EA. PILE
(2 BOLTS AT SPLICES)

INSTALL WASHER (TYP.)

TYPE ST12 STRAPS
PER SIMPSON CO.
AT 4'-0" O/C WITH
TYPE PS418 BOLTED
PLATES AT ALL 6
CORNERS (CONNECTED
TO PILES)

WOOD PILE (TRTD.) -
SEE PLAN FOR LOCATIONS

PERIMETER SECTION

RELOCATED BUILDING TO NEW PILE SUPPORTS
(DETAIL T6-PS1)

NOT TO SCALE

416

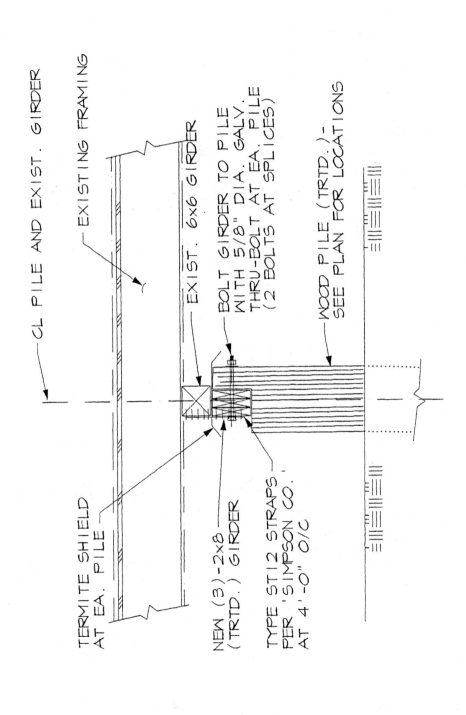

CL PILE AND EXIST. GIRDER

EXISTING FRAMING

EXIST. 6×6 GIRDER

BOLT GIRDER TO PILE
WITH 5/8" DIA. GALV.
THRU-BOLT AT EA. PILE
(2 BOLTS AT SPLICES)

WOOD PILE (TRTD.) -
SEE PLAN FOR LOCATIONS

TERMITE SHIELD
AT EA. PILE

NEW (3)-2×8
(TRTD.) GIRDER

TYPE ST12 STRAPS
PER 'SIMPSON CO.'
AT 4'-0" O/C

INTERIOR SECTION

RELOCATED BUILDING TO NEW PILE SUPPORTS
(DETAIL T6-PS2)

NOT TO SCALE

417

Commentary—Wood Details

2× FRAMING

TYPICAL 2× JOIST PERIMETER BEARING (DETAIL T6-2×1)

This detail shows a typical perimeter 2× floor joist bearing condition. The floor joists are spaced at 16″ on center. The joist size is noted as being shown on the plan; however, it could be scheduled or shown on the detail. The joists are supported by 2 × 6 studs spaced to match the joists. The stud spacing at the upper level switches to 24″ because roof framing occurs above. The floor sheathing is ¾″ tongue and groove Sturd-I-Floor plywood. A continuous 2× rim joist is at the outer edge to help transfer the loads from the upper studs to the lower studs. It also prevents the joists from rolling over. The wall is framed with a double 2 × 6 top plate and single 2 × 6 sill plate. The exterior structural sheathing element is ½″ plywood, the requirements of which should be contained in a shearwall schedule (or notes and specifications).

TYPICAL 2× JOIST PERIMETER NONBEARING (DETAIL T6-2×2)

This detail shows a typical perimeter nonbearing condition. The 2× floor joists are spaced at 16″ on center running in the same direction as the wall. A suspended ceiling is indicated so 2× outlookers are required to brace the top of the wall. All studs are shown spaced at 24″ on center.

TYPICAL 2× JOIST INTERIOR BEARING (DETAIL T6-2×3)

This detail shows a typical interior bearing condition. The 2× joists are spaced at 16″ on center and bear on 2 × 4 studs. The joists are lapped over the 2 × 4 stud bearing wall. Blocking is shown between the joists to help transfer loads from the level above and to prevent the joists from rolling over. In this detail, the wall is not considered a major shearwall (the architectural drawings are being referred to for the sheathing requirements).

TYPICAL 2× CANTILEVERED DECK JOISTS (DETAIL T6-2×4)

This detail shows a cantilevered deck condition. The 2× deck joists are lapped back with the main floor joists. The joists are supported by 2 × 6 studs spaced to match the joists. The deck joists need to be properly nailed to the main floor joists. All deck framing should be pressure treated. Blocking is shown between the joists. A vertical drop is provided between the deck and main floor level to provide better weatherproofing.

TYPICAL 2× CANTILEVERED JOIST DETAIL (DETAIL T6-2×5)

A short floor joist cantilever condition is shown with 2× joists spaced at 16" on center. The joists are supported by 2 × 6 studs spaced to match the joists. A continuous rim joist helps transfer loads from the level above. Blocking is shown between the joists directly over the bearing wall.

TYPICAL DECK PERIMETER FRAMING DETAIL (DETAIL T6-2×6)

This detail shows a perimeter deck condition with joists, decking, beams, and posts. All deck framing should be pressure treated. The architectural drawings are referred to for the railing and top of deck level. Galvanized thru-bolts connect the wood beams to the wood posts. Due to high loading requirements, wood chocks are provided on the posts to help support the wood beams.

TYPICAL 2× RAFTER BEARING ON STUDWALL (DETAIL T6-R2×1)

With a typical rafter bearing condition, the 2× rafters and ceiling joists are spaced at 24" on center. The framing sizes can be scheduled, called out on the plan, or shown on the detail. The rafters are supported by 2 × 6 studs spaced to match the rafters. The roof sheathing is ⅝" plywood. Blocking is provided at the alternate spaces between the framing to help prevent the framing from rolling over. The wall is framed with a double 2 × 6 top plate. The exterior structural sheathing element is ½" plywood, the requirements of which should be contained in a shearwall schedule (or notes and specifications). The architectural drawings are referred to for the exterior finish system, soffit, and overhang. Hurricane clips are indicated for each rafter. The required uplift capacity of the hurricane clips should also be noted or specified.

TYPICAL 2× RAFTER BEARING ON STUDWALL (DETAIL T6-R2×2)

This detail is similar to the previous one, but there are no ceiling joists, providing a cathedral ceiling. This system needs to be investigated very carefully. Horizontal thrusts can develop at the rafter bearings causing the walls to push out. This could result in a roof collapse. The rafters are supported by the wall, but they also need to be supported at the ridge, such as with a wood beam. Another option is to add *collar ties* several feet below the ridge, essentially allowing the system to span from wall to wall without requiring a support at the ridge.

TYPICAL 2× RAFTER BEARING ON STUDWALL (DETAIL T6-R2×3)

This detail is similar to the previous two, with the rafters bearing on a high wall condition and without an overhang.

MISCELLANEOUS 2× RAFTER BEARING DETAILS (DETAIL T6-R2×4)

Two bearing conditions are shown: one at an interior steel beam and the other at an exterior wood beam. A continuous 2× plate is provided on the top of the steel beam. This plate is shown as being attached with power actuated fasteners; however, it could also be bolted. The perimeter wood beam is shown being attached to the wood posts with a prefabricated metal cap connector.

TYPICAL 2× RAFTER NONBEARING DETAIL (DETAIL T6-R2×5)

This detail shows a typical perimeter nonbearing rafter condition. The 2× rafters are spaced at 24″ on center running in the same direction as the wall. The studs are shown spaced at 24″ on center continuous to the top of wall. The diaphragm loads are transferred to the wall system by connecting the end rafter to the studs with lag screws.

TYPICAL 2× RAFTER NONBEARING DETAIL (DETAIL T6-R2×6)

This detail is a variation of the previous one, with a roof overhang in lieu of a parapet wall. The 2× outlookers are used to frame the overhang. A hurricane clip is indicated at each outlooker. Hurricane clips may not be actually required depending on the toe-nailing capacity of the outlooker connection and the actual wind uplift force.

TYPICAL 2× RAFTER NONBEARING DETAIL (DETAIL T6-R2×7)

This detail is similar to the previous one, but there are ceiling joists and the wall is indicated as being constructed using the platform method. The platform construction method results in the gable end walls having a potential wall hinge. This hinge needs to be properly braced off, as indicated by the 2 × 6 diagonal braces. The ceiling diaphragm may also provide some bracing capability. The 2× outlookers are used to frame the overhang. A hurricane clip is indicated at each outlooker. Hurricane clips may not be actually required depending on the toe-nailing capacity of the outlooker connection and the actual wind uplift force.

2× RAFTER FRAMING DETAIL AT SKYLIGHT (DETAIL T6-R2×8)

A header is shown at the top, which spans between hip rafters. The architectural drawings are referred to for the curb, skylight, and clear opening.

TYPICAL 2× FRAMING DETAILS AT RIDGE (DETAIL T6-R2×9)

Two details are shown. One shows the rafters supported at the ridge by a wood beam. The other shows the rafter condition at a 2× ridge, where collar ties are needed to maintain continuity and prevent horizontal thrusts from being developed at the lower rafter bearing support (see detail T6-R2×2).

MISCELLANEOUS 2× RAFTER FRAMING DETAIL (DETAIL T6-R2×10)

A cantilevered rafter overhang condition is shown. A short kneewall runs from the floor level to the rafter bearing elevation. 2 × 6 studs are indicated. The floor framing runs parallel to the wall.

MISCELLANEOUS 2× RAFTER FRAMING DETAIL (DETAIL T6-R2×11)

A load-bearing condition is shown. The rafters are supported on the interior by a short kneewall. The ceiling joist framing bears on the 2 × 6 studwall. A beam supports the kneewall and ceiling joists. This condition shows an attic floor level.

PREMANUFACTURED WOOD I-JOIST FRAMING

TYPICAL I-JOIST BEARING ON STUDWALL (DETAIL T6-IJ1)

This detail shows a typical bearing condition. The floor I-joists are spaced at 16″ on center. The joist size can be scheduled, called out on the plan, or shown on the detail. The joists are supported by 2 × 6 studs spaced to match the joists. The stud spacing at the upper level switches to 24″, assuming that roof framing occurs above. The floor sheathing is ¾″ tongue and groove Sturd-I-Floor plywood. A continuous rim joist is at the outer edge to help transfer the loads from the upper studs to the lower studs. It also prevents the joists from rolling over. The wall is framed with a double 2 × 6 top plate and single 2 × 6 sill plate. The exterior structural sheathing element is ½″ plywood, the requirements of which should be contained in a shearwall schedule (or notes and specifications). The I-joist manufacturer is required to investigate whether web stiffeners are required.

TYPICAL I-JOIST PERIMETER FRAMING DETAIL (DETAIL T6-IJ2)

This detail shows a typical perimeter nonbearing condition. The floor I-joists are spaced at 16″ on center running in the same direction as the wall. A suspended ceiling is indicated, so I-joist outlookers are required to brace the top of the wall. All studs are shown spaced at 24″ on center. Joist bridging is recommended to help distribute concentrated loads and brace the joists during construction.

TYPICAL I-JOIST INTERIOR BEARING DETAILS (DETAIL T6-IJ3)

Two interior bearing details are shown. One detail is at a wood beam. Joist hangers are required. The other detail is at an interior studwall where the I-joists are spaced at 16″ on center and bear on 2 × 4 studs. The joists are lapped over the 2 × 4 stud bearing wall. I-joist blocking is shown between the joists to prevent the joists from rolling over. In this detail, the wall is not considered a major shearwall because the architectural drawings are referred to for the sheathing requirements.

TYPICAL FRAMING DETAILS AT STAIRS (DETAIL T6-IJ4)

Two stair-area-related details are shown. One detail is at an exterior stairwell wall and may also be applicable for regular 2× framed buildings. The exterior wall is shown as being constructed using the platform method. Additional 2 × 6 wall plates are provided to span the width of the stairwell opening. This prevents a wall system hinge. The other detail is at a stairwell wall where the I-joist outlookers are provided to brace the wall. The architectural drawings are referred to for the wood stair framing.

TYPICAL I-JOIST AND EXTERIOR DECK DETAIL (DETAIL T6-IJ5)

This detail is similar to the typical bearing condition, but there is a lower level deck supported by the perimeter wall. 2 × 10s are shown for the deck framing.

TYPICAL I-JOIST BEARING ON CMU WALL (DETAIL T6-IJ6)

This detail is similar to the typical bearing condition, but the supporting wall is masonry. Since the 2 × 8 plate is only anchored with a nominal connection to the masonry bond beam, this detail provides nominal uplift resistance. All 2× framing in contact with masonry or concrete should be treated.

TYPICAL I-JOIST BEARING ON INTERIOR CMU (DETAIL T6-IJ7)

This detail is similar to the previous one, but it is at an interior fire rated wall condition.

TYPICAL I-JOIST INTERIOR FRAMING DETAILS (DETAIL T6-IJ8)

Two interior conditions are shown: one at an interior wood beam and the other at a railing condition. The architectural drawings are referred to for the rail.

TYPICAL I-JOIST CONNECTIONS TO STEEL (DETAIL T6-IJ9)

Two bearing conditions are shown: one at a floor level and the other at a roof level. The joists are shown framing into the steel beam due to headroom considerations. Otherwise, it is more economical to run the joists continuous over the beams. A continuous nailer is provided at the top of the steel beams. The width of this nailer needs to be ripped to match the beam top flange width (or to exceed the top flange width). Joist hangers are provided on each side of the beams.

TYPICAL I-JOIST BEARING ON STUDWALL (DETAIL T6-IJ10)

This detail is similar to the typical bearing condition, but it has a brick façade.

TYPICAL I-JOIST PERIMETER FRAMING DETAIL (DETAIL T6-IJ11)

This detail is similar to the typical perimeter framing condition, but it has a brick façade.

TYPICAL I-JOIST PARALLEL TO CMU WALL (DETAIL T6-IJ12)

This detail is similar to the typical perimeter framing condition, but the exterior wall is concrete masonry. The masonry wall is a shearwall. Lateral loads are transferred to the shearwall by bolting the continuous 2 × 12 ledger to the shearwall.

TYPICAL I-JOIST INTERIOR FRAMING DETAILS (DETAIL T6-IJ13)

Two interior framing conditions are shown: one at a floor joist direction change and another at an interior shearwall.

TYPICAL I-JOIST FRAMING AT STAIRS (DETAIL T6-IJ14)

Two stair framing conditions are shown: one at the main floor level and another at an intermediate landing level.

TYPICAL I-JOIST PERIMETER FRAMING DETAILS (DETAIL T6-IJ15)

Two interior framing conditions are shown: one where the joists are parallel to a steel beam and another where the floor joists cantilever over a lower studwall.

MISCELLANEOUS I-JOIST FRAMING DETAILS (DETAIL T6-IJ16)

Two interior framing conditions are shown: one where the joists support an interior shearwall and another at an interior mechanical duct chase.

MISCELLANEOUS I-JOIST CONNECTIONS TO STEEL (DETAIL T6-IJ17)

Some buildings require the main second floor structure to be framed with steel to maximize the open area (such as for lower level parking). Framing above the second floor level is wood. This detail shows the typical perimeter condition, where the I-joists frame into the steel beams.

MISCELLANEOUS I-JOIST CONNECTIONS TO STEEL (DETAIL T6-IJ18)

This detail is similar to the previous one, but the I-joists run in the same direction as the wall.

MISCELLANEOUS I-JOIST CONNECTIONS TO STEEL (DETAIL T6-IJ19)

This detail is similar to the previous one, but it is at an interior load-bearing wall condition.

TYPICAL CANTILEVERED DECK JOIST DETAIL (DETAIL T6-IJ20)

This detail is similar to the previous one, but it is at an exterior deck condition. Since the deck joists lap back with the main floor I-joists, the steel beam has been lowered. It is preferable that the deck joist depth be such that the deck joists fit snug next to the I-joist webs.

TYPICAL CANTILEVERED DECK JOIST DETAIL (DETAIL T6-IJ21)

This detail is similar to the previous one, but it occurs where there is a 2 × 6 studwall above and below the joists.

TYPICAL CANTILEVERED WOOD I-JOIST DETAIL (DETAIL T6-IJ22)

This detail is similar to detail T6-IJ20, but it involves an I-joist cantilever. A steel beam supports the joists.

WOOD I-JOIST LOADING DIAGRAMS (DETAIL T6-FJOIS) .

Any special I-joist loading requirements should be shown on the drawings.

TYPICAL ROOF I-JOIST BEARING DETAIL (DETAIL T6-RJ1)

This detail shows a typical roof I-joist bearing condition. The joists are spaced at 24″ on center and bear on a 2 × 6 studwall. The parapet wall is framed by lapping 2 × 6 studs with the I-joists. The joist size can be scheduled, called out on the plan, or shown on the detail. The roof sheathing is ⅝″ plywood. The exterior structural sheathing element is ½″ plywood, the requirements of which should be contained in a shearwall schedule (or notes and specifications). The I-joist manufacturer is required to investigate whether web stiffeners are required. Double hurricane clips are indicated. The architectural drawings are referred to for the top of parapet wall elevation.

TYPICAL ROOF I-JOIST PARALLEL TO WALL (DETAIL T6-RJ2)

This detail shows a typical perimeter nonbearing condition. The roof I-joists are spaced at 24″ on center running in the same direction as the wall. A suspended ceiling is indicated, so 2 × 12 outlookers are required to brace the top of the wall. All studs are shown continuous to the roof without splices.

TYPICAL ROOF JOIST BEARING DETAIL (DETAIL T6-RJ3)

This detail is similar to detail T6-RJ1, but it involves a different type of wood joist. This wood joist has steel webs and wood chords. It also bears on the top chord. The architectural drawings are referred to for the brick façade and mansard.

TYPICAL ROOF JOIST PARALLEL TO WALL (DETAIL T6-RJ4)

This detail is similar to the previous one, but the roof joists are running parallel to the wall.

ROOF I-JOIST DETAIL AT ELEVATOR WALL (DETAIL T6-RJ5)

This detail is similar to detail T6-IJ12, but the cmu wall is at an elevator shaft.

MISCELLANEOUS ROOF I-JOIST DETAILS (DETAIL T6-RJ6)

Two interior framing conditions are shown: one where the roof I-joists are supported by a wood beam (next to an attic) and another at an interior shearwall.

PRE-ENGINEERED WOOD TRUSS FRAMING

TYPICAL ROOF TRUSS BEARING DETAIL (DETAIL T6-TR1)

This detail shows a roof truss bearing condition. The trusses are spaced at 24″ on center, and in this detail, the trusses are shown bearing on an 8″ masonry wall. A masonry bond beam is provided at the top of the wall. The overall truss geometry should be shown elsewhere (see detail T6-TEL1). The roof sheathing is ⅝" plywood. 2 × 8 blocking is provided to help restrain the trusses from rolling over. The wind uplift condition must be carefully investigated. In this detail, the wind uplift is transferred from the trusses to the double top plate (at the top of the masonry wall) via the double hurricane clips at each truss. The required uplift capacity of the hurricane clips should be called out in the detail (or elsewhere). The double top plates span between the ⅝" diameter all-thread rods, which are spaced at 4' on center. The rods are embedded 24" into the masonry wall to satisfy the minimum development length. The top 24" of the masonry wall is grouted solid, forming a continuous 8 × 24" deep masonry beam. A head is provided at the bottom of the all-thread rods, increasing their pull-out capacity. The uplift forces from the all-thread rods are transferred to the masonry wall reinforcing, which should be spaced no more than the all-thread rods (4' on center). The 8 × 24" masonry beam serves a structural purpose during uplift. It transfers the force from the all-thread rods to the vertical reinforcing bars. It also provides more wall mass at the top of the wall to resist wind uplift.

TYPICAL ROOF TRUSS ENDWALL DETAIL (DETAIL T6-TR2)

A gable endwall condition can be difficult to brace. In this detail, it was decided to brace the top of the masonry endwall by creating a mechanical equipment attic. The forces from the roof diaphragm are transferred to the plywood shearwall, which has blocked edges. The top 16" of the masonry wall is grouted solid (compared to the top 24" of the bearing condition per detail T6-TR1) because the wind uplift is not as significant. The all-thread rods are spaced more closely due to in-plane shear transfer.

CANOPY ROOF TRUSS BEARING DETAIL (DETAIL T6-TR3)

This detail is similar to detail T6-TR1, but it involves a steel beam support in lieu of the masonry wall. A double 2 × 6 plate is bolted to the top of the steel beam. The soffit is framed with 2 × 4s at 24" on center.

ROOF TRUSS ENDWALL DETAIL AT DOOR (DETAIL T6-TR4)

This detail is similar to detail T6-TR2, but it is shown at an overhead roll-up door.

TYPICAL CANOPY ENDWALL DETAIL (DETAIL T6-TR5)

This detail is similar to detail T6-TR3, but it shows an endwall condition. A double 2 × 6 plate is bolted to the top of the steel beam. The outer studwall is run continuous, supported vertically off of the framing above the steel beam. Transverse wind loads are transferred to the roof diaphragm with the 2 × 4 diagonal braces. These braces also restrain the top (compression) flange of the steel beam. The ceiling joists are indicated as 2 × 6s at 24″ on center.

TYPICAL MECHANICAL CHASE DETAIL (DETAIL T6-TR6)

This detail is similar to detail T6-TR2, but it shows an interior mechanical attic condition. Multitrusses are on each side of the attic, and 2× framing spans between the trusses.

TYPICAL TRUSS BEARING ON HEADER (DETAIL T6-TR7)

This detail is similar to detail T6-TR3, but a wood header supports the trusses.

TYPICAL ATTIC/BALCONY DETAIL (DETAIL T6-TR8)

An attic and balcony detail is shown with roof trusses and 2 × framing. The masonry wall is used as a shearwall.

TYPICAL TRUSS BEARING ON STUDWALL (DETAIL T6-TR9)

This detail is similar to detail T6-TR1, but a 2 × 6 studwall supports the trusses. The 2 × 6 studs are spaced at 24″ on center to align with the trusses. It is recommended that the outer brick face be at least 6″ from the face of the stud. This allows an approximate clear cavity of 2″. In this detail, the wind uplift is transferred from the trusses to the double top plate (at the top of the wall) via the double hurricane clips at each truss. The required uplift capacity of the hurricane clips should be called out in the detail (or elsewhere). The force from the double top plates gets transferred to the 2 × 6 studs through the ½″ plywood sheathing, which must run continuous to the top of the double 2 × 6 plates.

TYPICAL ENDWALL FRAMING DETAIL (DETAIL T6-TR10)

This detail is similar to detail T6-R2×7, but roof trusses are used in lieu of rafters and ceiling joists. The 2 × 6 studs are spaced at 24″ on center. Outlookers consist of 2 × 6s at 24″ on center. Hurricane clips are shown at 48″ on center. The gable endwall creates a wall hinge, which in this case is braced off by the 2 × 6 diagonals. A brick façade is indicated.

ENDWALL FRAMING DETAIL AT ENTRY
(DETAIL T6-TR11)

This detail is similar to detail T6-TR10, but it occurs over an entry and a lower header is provided at the soffit level. A brick façade is indicated, supported by an angle that is bolted to the header. The brick weight induces a torsional (twisting) force on the header, which is resisted by the 2 × 6 diagonal braces and 2 × 4 soffit framing.

DETAIL AT TRUSS DIRECTION CHANGE
(DETAIL T6-TR12)

This detail shows one series of trusses bearing on an interior 2 × 4 studwall. The other trusses are running in the same direction as the wall. The roof framing is built-up over the trusses that bear on the studwall.

TYPICAL INTERIOR TRUSS BEARING DETAIL
(DETAIL T6-TR13)

Truss manufacturers generally prefer to design trusses to *clear span* without bearing on the interior walls. In most cases, their engineering analyses are simplified by clear spanning. However, interior bearings may be considered for longer span trusses or to reduce the wind uplift at the end bearings. Temporary diagonal truss bracing should be provided over the bearing wall.

TYPICAL INTERIOR TRUSS BEARING DETAIL
(DETAIL T6-TR14)

This detail is similar to the previous one, but the trusses are in two pieces. Diagonal braces are shown over the interior 2 × 6 studwall. This prevents the trusses from rolling over and also helps transfer diaphragm shear to the shearwall.

TYPICAL DETAIL AT ROOF 'STEP'
(DETAIL T6-TR15)

This detail shows one way to provide a roof step using roof truss construction.

ENGINEERED WOOD TRUSS ELEVATIONS (DETAIL T6-TEL1)

The truss manufacturer's work is greatly simplified if the different types of trusses are shown, including the critical dimensions, elevations, roof pitches, and so on. Another option, especially for projects that involve a multitude of trusses, is to show them using single line diagrams. All dimensions are shown as being +/− to be verified by the manufacturer.

WOOD SCISSOR TRUSS REQUIREMENTS (DETAIL T6-SCISS)

Wood scissor trusses present several potential design problems. This detail provides the truss manufacturer with a diagram and set of notes specific to scissor trusses.

HOLD-DOWN ANCHORS

TYPICAL HOLD-DOWN ANCHOR DETAILS (DETAIL T6-HDA)

Hold-down anchors are typically used at the ends of shearwalls, building corners, and so on. This detail shows a typical foundation level condition with a continuous wall footing and masonry foundation wall. Double 2× studs are shown connected to the hold-down anchor; however, other stud configurations may be needed depending on the required hold-down anchor capacity. The force from the hold-down anchor is transmitted to the foundation through the two #3 vertical reinforcing bars.

TYPICAL HOLD-DOWN ANCHOR DETAILS (DETAIL T6-HDA2)

This detail is similar to the previous one, but the foundation is a continuous turned down concrete slab edge.

TYPICAL HOLD-DOWN ANCHOR DETAILS (DETAIL T6-EHD)

This detail is similar to the previous two, but a condition at an elevated wood framed floor level is shown.

TYPICAL HOLD-DOWN ANCHOR DETAILS (DETAIL T6-EHD2)

This detail is similar to the previous one, but a condition at an elevated steel framed floor level is shown.

TYPICAL HOLD-DOWN ANCHOR DETAILS (DETAIL T6-EHD3)

This detail is similar to the previous one, but a condition at a masonry wall level is shown.

DIAPHRAGMS

TYPICAL ROOF/FLOOR DIAPHRAGM DETAIL (DETAIL T6-RDIA)

Horizontal roof and floor diaphragms typically consist of plywood sheathing. The plywood is laid in a staggered joint pattern. Blocking is required if noted on the plan. A ⅛" gap is typically provided between the plywood sheets. Separate notes address the nailing requirements. An option in lieu of plywood is to use oriented strand board, which provides better weather resistance (during construction).

TYPICAL SHEARWALL ELEVATION DETAIL (DETAIL T6-SWE)

Vertical wall diaphragms typically consist of ½" plywood sheathing. The plywood is laid in a staggered joint pattern. Blocking is required if noted in the shearwall schedule. A ⅛" gap is typically provided between the plywood sheets. The shearwall schedule addresses the nailing requirements. An option in lieu of plywood is to use oriented strand board, which provides better weather resistance (during construction). A hold-down anchor is shown at the end of the shearwall.

MISCELLANEOUS DETAILS

TYPICAL FRAMED OPENING DETAIL (DETAIL T6-HDR)

Wall openings need to be carefully investigated. Headers should be sized for the specific loading conditions. The jamb conditions need to have full-height studs and jack studs as required for the transverse wind loading requirements. The Header Schedule contains a description of the header and jamb stud requirements.

TYPICAL DORMER ATTACHMENT DETAIL (DETAIL T6-DORMR)

Two details are shown: one for connecting a dormer to the main roof framing and another for a typical condition at a ridge.

TYPICAL BUILT-UP ROOF FRAMING DETAILS (DETAIL T6-BURF)

This detail shows a way to provide framing that is built-up over the main roof level.

TYPICAL PARTITION WALL BRACING DETAILS
(DETAIL T6-PWALL)

For buildings with suspended ceilings, the trusses should not bear on the interior partition walls. This detail shows a way to brace the interior partition wall and provide for vertical movement of the roof trusses so that vertical load is not transferred to the interior partition.

TYPICAL DIAGONAL BRACE DETAIL
(DETAIL T6-DBR)

This detail shows a way to brace the lower level of a wood framed building that uses 6 × 6 posts for vertical support.

TYPICAL FLOOR GIRDER CONNECTION DETAILS
(DETAIL T6-GRD1)

This condition needs to be investigated for the actual loading requirements. Wood chocks are provided to help transfer the force from the floor girder to the 6 × 6 wood post.

RELOCATED BUILDING TO NEW PILE SUPPORTS
(DETAIL T6-PS1)

A special building condition occurs when an existing building is to be relocated to a new pile-supported foundation. This detail shows a typical perimeter condition.

RELOCATED BUILDING TO NEW PILE SUPPORTS
(DETAIL T6-PS2)

This detail is similar to the previous one, but it shows a typical interior condition.

Appendix A

Structural Notes

GENERAL STRUCTURAL NOTES:

1. THE STRUCTURAL DRAWINGS SHALL BE USED IN CONJUNCTION WITH THE DRAWINGS OF ALL OTHER DISCIPLINES AND THE SPECIFICATIONS. THE CONTRACTOR SHALL VERIFY THE REQUIREMENTS OF OTHER TRADES AS TO SLEEVES, CHASES, HANGERS, INSERTS, ANCHORS, HOLES AND OTHER ITEMS TO BE PLACED OR SET IN THE STRUCTURAL WORK.

2. THE CONTRACTOR SHALL BE RESPONSIBLE FOR COMPLYING WITH ALL SAFETY PRECAUTIONS AND REGULATIONS DURING THE WORK. THE ENGINEER WILL NOT ADVISE ON NOR ISSUE DIRECTION AS TO SAFETY PRECAUTIONS AND PROGRAMS.

3. THE STRUCTURAL DRAWINGS HEREIN REPRESENT THE FINISHED STRUCTURE. THE CONTRACTOR SHALL PROVIDE ALL TEMPORARY GUYING AND BRACING REQUIRED TO ERECT AND HOLD THE STRUCTURE IN PROPER ALIGNMENT UNTIL ALL STRUCTURAL WORK AND CONNECTIONS HAVE BEEN COMPLETED. THE INVESTIGATION, DESIGN, SAFETY, ADEQUACY AND INSPECTION OF ERECTION BRACING, SHORING, TEMPORARY SUPPORTS, ETC. IS THE SOLE RESPONSIBILITY OF THE CONTRACTOR.

4. THE ENGINEER SHALL NOT BE RESPONSIBLE FOR THE METHODS, TECHNIQUES AND SEQUENCES OF PROCEDURES TO PERFORM THE WORK. THE SUPERVISION OF THE WORK IS THE SOLE RESPONSIBILITY OF THE CONTRACTOR.

5. DRAWINGS INDICATE GENERAL AND TYPICAL DETAILS OF CONSTRUCTION. WHERE CONDITIONS ARE NOT SPECIFICALLY SHOWN, SIMILAR DETAILS OF CONSTRUCTION SHALL BE USED, SUBJECT TO APPROVAL BY THE ENGINEER.

6. ALL STRUCTURAL SYSTEMS WHICH ARE TO BE COMPOSED OF COMPONENTS TO BE FIELD ERECTED SHALL BE SUPERVISED BY THE SUPPLIER DURING MANU-FACTURING, DELIVERY, HANDLING, STORAGE AND ERECTION IN ACCORDANCE WITH THE SUPPLIER'S INSTRUCTIONS AND REQUIREMENTS.

7. LOADING APPLIED TO THE STRUCTURE DURING THE PROCESS OF CONSTRUCT-ION SHALL NOT EXCEED THE SAFE LOAD-CARRYING CAPACITY OF THE STRUCTURAL MEMBERS. THE LIVE LOADINGS USED IN THE DESIGN OF THIS STRUCTURE ARE INDICATED IN THE "DESIGN CRITERIA NOTES". DO NOT APPLY ANY CONSTRUCTION LOADS UNTIL STRUCTURAL FRAMING IS PROPERLY CONNECTED TOGETHER AND UNTIL ALL TEMPORARY BRACING IS IN PLACE.

8. ALL ASTM AND OTHER REFERENCES ARE PER THE LATEST EDITIONS OF THESE STANDARDS, UNLESS OTHERWISE NOTED.

9. SHOP DRAWINGS AND OTHER ITEMS SHALL BE SUBMITTED TO THE ENGINEER FOR REVIEW PRIOR TO FABRICATION. ALL SHOP DRAWINGS SHALL BE REVIEWED BY THE GENERAL CONTRACTOR BEFORE SUBMITTAL. THE ENGINEER'S REVIEW IS TO BE FOR CONFORMANCE WITH THE DESIGN CONCEPT AND GENERAL COMPLIANCE WITH THE RELEVANT CONTRACT DOCUMENTS. THE ENGINEER'S REVIEW DOES NOT RELIEVE THE CONTRACTOR OF THE SOLE RESPONSIBILITY TO REVIEW, CHECK AND COORDINATE THE SHOP DRAWINGS PRIOR TO SUBMISSION. THE CONTRACTOR REMAINS SOLEY RESPONSIBLE FOR ERRORS AND OMISSIONS ASSOCIATED WITH THE PREPARATION OF SHOP DRAWINGS AS THEY PERTAIN TO MEMBER SIZES, DETAILS, DIMENSIONS, ETC.

10. SUBMIT SHOP DRAWINGS IN THE FORM OF TWO BLUELINE PRINTS AND ONE SEPIA. IN NO CASE SHALL REPRODUCTION OF THE CONTRACT DRAWINGS BE USED AS SHOP DRAWINGS. AS A MINIMUM, SUBMIT THE FOLLOWING ITEMS FOR REVIEW:

    A. CONCRETE MIX DESIGN(S).
    B. REINFORCING STEEL SHOP DRAWINGS.
    C. STRUCTURAL STEEL SHOP DRAWINGS.
    D. STEEL JOIST / GIRDER SHOP DRAWINGS.
    E. METAL DECKING SHOP DRAWINGS.
    F. PRE-MANUFACTURED WOOD SYSTEM / TRUSS SHOP DRAWINGS (SEE NOTES).
    G. PRE-ENGINEERED METAL BUILDING SYSTEM (SEE NOTES).

    OTHER SUBMITTALS MAY BE REQUIRED PER THE "SCHEDULE OF SPECIAL INSPECTIONS" OR THE SEPARATE NOTES CONTAINED HEREIN.

11. IN ACCORDANCE WITH SECTION 1705 OF BOCA 1996, SPECIAL INSPECTIONS WILL BE REQUIRED FOR THIS PROJECT. SPECIAL INSPECTIONS SHALL BE PERFORMED IN ACCORDANCE WITH THE "SCHEDULE OF SPECIAL INSPECTIONS". ALL FABRICATORS SHALL SATISFY THE "EXCEPTION" NOTED IN SECTION 1705.2.2, WHICH REQUIRES THE FABRICATOR TO MAINTAIN AN AGREEMENT WITH AN APPROVED INDEPENDENT INSPECTION OR QUALITY CONTROL AGENCY. THE CONTRACTOR SHALL NOTIFY THE SPECIAL INSPECTOR AT LEAST 48 HOURS IN ADVANCE FOR WORK THAT WILL REQUIRE INSPECTION OR TESTING.

12. UNLESS OTHERWISE INDICATED, ALL ITEMS NOTED TO BE DEMOLISHED SHALL BECOME THE CONTRACTOR'S PROPERTY AND BE REMOVED FROM THE SITE.

13. CONTRACTORS SHALL VISIT THE SITE PRIOR TO BID TO ASCERTAIN CONDITIONS WHICH MAY ADVERSELY AFFECT THE WORK OR COST THEREOF.

DESIGN CRITERIA NOTES:

1. THE INTENDED DESIGN STANDARDS AND/OR CRITERIA ARE AS FOLLOWS:

GENERAL	UNIFORM STATEWIDE BLDG CODE (BOCA 1996, AS AMENDED)
CONCRETE	ACI 318-95
MASONRY	ACI 530-95
STRUCTURAL STEEL	AISC L.R.F.D.-93
STEEL JOISTS/GIRDERS	SJI-94 (& 50-YEAR JOIST DIGEST)
METAL DECK	SDI-92
COLD-FORMED METAL	AISI, ASD-86 (REV. 89)
WOOD	NDS-91
WOOD TRUSSES	TPI 1-95
FOUNDATIONS	SOILS INVESTIGATION AND REPORT PERFORMED BY xxx

2. DESIGN GRAVITY <u>DEAD LOADS</u> USED IN THE DESIGN OF THIS STRUCTURE ARE AS FOLLOWS:

ROOF	20 PSF MAX. 10 PSF MIN.
FLOORS - TYPICAL	X PSF
- MEZZANINES	X PSF
PARTITION ALLOWANCE	20 PSF
ALL OTHER	ACTUAL WEIGHT

3. DESIGN GRAVITY <u>LIVE LOADS</u> USED IN THE DESIGN OF THIS STRUCTURE ARE AS FOLLOWS:

ROOF, TYPICAL	20 PSF MIN.	ROOF, OVERHANGS	60 PSF
FIRST FLOOR	150 PSF	DWELLING UNITS	40 PSF
LOBBIES & STAIRS	100 PSF	BALCONIES	100 PSF
MEZZANINE	150 PSF	LIGHT STORAGE	125 PSF
OFFICES	50 PSF	HEAVY STORAGE	250 PSF
CORRIDORS	80 PSF	OVERHEAD CRANES	X TONS
MECHANICAL SPACE	125 PSF	ELEVATOR EQUIP. RM.	125 PSF

   FLOOR LIVE LOAD REDUCTION PER BOCA 96, 1606.7 HAS BEEN UTILIZED.
   ROOF LIVE LOAD REDUCTION PER BOCA 96, 1607.3 HAS BEEN UTILIZED.

4. DESIGN LATERAL LIVE LOADS USED IN THE DESIGN OF THIS STRUCTURE ARE AS FOLLOWS:

   WIND, MAIN SYSTEM: 100 MPH, EXP. 'B', I=1.1, P=21.9 PSF
   WIND, COMPONENTS: 100 MPH, EXP. 'C', I=1.1, GCpi=+0.75 & -0.25, Kh=0.93
   SEISMIC   Av< 0.05, Aa<0.05 (NONGOVERNING COMPARED TO WIND)

5. DESIGN SNOW LOADS USED IN THE DESIGN OF THIS STRUCTURE ARE AS FOLLOWS:
   Pg (GROUND SNOW LOAD) =10 PSF, Ce=0.9, I=1.0, Pf= 9 PSF

6. THE LATERAL LOAD RESISTING SYSTEM OF THIS BUILDING CONSISTS OF:
   PERIMETER- BRACED BAYS WITH STEEL RODS
   TYP. FRAME- RIGID FRAME ACTION (WITH "PIN-BASED" COLUMNS)

7. THIS STRUCTURE HAS BEEN DESIGNED WITH "SAFETY FACTORS" IN ACCORDANCE WITH GENERALLY ACCEPTED PRINCIPLES OF STRUCTURAL ENGINEERING. THE FUNDAMENTAL NATURE OF THE "SAFETY FACTOR" IS TO COMPENSATE FOR UNCERTAINTIES IN THE DESIGN, FABRICATION AND ERECTION OF STRUCTURAL BUILDING COMPONENTS. IT IS INTENDED THAT "SAFETY FACTORS" BE USED SO THAT THE LOAD CARRYING CAPACITY OF THE STRUCTURE DOES NOT FALL BELOW THE DESIGN LOAD AND THAT THE BUILDING WILL PERFORM UNDER DESIGN LOAD WITHOUT DISTRESS. WHILE THE USE OF "SAFETY FACTORS" IMPLIES SOME EXCESS CAPACITY BEYOND DESIGN LOAD, SUCH EXCESS CAPACITY CANNOT BE ADEQUATELY PREDICTED AND <u>SHALL NOT BE RELIED UPON</u>.

8. THIS STRUCTURE IS LOCATED IN FLOOD PLAIN X AND HAS BEEN DESIGNED FOR HYDRAULIC FORCES IN ACCORDANCE WITH BOCA 1996.

EXISTING CONSTRUCTION NOTES:

1. BEFORE PROCEEDING WITH ANY WORK WITHIN THE EXISTING FACILITY, THE CONTRACTOR SHALL FAMILIARIZE HIMSELF WITH EXISTING STRUCTURAL AND OTHER CONDITIONS. IT SHALL BE THE CONTRACTOR'S RESPONSIBILITY TO PROVIDE ALL NECESSARY BRACING, SHORING AND OTHER SAFEGUARDS TO MAINTAIN ALL PARTS OF THE EXISTING WORK IN A SAFE CONDITION DURING THE PROCESS OF DEMOLITION AND CONSTRUCTION AND TO PROTECT FROM DAMAGE THOSE PORTIONS OF THE EXISTING WORK WHICH ARE TO REMAIN.

2. THE CONTRACTOR SHALL FIELD VERIFY THE DIMENSIONS, ELEVATIONS, ETC. NECESSARY FOR THE PROPER CONSTRUCTION AND ALIGNMENT OF THE NEW PORTIONS OF THE WORK TO THE EXISTING WORK. THE CONTRACTOR SHALL MAKE ALL MEASUREMENTS NECESSARY FOR FABRICATION AND ERECTION OF STRUCTURAL MEMBERS. ANY DISCREPANCY SHALL BE IMMEDIATELY BROUGHT TO THE ATTENTION OF THE ENGINEER.

3. WELDING TO AND WITHIN AN EXISTING FACILITY PRESENTS POTENTIAL HAZARDS, INCLUDING:

   A. FIRE HAZARD - DUE TO THE EXISTING CONSTRUCTION AND BUILDING CONTENTS.

   B. STRUCTURAL LIQUEFACTION - DUE TO WELDING ACROSS THE FULL SECTION OF STRUCTURAL STEEL MEMBERS.

   RECOMMENDATIONS TO PREVENT THESE HAZARDS INCLUDE:

   A. FIRE HAZARD - PROTECT EXISTING COMBUSTIBLES PRIOR TO WELDING. KEEP A SEPARATE WATCHMAN AND SEVERAL FIRE EXTINGUISHERS ON HAND.

   B. STRUCTURAL LIQUEFACTION - WELD IN SMALL INCREMENTS. ALLOW WELDS TO TO HARDEN BEFORE CONINTUING TO THE NEXT INCREMENT.

   C. DO NOT LEAVE THE SITE UNTIL SATISFIED THAT NO FIRE HAZARD EXISTS.

4. INFORMATION USED IN PREPARING THESE DRAWINGS WAS TAKEN FROM DRAWINGS PREPARED BY THE FIRM OF [   ], DATED [   ].

5. THE CONTRACTOR SHALL BE RESPONSIBLE FOR THE DESIGN AND ERECTION OF ALL SHORING NECESSARY TO SAFEGUARD THE EXISTING STRUCTURE. THE SHORING SHOWN IS A PARTIAL AND SCHEMATIC REPRESENTATION OF THAT REQUIRED. THE CONTRACTOR SHALL SUBMIT A DETAILED PLAN FOR SHORING, BRACING AND PROTECTION OF THE EXISTING CONSTRUCTION. THE PLAN SHALL INCLUDE A CONSTRUCTION SEQUENCE, BEAR THE SEAL OF A PROFESSIONAL ENGINEER REGISTERED IN THE COMMONWEALTH OF VIRGINIA AND BE SUBMITTED TO THE ENGINEER FOR REVIEW PRIOR TO BEGINNING THE WORK.

FUTURE CONSTRUCTION NOTES:

1. THE FOUNDATIONS OF SHOP "G" HAVE BEEN DESIGNED FOR THE FUTURE PROVISION OF A MEZZANINE WITH A JOIST BRG. EL. OF (+12'-0").

2. THE STRUCTURAL SYSTEMS AT SHOPS "B" THRU "F" HAVE BEEN DESIGNED FOR A FUTURE FLOOR, ASSUMING A 3" CONCRETE FLOOR DECK (ON TOP OF THE COMPOSITE METAL DECKING BEING PROVIDED NOW). COLUMNS ON LINES 5, 7 AND 8 CAN BE EXENDED UP FOR FUTURE ROOF FRAMING SUPPORT. THE TOTAL HEIGHT OF THE FUTURE BUILDING ADDITION IS ASSUMED TO BE 10' ABOVE THE CURRENT CONSTRUCTION (FOR WIND LOAD PURPOSES).

PRE-ENGINEERED METAL BUILDING NOTES:

1. THE ENTIRE PRE-ENGINEERED METAL BUILDING SYSTEM SHALL BE DESIGNED BY THE METAL BUILDING MANUFACTURER IN CONFORMANCE TO THE PROVISIONS OF THE "VIRGINIA UNIFORM STATEWIDE BUILDING CODE" (BOCA 1996, AS AMENDED BY THE STATE) AND THE "LOW-RISE BUILDING SYSTEMS MANUAL" AS PUBLISHED BY THE METAL BUILDING MANUFACTURER'S ASSOCIATION. WHERE THESE CRITERIA CONFLICT, THE MORE STRINGENT CRITERIA SHALL APPLY.

2. IT IS THE PRE-ENGINEERED METAL BUILDING MANUFACTURER'S RESPONSIBILITY TO DESIGN THE COMPLETE BUILDING SYSTEM (STEEL FRAMING, ANCHOR BOLTS, PURLINS, GIRTS, BRACINGS, CONNECTIONS, ROOFING, WALL PANELS, COMPONENTS, ATTACHMENTS, ETC.). THE MANUFACTURER SHALL SUBMIT A CERTIFICATION LETTER BEARING THE SEAL OF A PROFESSIONAL ENGINEER REGISTERED IN VIRGINIA STATING THAT THE BUILDING SYSTEM DESIGN MEETS THE INDICATED CODE, PERFORMANCE AND LOADING REQUIREMENTS.

3. THE PRE-ENGINEERED METAL BUILDING MANUFACTURER SHALL BE CERTIFIED BY THE AMERICAN INSTITUTE OF STEEL CONSTRUCTION CATEGORY MB. THE MANUFACTURER SHALL MEET THE "EXCEPTION" NOTED PER BOCA 96, SECTION 1705.2.2, WHICH SPECIFIES QUALITY CONTROL REQUIREMENTS OF THE MANUFACTURER PERTAINING TO "SPECIAL INSPECTIONS".

4. THE CONTRACTOR SHALL SUBMIT SHOP DRAWINGS OF THE ENTIRE METAL BUILDING SYSTEM FOR REVIEW. THE CONTRACTOR SHALL ALSO SUBMIT A COMPLETE STRUCTURAL DESIGN ANALYSIS OF THE BUILDING SYSTEM (FOR RECORD PURPOSES ONLY). THE SHOP DRAWING SUBMITTAL SHALL INCLUDE ALL ANCHOR BOLT REQUIREMENTS AND FOUNDATION REACTIONS. ALL SHOP DRAWING AND CALCULATION SUBMITTALS SHALL BEAR THE SEAL OF A PROFESSIONAL ENGINEER REGISTERED IN x.

5. DESIGN LOADS TO BE USED IN CONNECTION WITH THE METAL BUILDING DESIGN ARE PER THE "DESIGN CRITERIA NOTES". IN ADDITION TO THE ACTUAL DEAD LOAD, AN ADDITIONAL COLLATERAL ROOF FRAMING DEAD LOAD OF 5 PSF SHALL BE INCLUDED. COORDINATE ANY EQUIPMENT LOADS WITH THE MECHANICAL AND ARCHITECTURAL DRAWINGS. PAY PARTICULAR ATTENTION TO THE CODE REQUIRED WIND LOADING REQUIREMENTS (WIND EXPOSURE CATEGORY, IMPORTANCE FACTOR (I), ETC.) NOTE THAT DUE TO THE ACTUAL WALL OPENINGS (STOREFRONT) CONDITION, THE INTERNAL WIND PRESSURE COEFFICIENTS (GCpi) ARE +0.75 AND -0.25 (SEE BOCA 96, TABLE 1609.7(6)).

6. CALCULATIONS FOR FRAME DEFLECTIONS SHALL BE DONE USING ONLY THE BARE FRAME METHOD. REDUCTIONS BASED ON ENGINEERING JUDGEMENT USING THE ASSUMED COMPOSITE STIFFNESS OF THE BUILDING ENVELOPE SHALL NOT BE PERMITTED. DRIFT SHALL FOLLOW AISC'S "SERVICEABILITY DESIGN CONSIDERATIONS FOR LOW-RISE BUILDINGS." CALCULATIONS SHALL BE SUBMITTED VERIFYING THAT THE ACTUAL DRIFT UNDER CODE REQUIRED LOADINGS DOES NOT EXCEED THE ALLOWABLE.

7. THE PRE-ENGINEERED MANUFACTURER SHALL PROVIDE ALL GIRTS, PURLINS, AND OTHER COMPONENTS REQUIRED FOR A COMPLETE SYSTEM. ALL WALL SYSTEMS, SUCH AS METAL STUDS, STOREFRONTS, ETC. SHALL BE PROPERLY SUPPORTED BY THE METAL BUILDING SYSTEM. ALLOWABLE DEFLECTIONS OF COMPONENTS SHALL BE IN ACCORDANCE WITH BOCA 1996.

8. THE FOUNDATION DESIGN IS BASED UPON THE "VACO-FRUDEN" BUILDING SYSTEM. THE CONTRACTOR SHALL BE RESPONSIBLE FOR COORDINATION OF ANY REVISIONS REQUIRED AS A RESULT OF A CHANGE IN THE BUILDING MANUFACTURER, INCLUDING REDESIGN OF THE FOUNDATIONS.

9. THE SIZE, NUMBER AND PLACEMENT PATTERN OF ALL ANCHOR BOLTS SHALL BE DETERMINED BY THE PRE-ENGINEERED BUILDING MANUFACTURER. ANCHOR BOLT EMBEDMENTS ARE INDICATED ON THE DRAWINGS.

10. THE PRE-ENGINEERED METAL BUILDING SHALL BE DESIGNED BY THE MANUFACTURER TO RESIST LATERAL LOADS AS FOLLOWS:
    INTERIOR FRAME LINES - RIGID FRAMES (PINNED-BASED COLUMNS)
    PERIMETER WALL LINES - BRACED BAYS OR PORTAL FRAMES

11. THE METAL BUILDING ERECTOR SHALL PROVIDE ALL TEMPORARY GUYING AND BRACING (SEE 'GENERAL STRUCTURAL NOTES').

12. UNLESS OTHERWISE NOTED OR SPECIFIED, ALL STEEL MEMBERS SHALL BE CLEANED AND PAINTED IN ACCORDANCE WITH MANUFACTURER'S STANDARD PROCEDURES.

13. DESIGN COLUMNS FOR REACTIONS FROM THE MEZZANINE STRUCTURE INDICATED BY XX.X KIPS ON THE PLANS.

14. THE FOUNDATIONS HAVE BEEN DESIGNED FOR THE REACTIONS INDICATED. THESE ARE BASED ON PINNED COLUMN BASES. NO "FIXED BASE" COLUMNS ARE PERMITTED WITHOUT THE ENGINEER'S WRITTEN APPROVAL.

    TYP. FRAME INTERIOR COLUMN: 25k (DOWN), 15k (UPLIFT), 0k (HORIZ.)
    TYP. FRAME END COLUMN: 15k (DOWN), 10k (UPLIFT), 5k (HORIZ. DL + 11)
    TYP. ENDWALL COLUMN: 10k (DOWN), 5k (UPLIFT), 0k (HORIZ. DL + 11)

FOUNDATION NOTES:

1. ALL FOOTINGS SHALL BEAR ON UNDISTURBED, FIRM NATURAL SOIL OR COMPACTED FILL CAPABLE OF SUPPORTING A DESIGN BEARING PRESSURE OF 2,000 PSF. ALL FOUNDATION EXCAVATIONS SHALL BE EVALUATED BY THE GEOTECHNICAL ENGINEER / TESTING AGENCY PRIOR TO POURING FOUNDATION CONCRETE.

1. ALL FOOTINGS HAVE BEEN DESIGNED BASED UPON AN ASSUMED SOIL BEARING PRESSURE OF 1,500 PSF. ALL FOOTINGS SHALL BEAR ON UNDISTURBED, FIRM NATURAL SOIL OR COMPACTED FILL. ALL FOUNDATION EXCAVATIONS SHALL BE EVALUATED BY THE GEOTECHNICAL ENGINEER / TESTING AGENCY PRIOR TO POURING FOUNDATION CONCRETE.

2. TOP OF FOOTING ELEVATION SHALL BE AS SHOWN ON THE FOUNDATION PLAN. THESE ELEVATIONS ARE A MAXIMUM AND SHALL BE LOWERED AS REQUIRED TO OBTAIN THE REQUIRED DESIGN BEARING PRESSURE.

3. ALL FOUNDATION CONCRETE SHALL OBTAIN A 28 DAY COMPRESSIVE STRENGTH OF 3000 PSI. ALL CONCRETE TO BE PERMANENTLY EXPOSED TO WEATHER SHALL BE AIR ENTRAINED TO 5% (± 1%) WITH AN ADMIXTURE THAT CONFORMS TO ASTM C-260.

4. ALL CONCRETE WORK SHALL CONFORM TO THE REQUIREMENTS OF ACI 301, "SPECIFICATION FOR STRUCTURAL CONCRETE BUILDINGS". HOT WEATHER CONCRETING SHALL BE IN ACCORDANCE WITH ACI 305. COLD WEATHER CONCRETING SHALL BE IN ACCORDANCE WITH ACI 306.

5. ALL REINFORCING STEEL SHALL CONFORM TO ASTM A-615, GRADE 60.

6. UNLESS OTHERWISE NOTED, THE FOLLOWING MINIMUM CONCRETE COVER SHALL BE PROVIDED FOR REINFORCEMENT:

    A) CONCRETE CAST AGAINST & PERMANENTLY EXPOSED TO EARTH - 3"

    B) CONCRETE EXPOSED TO EARTH OR WEATHER:
    #6 THROUGH #18 BARS   - 2"
    #5 BAR, W31 OR D31 WIRE & SMALLER - 1 1/2"

7. ALL REINFORCING MARKED CONTINUOUS (CONT.) ON THE PLANS AND DETAILS SHALL BE LAPPED 36xBAR DIAMETERS AT SPLICES UNLESS OTHERWISE NOTED.

8. NO UNBALANCED BACKFILLING SHALL BE DONE AGAINST FOUNDATION WALLS UNLESS WALLS ARE SECURELY BRACED AGAINST OVERTURNING, EITHER BY TEMPORARY BRACING OR BY PERMANENT CONSTRUCTION.

9. PRIOR TO COMMENCING ANY FOUNDATION WORK, COORDINATE WORK WITH ANY EXISTING UTILITIES. FOUNDATIONS SHALL BE LOWERED WHERE REQUIRED TO AVOID UTILITIES.

10. UNLESS OTHERWISE NOTED, THE CENTERLINES OF COLUMN FOUNDATIONS [PILE CAPS] SHALL BE LOCATED ON COLUMN CENTERLINES.

11. ALL RETAINING WALLS SHALL HAVE AT LEAST 12" OF FREE-DRAINING GRANULAR BACKFILL, FULL HEIGHT OF WALL. PROVIDE CONTROL JOINTS IN RETAINING WALLS AT APPROXIMATELY EQUAL INTERVALS NOT TO EXCEED 25 FEET NOR 3 TIMES THE WALL HEIGHT. PROVIDE EXPANSION JOINTS AT EVERY FOURTH CONTROL JOINT, UNLESS OTHERWISE INDICATED.

SITE PREPARATION NOTES:

1. WITHIN AN AREA A MINIMUM OF 5 FEET BEYOND THE BUILDING LIMITS, EXCAVATE A MINIMUM OF 4" OF EXISTING SOIL. REMOVE ALL ORGANICS, PAVEMENT, ROOTS, DEBRIS AND OTHERWISE UNSUITABLE MATERIAL.

2. THE SURFACE OF THE EXPOSED SUBGRADE SHALL BE INSPECTED BY PROBING OR TESTING TO CHECK FOR POCKETS OF SOFT OR UNSUITABLE MATERIAL. EXCAVATE UNSUITABLE SOIL AS DIRECTED BY THE GEOTECHNICAL ENGINEER / TESTING AGENCY.

3. PROOFROLL THE SURFACE OF THE EXPOSED SUBGRADE WITH A LOADED TANDEM AXLE DUMP TRUCK. REMOVE ALL SOILS WHICH PUMP OR DO NOT COMPACT PROPERLY AS DIRECTED BY THE GEOTECHNICAL ENGINEER / TESTING AGENCY.

4. FILL ALL EXCAVATED AREAS WITH APPROVED CONTROLLED FILL. PLACE IN 8 INCH LOOSE LIFTS AND COMPACT TO A MINIMUM OF 95% OF THE MAXIMUM DRY DENSITY IN ACCORDANCE WITH ASTM D-698.

5. ALL CONTROLLED FILL MATERIAL SHALL BE A SELECT GRANULAR MATERIAL FREE FROM ALL ORGANICS OR OTHERWISE DELETERIOUS MATERIAL WITH NOT MORE THAN 20% BY WEIGHT PASSING A NO. 200 SIEVE (CLASSIFIED AS SC, SM, SP OR BETTER IN ACCORDANCE WITH THE UNIFIED SOIL CLASSIFICATION SYSTEM) AND WITH A PLASTICITY INDEX NOT EXCEEDING 6%.

6. PROVIDE FIELD DENSITY TESTS FOR EACH 3,000 S.F. OF BUILDING AREA FOR EACH LIFT OF CONTROLLED FILL.

PILING NOTES:

1. ALL PILING SHALL BE SOUTHERN PINE OR DOUGLAS FIR CONFORMING TO ASTM D25, PRESSURE TREATED IN ACCORDANCE WITH AWPA STANDARD C-3 (6" MINIMUM TIP).

   [ALL PILING SHALL BE 10" SQUARE PRECAST, PRESTRESSED CONCRETE.]

2. DESIGN PILE CAPACITY IS AS FOLLOWS:

   COMPRESSION- 25 TONS
   TENSION- 10 TONS (TENSION PILES ONLY)
   LATERAL LOAD- 2 TONS (AT TOP OF PILE)

3. DRIVE SIX (6) INDICATOR TEST PILES IN LOCATIONS SHOWN THUS (TP-) ON THE FOUNDATION PLAN TO AN ACTUAL TIP ELEVATION OF ____. TOP ELEVATION SHALL BE 5 FT. MIN. ABOVE CUT-OFF ELEVATION.

4. BASE BID TIP ELEVATION FOR ALL PRODUCTION PILING SHALL BE THE SAME AS FOR TEST PILES. TOP ELEVATION OF PRODUCTION PILING SHALL BE AT CUT-OFF ELEVATION.

5. PRIOR TO COMMENCING PILE OPERATIONS, THE CONTRACTOR SHALL SUBMIT A PILE LOCATION PLAN SHOWING THE LOCATION & DESIGNATION OF ALL PILES. ALL DETAIL RECORDS FOR INDIVIDUAL PILES SHALL BEAR AN IDENTIFICATION. PRIOR TO PILE DRIVING, SUBMIT DATA PERTAINING TO THE PILE DRIVING HAMMER AND RIG.

6. THE BASIS FOR ACCEPTANCE OF THE PRODUCTION PILING SHALL BE:
   A. THAT THE BLOW COUNT FOR THE LAST 10 FEET OF DRIVING SHALL BE NOT LESS THAN THE BLOW COUNT RECORDED FOR THE LAST 10 FEET OF DRIVING OF THE SUCCESSFULLY DRIVEN[LOADED] TEST PILE WITH THE LEAST DRIVING RESISTANCE.

   B. THAT THE PILING MEETS THE MINIMUM PENETRATION AND ANY OTHER PRODUCTION PILE DRIVING CRITERIA.

7. ALL PILING SHALL BE DRIVEN TO A MAXIMUM TOLERANCE IN ANY DIRECTION OF THREE (3) INCHES PER PILE. WHERE AN INDIVIDUAL PILE IS DRIVEN OUT OF POSITION MORE THAN THREE (3) INCHES IN ANY DIRECTION AND/OR WHERE THE CENTER OF GRAVITY OF A PILE GROUP IS OUT OF POSITION MORE THAN TWO (2) INCHES, THE CONTRACTOR MAY BE REQUIRED TO DRIVE AN ADDITIONAL PILE OR PILES TO COMPENSATE FOR THE ECCENTRICITY OF THE PILE AND/OR PILE GROUP.

8. ALL PILE OPERATIONS, INCLUDING TEST PILES,[LOAD TEST] AND PRODUCTION PILES SHALL BE DONE UNDER THE SUPERVISION OF AN INDEPENDENT TESTING LABORATORY, DIRECTED BY A PROFESSIONAL ENGINEER REGISTERED IN THE STATE OF VIRGINIA.

9. BASED UPON THE DRIVING RESULTS OF THE TEST PILES THE ENGINEER WILL SELECT ONE (1) OF THE TEST PILES TO BE SUBJECTED TO A LOAD TEST. COMPRESSION LOAD TEST SHALL BE IN ACCORDANCE WITH ASTM D 1.143. TENSION LOAD TEST SHALL BE IN ACCORDANCE ASTM D 3689.

10. BASED UPON THE RESULTS OF THE TEST PILE DRIVING AND IN CONJUNCTION WITH THE PILE LOAD TEST RESULTS, THE ENGINEER WILL ESTABLISH THE PRODUCTION PILE DRIVING AND ACCEPTANCE CRITERIA.

11. THE TESTING AGENCY SHALL RECORD RESULTS OF ALL PILES DRIVEN, GIVING PILE HAMMER USED, PILE SIZE, LENGTH AND DRIVING RESISTANCE FOR THE ENTIRE LENGTH OF PILE, RECORDED IN BLOWS PER FOOT. DRIVING RESULTS SHALL BE REPORTED TO THE ENGINEER ON A DAILY BASIS.

12. UPON COMPLETION OF ALL PILE DRIVING, THE CONTRACTOR SHALL FURNISH THE ENGINEER A SURVEY OF AS-DRIVEN PILE LOCATIONS. THE SURVEY SHALL INDICATE THE MISALIGNMENT OF EACH PILE IN TWO PERPENDICULAR DIRECTIONS, GIVEN IN INCHES, AND THE ACTUAL CUT-OFF ELEVATION OF EACH PILE.

SLAB ON GRADE NOTES:

1. PROVIDE CONCRETE SLABS OVER A 6 MIL POLYETHYLENE VAPOR BARRIER AND 4" OF POROUS FILL AS FOLLOWS:

   MANUFACTURING AREAS- 6" SLAB REINFORCED WITH 6x6- W2.9xW2.9 WELDED WIRE FABRIC AND WITH 4000 PSI MIX CONCRETE.

   ALL OTHER AREAS- 4" SLAB REINFORCED WITH 6x6- W1.4xW1.4 WELDED WIRE FABRIC AND WITH 3000 PSI MIX CONCRETE.

   MAXIMUM SLUMP FOR ALL CONCRETE SLABS SHALL BE 5 INCHES, USING TYPE II CEMENT.

2. ALL WELDED WIRE FABRIC SHALL BE IN ACCORDANCE WITH ASTM A-185. LAP ADJOINING PIECES AT LEAST ONE FULL MESH.

3. ALL POROUS FILL MATERIAL SHALL BE A CLEAN GRANULAR MATERIAL WITH 100% PASSING A 1-1/2" SIEVE AND NO MORE THAN 5% PASSING A NO. 4 SIEVE. POROUS FILL SHALL BE COMPACTED TO 95% MAX. DRY DENSITY PER ASTM D-698.

4. SLAB JOINTS SHALL BE FILLED WITH APPROVED MATERIAL. THIS SHOULD TAKE PLACE AS LATE AS POSSIBLE, PREFERABLY 4 TO 6 WEEKS AFTER THE SLAB HAS BEEN CAST. PRIOR TO FILLING, REMOVE ALL DEBRIS FROM THE SLAB JOINTS, THEN FILL IN ACCORDANCE WITH THE MANUFACTURER'S RECOMMENDATIONS AS FOLLOWS:

   6" SLABS - FILL WITH EPOXY RESIN
   OTHER SLABS - FILL WITH FIELD MOLDED OR ELASTOMERIC SEALANT

5. UNLESS OTHERWISE APPROVED, ALL WELDED WIRE FABRIC SHALL BE BLOCKED INTO THE POSITION INDICATED WITH PRECAST CONCRETE BLOCKS HAVING A COMPRESSIVE STRENGTH EQUAL TO THAT OF THE SLAB.

6. WALKWAYS AND OTHER EXTERIOR SLABS ARE NOT INDICATED ON THE STRUCTURAL DRAWINGS. SEE THE SITE PLAN AND ARCHITECTURAL DRAWINGS FOR LOCATIONS, DIMENSIONS, ELEVATIONS, JOINTING DETAILS AND FINISH DETAILS. PROVIDE 4" WALKS REINFORCED WITH 6x6 - W1.4xW1.4 WWF UNLESS OTHERWISE NOTED.

7. SLABS TO BE PERMANENTLY EXPOSED TO WEATHER SHALL BE AIR ENTRAINED TO 5% (± 1%) WITH AN ADMIXTURE THAT CONFORMS TO ASTM C-260.

8. ALL CONCRETE WORK SHALL CONFORM TO THE REQUIREMENTS OF ACI 301, "SPECIFICATION FOR STRUCTURAL CONCRETE BUILDINGS". HOT WEATHER CONCRETING SHALL BE IN ACCORDANCE WITH ACI 305. COLD WEATHER CONCRETING SHALL BE IN ACCORDANCE WITH ACI 306.

9. IN ORDER TO AVOID CONCRETE SHRINKAGE CRACKING, PLACE CONCRETE SLABS IN AN ALTERNATING LANE (OR CHECKERBOARD) PATTERN. THE MAXIMUM LENGTH OF SLAB CAST IN ANY ONE CONTINUOUS POUR IS RECOMMENDED TO BE LESS THAN 100 FEET. THE MAXIMUM SPACING OF JOINTS SHALL BE 25'.

10. THE ALTERNATE WIRES OF THE WELDED WIRE FABRIC MUST BE PRECUT AT THE SLAB CONTRACTION JOINT LOCATIONS TO CREATE A "WEAKENED PLANE". WITHOUT CUTTING THE ALTERNATE WIRES, THE STRENGTH OF THE WIRE WILL PREVENT THE SLAB FROM CRACKING (SEPARATING) AT THE JOINT AND THE SLAB MAY BEGIN TO CRACK ELSEWHERE.

11. THE USE OF POLYPROPYLENE FIBERS (IN LIEU OF WELDED WIRE FABRIC) IS PROHIBITED WITHOUT THE WRITTEN AUTHORIZATION OF THE ENGINEER.

12. SEE THE ARCHITECTURAL DRAWINGS FOR EXACT LOCATIONS OF DEPRESSED SLAB AREAS AND DRAINS. SLOPE SLAB TO DRAINS WHERE SHOWN.

SLAB ON GRADE NOTES (CONTINUED):

13. SLABS HAVE BEEN DESIGNED IN ACCORDANCE WITH THE FOLLOWING:

    SLAB THICKNESS- "SLAB THICKNESS DESIGN FOR INDUSTRIAL CON-
    CRETE FLOORS ON GRADE", BY ROBERT PACKARD, PORTLAND CEMENT
    ASSOCIATION, 1976.

    REINFORCING- "DESIGN PROCEDURES FOR INDUSTRIAL SLABS", INTERIM
    REPORT (1973), THE WIRE REINFORCEMENT INSTITUTE.

14. SLABS HAVE BEEN DESIGNED BASED ON THE FOLLOWING CRITERIA:

    SUBGRADE MODULUS, K=200 pci (ASSUMED)
    UNIFORM LIVE LOADING = 350 PSF
    MAX. CONCENTRATED WHEEL LOADING = 3000 POUNDS
    MAX. CONCENTRATED POST LOADING = 10,000 POUNDS
    POST SPACING = XX
    POST BASEPLATE SIZE = XX

    THE ENGINEER SHALL NOT BE RESPONSIBLE FOR DIFFERENTIAL SETTLE-
    MENT, SLAB CRACKING OR OTHER FUTURE DEFECTS RESULTING FROM
    UNREPORTED CONDITIONS MITIGATING THE ABOVE ASSUMED CRITERIA.

15. THE FINISH TOLERANCE OF ALL SLABS SHALL BE IN ACCORDANCE WITH
    ACI 301, TYPE A.

16. SLABS SHALL BE CONSTRUCTED IN ACCORDANCE WITH THE FOLLOWING
    FLATNESS/LEVELNESS REQUIREMENTS:

SLAB CATEGORY	SPECIFIED	LOCAL MINIMUM
BULL-FLOATED	$F_F=15$, $F_L=13$	$F_F=13$, $F_L=10$
STRAIGHT EDGED	$F_F=20$, $F_L=15$	$F_F=15$, $F_L=10$
FLAT	$F_F=30$, $F_L=20$	$F_F=15$, $F_L=10$
VERY FLAT	$F_F=50$, $F_L=30$	$F_F=25$, $F_L=15$
SUPERFLAT	$F_F=100+$, $F_L=50+$	— —

    FLOOR FLATNESS AND LEVELNESS TESTS SHALL BE CONDUCTED BY THE
    OWNER IN ACCORDANCE WITH ASTM E 1155. RESULTS, INCLUDING
    ACCEPTANCE OR REJECTION OF THE WORK WILL BE PROVIDED TO THE
    CONTRACTOR WITHIN 48 HOURS AFTER DATA COLLECTION. REMEDIES FOR
    OUT OF TOLERANCE WORK SHALL BE IN ACCORDANCE WITH THE SPECIFICA-
    TIONS.

PRECAST/PRESTRESSED CONCRETE NOTES:

1. THE DESIGN, FABRICATION & ERECTION OF ALL PRECAST/PRESTRESSED CONCRETE (P/C) SHALL BE THE RESPONSIBILITY OF THE P/C MANUFACTURER.

2. P/C MEMBERS SHALL BE DESIGNED BY THE MANUFACTURER IN ACCORDANCE WITH ACI 318-95 AND PCI MNL-116, FOR THE LOADS INDICATED PER THE "DESIGN CRITERIA NOTES". AS WELL AS FOR ALL HANDLING AND ERECTION LOADINGS.

3. THE P/C MANUFACTURER SHALL BE A PCI CERTIFIED PLANT AND SHALL MAINTAIN DETAILED FABRICATION AND QUALITY CONTROL PROCEDURES AS REQUIRED TO SATISFY THE EXCEPTION IN BOCA 1996, SECTION 1705.2.2.

4. THE P/C MANUFACTURER SHALL SUBMIT CALCULATIONS AND SHOP DRAWINGS, BEARING THE SIGNED AND DATED SEAL OF A PROFESSIONAL ENGINEER REGISTERED IN THE STATE WHERE MANUFACTURED, FOR ALL P/C MEMBERS. THE SHOP DRAWINGS SHALL INCLUDE ALL REINFORCING, INSERTS, BEARING PADS, OPENINGS, BEARING PLATES AND ANCHORS.

5. ALL P/C CONCRETE SHALL HAVE A MINIMUM 28 DAY COMPRESSIVE STRENGTH OF 5000 PSI. MINIMUM COMPRESSIVE STRENGTH AT TRANSFER OF PRESTRESSING FORCE SHALL BE 3500 PSI. CONCRETE PERMANENTLY EXPOSED TO EARTH OR WEATHER SHALL BE AIR-ENTRAINED TO 5% (± 1%) WITH AN ADMIXTURE THAT CONFORMS TO ASTM C 260.

6. ALL NON-SHRINK GROUT SHALL HAVE A MINIMUM 28 DAY COMPRESSIVE STRENGTH OF 7500 PSI AND SHALL CONFORM TO ASTM C 1107.

7. ALL CONCRETE TOPPINGS SHALL HAVE A MINIMUM 28 DAY COMPRESSIVE STRENGTH OF 3000 PSI, UNLESS OTHERWISE REQUIRED BY THE P/C MANUFACTURER. CONCRETE TOPPING SHALL HAVE A MAXIMUM AGGREGATE SIZE OF 1/4".

8. THE WELDED-WIRE-FABRIC (W.W.F.) INDICATED ON THE DRAWINGS IS INTENDED FOR CRACK CONTROL PURPOSES ONLY. THE P/C MANUFACTURER SHALL PROVIDE ADDITIONAL W.W.F. OR REBARS IN THE TOPPING OR P/C MEMBER AS REQUIRED FOR STRUCTURAL LOADINGS.

9. UNLESS GREATER STRENGTH IS REQUIRED BY THE P/C MANUFACTURER, ALL REINFORCING MATERIALS SHALL CONFORM TO THE FOLLOWING:

REINFORCING	ASTM	MIN. STRENGTH
PRESTRESSING STRAND (UNCOATED 7 WIRE, STRESS RELIEVED)	A 416	250 KSI (ULT)
REINFORCING BARS	A 615	60 KSI
WELDED WIRE FABRIC	A 185	70 KSI

10. THE P/C MANUFACTURER SHALL PROVIDE MINIMUM CLEAR COVER TO REINFORCING IN ACCORDANCE WITH ACI 318.

11. CONNECTION DETAILS SHOWN ARE SCHEMATIC ONLY. ALL CONNECTION DETAILS SHALL BE DESIGNED BY THE P/C MANUFACTURER TO SUIT THE SPECIFIED LOADINGS. ALL CONNECTIONS SHALL ACCOUNT FOR THERMAL MOVEMENT, SHRINKAGE AND CREEP OF THE P/C MEMBERS.

12. ALL WELD PLATES, INSERTS, ANCHOR BOLTS, WELDING, LIFTING HARDWARE, GROUT SLEEVES, ETC. SHALL BE DESIGNED AND PROVIDED BY THE P/C MANUFACTURER. UNLESS OTHERWISE NOTED, ALL CONNECTIONS EXPOSED TO EARTH OR WEATHER SHALL BE HOT DIP GALVANIZED IN ACCORDANCE WITH ASTM A 153.

PRESTRESSED HOLLOW CORE PLANK NOTES:

1. THE DESIGN, FABRICATION & ERECTION OF ALL PRECAST/PRESTRESSED HOLLOW CORE CONCRETE SLABS (H/C) SHALL BE THE RESPONSIBILITY OF THE H/C MANUFACTURER.

2. H/C MEMBERS SHALL BE DESIGNED BY THE MANUFACTURER IN ACCORDANCE WITH ACI 318-95 AND PCI MNL-116, FOR THE LOADS INDICATED PER THE "DESIGN CRITERIA NOTES", AS WELL AS FOR ALL HANDLING AND ERECTION LOADINGS.

3. THE H/C MANUFACTURER SHALL BE A PCI CERTIFIED PLANT AND SHALL MAINTAIN DETAILED FABRICATION AND QUALITY CONTROL PROCEDURES AS REQUIRED TO SATISFY THE EXCEPTION IN BOCA 1996, SECTION 1705.2.2.

4. THE H/C MANUFACTURER SHALL SUBMIT CALCULATIONS AND SHOP DRAWINGS, BEARING THE SIGNED AND DATED SEAL OF A PROFESSIONAL ENGINEER REGISTERED IN THE STATE WHERE MANUFACTURED, FOR ALL H/C PLANKS, INSERTS, BEARING PADS/PLATES, OPENINGS AND ANCHORS.

5. ALL H/C CONCRETE SHALL HAVE A MINIMUM 28 DAY COMPRESSIVE STRENGTH OF 5000 PSI. MINIMUM COMPRESSIVE STRENGTH AT TRANSFER OF PRESTRESSING FORCE SHALL BE 3500 PSI. CONCRETE PERMANENTLY EXPOSED TO EARTH OR WEATHER SHALL BE AIR-ENTRAINED TO 5% (± 1%) WITH AN ADMIXTURE THAT CONFORMS TO ASTM C 260.

6. ALL NONSHRINK GROUT SHALL HAVE A MINIMUM 28 DAY COMPRESSIVE STRENGTH OF 7500 PSI AND SHALL CONFORM TO ASTM C 1107.

7. ALL CONCRETE TOPPINGS SHALL HAVE A MINIMUM 28 DAY COMPRESSIVE STRENGTH OF 3000 PSI, UNLESS OTHERWISE REQUIRED BY THE H/C MANUFACTURER. CONCRETE TOPPING SHALL HAVE A MAXIMUM AGGREGATE SIZE OF 1/4".

8. THE WELDED WIRE FABRIC (W.W.F.) INDICATED ON THE DRAWINGS IS INTENDED FOR CRACK CONTROL PURPOSES ONLY. THE H/C MANUFACTURER SHALL PROVIDE ADDITIONAL W.W.F. OR REBARS IN THE TOPPING OR P/C MEMBER AS REQUIRED FOR STRUCTURAL LOADINGS.

9. UNLESS GREATER STRENGTH IS REQUIRED BY THE H/C MANUFACTURER, ALL REINFORCING MATERIALS SHALL CONFORM TO THE FOLLOWING:

REINFORCING	ASTM	MIN. STRENGTH
PRESTRESSING STRAND (UNCOATED 7 WIRE, STRESS RELIEVED)	A 416	250 KSI (ULT)
REINFORCING BARS	A 615	60 KSI
WELDED WIRE FABRIC	A 185	70 KSI

10. THE H/C MANUFACTURER SHALL PROVIDE MINIMUM CLEAR COVER TO REINFORCING IN ACCORDANCE WITH ACI 318.

11. ALL WELD PLATES, INSERTS, ANCHOR BOLTS, WELDING, LIFTING HARDWARE, GROUT SLEEVES, ETC. SHALL BE DESIGNED AND PROVIDED BY THE H/C MANUFACTURER. UNLESS OTHERWISE NOTED, ALL CONNECTIONS EXPOSED TO EARTH OR WEATHER SHALL BE HOT DIP GALVANIZED IN ACCORDANCE WITH ASTM A 153.

12. OPENINGS FOR MECHANICAL AND ELECTRICAL ITEMS SHALL BE CORE DRILLED THROUGH HOLLOW CELLS ONLY, IN ACCORDANCE WITH THE H/C MANUFACTURER'S RECOMMENDATIONS. ADDITIONAL REINFORCEMENT SHALL BE PROVIDED AS REQUIRED BY THE H/C MANUFACTURER.

13. ALL H/C UNITS SUPPORTED BY MASONRY OR CONCRETE SHALL BEAR ON 1/8" CONTINUOUS STRIPS OF HARDBOARD TEMPERED MASONITE.

14. THE CONTRACTOR SHALL PROVIDE HOLES OR OTHER PROTECTIVE MEANS TO ALLOW THE HOLLOW CELLS TO DRAIN ANY WATER ACCUMULATION THAT MAY OCCUR DURING THE PROCESS OF CONSTRUCTION.

CAST-IN-PLACE CONCRETE NOTES:

1. CONCRETE MIXES SHALL BE DESIGNED PER ACI 301, USING PORTLAND CEMENT CONFORMING TO ASTM C-150 OR C-595, AGGREGATE CONFORMING TO ASTM C-33, AND ADMIXTURES CONFORMING TO ASTM C-494, C-1017, C-618, C-989 AND C-260. CONCRETE SHALL BE READY-MIXED IN ACCORDANCE WITH ASTM C-94.

2. CONCRETE SHALL CONFORM TO THE FOLLOWING COMPRESSIVE STRENGTH, SLUMP AND WATER/CEMENT RATIO REQUIREMENTS:

CONCRETE	MIN. F'c (28 DAYS)	SLUMP*	W/C RATIO
COLUMNS	4000 PSI	2" TO 4"	.46
ELEVATED SLABS	3000 PSI	2" TO 4"	.46
CONCRETE NOT NOTED	3000 PSI	2" TO 4"	.50
FOUNDATION	"SEE FDN NOTES"	2" TO 4"	.50
SLABS-ON-GRADE	SEE "SLAB-ON-GRADE NOTES"		.50

   * AT CONTRACTOR'S OPTION, AN APPROVED ADMIXTURE MAY BE USED TO PRODUCE FLOWABLE CONCRETE. MAXIMUM SLUMP SHALL NOT EXCEED 10 INCHES. THE CONTRACTOR SHALL SUBMIT TEST RESULTS OF THE PROPOSED CONCRETE MIXES ALONG WITH THE MANUFACTURER'S TECHNICAL DATA FOR APPROVAL PRIOR TO POURING CONCRETE.

3. ALL CONCRETE WORK SHALL CONFORM TO THE REQUIREMENTS OF ACI 301, "SPECIFICATION FOR STRUCTURAL CONCRETE BUILDINGS". HOT WEATHER CONCRETING SHALL BE IN ACCORDANCE WITH ACI 305. COLD WEATHER CONCRETING SHALL BE IN ACCORDANCE WITH ACI 306.

4. ALL REINFORCING STEEL SHALL CONFORM TO ASTM A-615, GRADE 60. ALL WELDING OF REINFORCING STEEL SHALL BE IN ACCORDANCE WITH AWS D1.4. EPOXY COATED REINFORCING SHALL CONFORM TO ASTM A-775.

5. ALL WELDED WIRE FABRIC (WWF) SHALL CONFORM TO ASTM A-185.

6. ALL REINFORCING STEEL SHALL BE SET AND TIED IN PLACE PRIOR TO POURING OF CONCRETE, EXCEPT THAT VERTICAL DOWELS FOR MASONRY WALL REINFORCING MAY BE "FLOATED" IN PLACE. DO NOT FIELD BEND BARS PARTIALLY EMBEDDED IN HARDENED CONCRETE UNLESS SPECIFICALLY INDICATED OR APPROVED BY THE ENGINEER.

7. REINFORCING STEEL, INCLUDING HOOKS AND BENDS, SHALL BE DETAILED IN ACCORDANCE WITH ACI 315. ALL REINFORCING STEEL INDICATED AS BEING CONTINUOUS (CONT) SHALL BE LAPPED WITH A TYPE 2 LAP SPLICE UNLESS OTHERWISE NOTED.

8. UNLESS OTHERWISE NOTED, THE FOLLOWING MINIMUM CONCRETE COVER SHALL BE PROVIDED FOR REINFORCEMENT:

   A) CONCRETE EXPOSED TO EARTH OR WEATHER:
      #6 THROUGH #18 BARS   - 2"
      #5 BAR, W31 OR D31 WIRE & SMALLER - 1 1/2"

   B) CONCRETE NOT EXPOSED TO EARTH OR WEATHER:
      WALLS, ELEVATED SLABS [& JOISTS] - 3/4"
      BEAMS AND COLUMNS   - 1 1/2"

   C) FOUNDATION CONCRETE (SEE "FOUNDATION NOTES")

9. BAR SUPPORTS AND HOLDING BARS SHALL BE PROVIDED FOR ALL REINFORCING STEEL TO INSURE MINIMUM CONCRETE COVER. BAR SUPPORTS SHALL BE PLASTIC TIPPED OR STAINLESS STEEL.

10. THE CONTRACTOR SHALL ALLOW IN THE BID AN ADDITIONAL ONE (1) TON OF REINFORCING STEEL TO BE PLACED IN THE FIELD AT THE DIRECTION OF THE ENGINEER. ANY UNUSED PORTION OF THIS ALLOWANCE SHALL BE CREDITED TO THE OWNER.

CAST-IN-PLACE CONCRETE NOTES (CONTINUED):

11. UNLESS OTHERWISE NOTED, ALL ONE WAY SLABS SHALL BE REIN-
    FORCED AS FOLLOWS:

BOTTOM REINFORCING	#4 @ 16" O.C.	(BETWEEN SUPPORTS)
TOP REINFORCING	#4 @ 12" O.C.	(CENTERED ON SUPPORTS)
TEMP. REINFORCING	#4 @ 18" O.C.	(TRANSVERSE BOTTOM)

12. UNLESS OTHERWISE NOTED, ALL CONCRETE WALLS (OTHER THAN
    RETAINING WALLS) SHALL BE REINFORCED AS FOLLOWS:

WALL THICKNESS	HORIZONTAL	VERTICAL	LOCATION
4" TO 6"	#4 @ 16" O.C.	#4 @ 18" O.C.	CENTERED
8"	#4 @ 12" O.C.	#4 @ 16" O.C.	CENTERED
10"	#4 @ 18" O.C.	#4 @ 18" O.C.	EACH FACE
12"	#4 @ 16" O.C.	#4 @ 18" O.C.	EACH FACE

13. ALL EDGES OF PERMANENTLY EXPOSED CONCRETE SURFACES SHALL BE
    CHAMFERED 3/4" UNLESS OTHERWISE NOTED.

14. THE CONTRACTOR SHALL PROVIDE THE ENGINEER WITH DOCUMENTA-
    TION THAT ALL MATERIALS CONFORM TO THE QUALITY STANDARDS
    SPECIFIED IN BOCA 1996.

15. IN ACCORDANCE WITH BOCA 1996, SPECIAL INSPECTIONS ARE
    REQUIRED FOR THE CONCRETE WORK. THE OWNER WILL HIRE THE
    SPECIAL INSPECTOR TO PERFORM ALL REQUIRED SPECIAL INSPECTIONS.

16. IN ORDER TO AVOID CONCRETE SHRINKAGE CRACKING, PLACE CONCRETE
    SLABS IN AN ALTERNATING LANE PATTERN. THE MAXIMUM LENGTH OF
    SLAB CAST IN ANY ONE CONTINUOUS POUR SHALL BE LIMITED TO 80 FEET.

17. FORMWORK SHALL REMAIN IN PLACE UNTIL CONCRETE HAS OBTAINED AT
    LEAST 90% OF ITS 28 DAY COMPRESSIVE STRENGTH. THE CONTRACTOR
    SHALL PROVIDE ALL SHORING AND RESHORING.

MASONRY NOTES:

1. MASONRY CONSTRUCTION SHALL CONFORM TO THE REQUIREMENTS OF THE "SPECIFICATIONS FOR MASONRY STRUCTURES (ACI 530.1-95)", PUBLISHED BY THE AMERICAN CONCRETE INSTITUTE, DETROIT, MICHIGAN.

2. HOLLOW LOAD-BEARING MASONRY UNITS SHALL CONFORM TO ASTM C-90, GRADE N-I AND BE MADE WITH LIGHTWEIGHT AGGREGATE. THE MINIMUM PRISM COMPRESSIVE STRENGTH ($f'm$) SHALL BE 1,550 PSI AT AN AGE OF 28 DAYS, AS DETERMINED BY THE UNIT STRENGTH METHOD OF ACI 530.1.

3. FILL ALL BOND BEAMS AND REINFORCED CELLS SOLIDLY WITH GROUT. GROUT SHALL CONFORM TO ASTM C-476 AND SHALL OBTAIN A MIN. 28 DAY COMPRESSIVE STRENGTH OF 2,500 PSI.

4. REINFORCING STEEL SHALL BE IN ACCORDANCE WITH ASTM A-615, GRADE 60. SHOP FABRICATE REINFORCING BARS WHICH ARE SHOWN TO BE HOOKED OR BENT. PROVIDE A MINIMUM LAP OF 48 x BAR DIAMETERS AT ALL SPLICES, UNLESS INDICATED OTHERWISE.

5. THE USE OF MASONRY-CEMENT MORTAR IS STRICTLY PROHIBITED. MORTAR SHALL CONFORM TO ASTM C-270, TYPE S. ALL MORTAR SHALL MEET THE "PROPORTION SPECIFICATION" OF ASTM C-270 AND BE MADE WITH PORTLAND CEMENT/LIME (NON AIR-ENTRAINED).

6. UNLESS OTHERWISE INDICATED, ALL WALLS SHALL BE LAID IN RUNNING BOND. BOND CORNERS AND INTERSECTIONS OF LOAD-BEARING WALLS.

7. PROVIDE VERTICAL REINFORCING BARS OF THE GIVEN SIZE AND SPACING AS INDICATED. PROVIDE BARS AT ALL WALL CORNERS, INTERSECTIONS AND OPENING EDGES. MASONRY WALLS SHALL BE CONSTRUCTED IN ACCORDANCE WITH THE "LOW-LIFT" OR "HIGH-LIFT" DETAILS INDICATED. "HIGH-LIFT" MASONRY CONSTRUCTION IS LIMITED TO SPECIALLY QUALIFIED CONTRACTORS MEETING THE FOLLOWING MINIMUM REQUIREMENTS:
    A. SUCCESSFUL COMPLETION OF AT LEAST 3 PREVIOUS PROJECTS THAT UTILIIZED "HIGH-LIFT" WALL CONSTRUCTION.
    B. CONTRACTOR SHALL SUBMIT A DETAILED "HIGH-LIFT" WALL CONSTRUCTION PROCEDURE FOR APPROVAL, INCLUDING THE DOCUMENTATION OF ALL PERSONNEL WHO HAVE SUCCESSFULLY BEEN TRAINED IN "HIGH-LIFT" MASONRY CONSTRUCTION.

8. PROVIDE REBAR DOWELS FROM FOUNDATIONS TO MATCH VERTICAL REINFORCING SIZE AND SPACING. DOWELS SHALL HAVE STANDARD 90 DEGREE HOOKS AND LAP WITH THE FIRST LIFT OF REINFORCING.

9. PROVIDE HORIZONTAL BOND BEAMS WITH CONTINUOUS REINFORCING AS INDICATED. DISCONTINUE ALL HORIZONTAL REINFORCING AT CONTROL JOINTS EXCEPT FOR THE BOND BEAMS AT BEARING ELEVATIONS. INTERMEDIATE BOND BEAMS SHALL BE PROVIDED PER THE SCHEDULE ON SHEET S-X.

10. PROVIDE STANDARD 9 GAUGE HORIZONTAL JOINT REINFORCING AT 16" ON CENTER IN ALL WALLS. PROVIDE TRUSS TYPE JOINT REINFORCING FOR ALL CONCRETE MASONRY. COORDINATE BRICK TIE BACK REQUIREMENTS WITH THE ARCHITECTURAL DRAWINGS. UNLESS OTHERWISE NOTED, STOP ALL HORIZONTAL JOINT REINFORCING AT CONTROL JOINTS.

11. PROVIDE BOND BEAM LINTELS AND BRICK SHELF ANGLES ABOVE ALL WALL OPENINGS PER TYPICAL DETAILS ON SHEET S- . SEE THE ARCHITECTURAL DRAWINGS FOR LOCATIONS OF ALL DOOR AND WINDOW OPENINGS.

12. PROVIDE STEEL JOIST AND BEAM BEARING PLATES AND OTHER ACCESSORIES AS INDICATED. PROVIDE 3 COURSES OF SOLIDLY GROUTED CMU BELOW ALL BEAM BEARINGS OVER A WIDTH OF 2'-8", CENTERED ON THE WALL, PER TYP. BEAM BEARING DETAIL ON SHT. S-

13. PROVIDE CMU CONTROL JOINTS AS INDICATED, WITH ADDITIONAL JOINTS SUCH THAT THE SPACING BETWEEN JOINTS DOES NOT EXCEED A SPACING OF 3 x WALL HEIGHT (35 FEET MAXIMUM). WHERE BEAMS OR LINTELS BEAR AT CMU CONTROL JOINTS, OFFSET & LAP THE VERTICAL RINFORCING AS INDICATED.

14. THE MASONRY CONTRACTOR SHALL PROVIDE ALL REQUIRED TEMPORARY WALL BRACING DURING CONSTRUCTION (SEE 'GENERAL STRUCTURAL NOTES').

LIGHT GAUGE METAL PURLIN FRAMING NOTES:

1. ALL COLD FORMED STEEL FRAMING MEMBERS, THEIR DESIGN, FABRICATION, AND ERECTION SHALL CONFORM TO THE "SPECIFICATION FOR THE DESIGN OF COLD-FORMED STEEL STRUCTURAL MEMBERS" OF THE A.I.S.I. (1986 ED.).

2. ALL FRAMING MEMBERS SHALL BE FORMED FROM STEEL CONFORMING TO ASTM A446 WITH A MINIMUM YIELD STRENGTH AS FOLLOWS:

   12, 14 & 16 GAUGE MEMBERS: FY=50 KSI (GRADE D)
   18 & 20 GAUGE MEMBERS: FY=33 KSI (GRADE A)

3. ALL FRAMING MEMBERS SHALL BE PAINTED PER THE MANUFACTURER'S REQUIREMENTS.

4. MEMBERS SHALL BE THE MANUFACTURER'S STANDARD "Z" SHAPED PURLINS. THE DEPTH (9 1/2") AND PRELIMINARY FLANGE WIDTH ARE INDICATED. THE FINAL PURLIN DESIGN (INCLUDING THE REQUIRED GAUGE) SHALL BE BY THE MANUFACTURER. THE REQUIRED WIND VELOCITY IS PER THE "DESIGN CRITERIA NOTES". WIND PRESSURE COEFFICIENTS SHALL BE PER SECTION 1611.8 OF BOCA 93, BASED ON EXPOSURE 'C'.

5. THE GAUGE OF ALL CONNECTIONS SHALL BE NO LIGHTER THAN THE PURLINS BEING CONNECTED. THE MANUFACTURER SHALL DESIGN ALL CONNECTIONS.

6. ALL WELDING SHALL BE IN CONFORMANCE WITH AMERICAN WELDING SOCIETY SPECIFICATION D1.3. ALL WELDS SHALL BE TOUCHED UP WITH ZINC RICH PAINT.

7. ALL STRUCTURAL MEMBERS SHALL BE PROPERLY CONNECTED TO EACH OTHER AND TO THE SUPPORTING FRAMING. FASTENINGS SHALL BE MADE WITH SELF TAPPING SCREWS OR WELDS OF SUFFICIENT SIZE TO INSURE THE CONNECTION STRENGTH.

8. THE CONTRACTOR SHALL SUBMIT THE FOLLOWING FOR APPROVAL:

   A. MANUFACTURER'S PRODUCT DATA AND LATEST TECHNICAL DATA.
   B. ERECTION DRAWINGS SHOWING THE NUMBER, TYPE, LOCATION AND SPACING OF ALL MEMBERS. ALL CONNECTIONS AND ATTACHMENTS SHALL BE CLEARLY SHOWN.
   C. THE PROPERTIES OF ALL FRAMING MEMBERS THAT ARE USED.
   D. STRUCTURAL CALCULATIONS FOR ALL CONNECTIONS NOT OTHERWISE DETAILED ON THE DRAWINGS.

LIGHT GAUGE METAL GIRT FRAMING NOTES:

1. ALL COLD FORMED STEEL FRAMING MEMBERS, THEIR DESIGN, FABRICATION, AND ERECTION SHALL CONFORM TO THE "SPECIFICATION FOR THE DESIGN OF COLD-FORMED STEEL STRUCTURAL MEMBERS" OF THE A.I.S.I. (1986 ED.).

2. ALL FRAMING MEMBERS SHALL BE FORMED FROM STEEL CONFORMING TO ASTM A446 WITH A MINIMUM YIELD STRENGTH AS FOLLOWS:

   12, 14 & 16 GAUGE MEMBERS: FY=50 KSI (GRADE D)
   18 & 20 GAUGE MEMBERS: FY=33 KSI (GRADE A)

3. ALL FRAMING MEMBERS SHALL BE PAINTED (SEE SPECIFICATIONS).

4. MEMBERS SHALL BE THE MANUFACTURER'S STANDARD "Z" SHAPED GIRTS. THE DEPTH (9 1/2") AND PRELIMINARY FLANGE WIDTH ARE INDICATED. THE FINAL GIRT DESIGN (INCLUDING THE REQUIRED GAUGE) SHALL BE BY THE GIRT MANUFACTURER. THE REQUIRED WIND CRITERIA IS PER THE "DESIGN CRITERIA NOTES". WIND DESIGN SHALL BE PER SECTION 1609.8 OF BOCA 96.

5. THE GAUGE OF ALL CONNECTIONS SHALL BE NO LIGHTER THAN THE GIRTS BEING CONNECTED. THE MANUFACTURER SHALL DESIGN ALL GIRT CONNECTIONS. SEE THE TYPICAL DETAILS ON SHEET S- .

6. ALL WELDING SHALL BE IN CONFORMANCE WITH AMERICAN WELDING SOCIETY SPECIFICATION D1.3. ALL WELDS SHALL BE TOUCHED UP WITH ZINC RICH PAINT.

7. ALL STRUCTURAL MEMBERS SHALL BE PROPERLY CONNECTED TO EACH OTHER AND TO THE SUPPORTING BACK-UP FRAMING. FASTENINGS SHALL BE MADE WITH SELF TAPPING SCREWS OR WELDS OF SUFFICIENT SIZE TO INSURE THE CONNECTION STRENGTH.

8. PROVIDE GIRTS WITH A SLIGHT DOWNWARD "BOW" AT MIDSPAN.

9. THE CONTRACTOR SHALL SUBMIT THE FOLLOWING FOR APPROVAL:

   A. MANUFACTURER'S PRODUCT DATA AND LATEST TECHNICAL DATA.
   B. ERECTION DRAWINGS SHOWING THE NUMBER, TYPE, LOCATION AND SPACING OF ALL MEMBERS. ALL CONNECTIONS AND ATTACHMENTS SHALL BE CLEARLY SHOWN.
   C. THE PROPERTIES OF ALL FRAMING MEMBERS THAT ARE USED.
   D. STRUCTURAL CALCULATIONS FOR ALL CONNECTIONS NOT OTHERWISE DETAILED ON THE DRAWINGS.

STRUCTURAL STEEL NOTES:

1. ALL STRUCTURAL STEEL SHALL CONFORM TO THE 1994 EDITION OF THE LOAD AND RESISTANCE FACTOR DESIGN (L.R.F.D) "MANUAL OF STEEL CONSTRUCTION" OF THE AISC.

2. UNLESS OTHERWISE NOTED, ALL MATERIALS SHALL BE IN ACCORDANCE WITH THE FOLLOWING ASTM SPECIFICATIONS:

MEMBER	ASTM	MIN. STRENGTH
STRUCTURAL TUBING	A500 (GRADE B)	46 KSI
STEEL PIPE	A53 (TYPE E, GR. B)	35 KSI
OTHER ROLLED PLATES/SHAPES	A36	36 KSI
CONNECTION BOLTS	A325	92 KSI
ANCHOR BOLTS	A307	—
THREADED RODS	A36	36 KSI
NONSHRINK GROUT	C1107	8000 PSI

3. ALL CONNECTIONS SHALL BE SHEAR TYPE CONNECTIONS AND DESIGNED BY THE FABRICATOR FOR THE FACTORED SHEAR FORCES INDICATED ON PLAN IN ACCORDANCE WITH THE AISC SPECIFICATIONS FOR LOAD AND RESISTANCE FACTOR DESIGN. MINIMUM BOLT DIAMETER SHALL BE 3/4". UNLESS OTHERWISE NOTED ALL BOLTS SHALL BE SHEAR/BEARING TYPE BOLTS AND BE "SNUG-TIGHT".

4. ALL WELDING SHALL BE IN ACCORDANCE WITH AWS D1.1 USING E70XX ELECTRODES. UNLESS OTHERWISE NOTED, PROVIDE CONT. MIN. SIZED FILLET WELDS PER AISC REQUIREMENTS. ALL FILLER MATERIAL SHALL HAVE A MINIMUM YIELD STRENGTH OF 58 KSI.

5. WHERE "CONTINUOUS CHORD" ANGLES ARE INDICATED, PROVIDE A CONTINUOUS BUTT WELD OR FULL PENETRATION WELD AT THE SPLICE POINTS. THE STEEL FABRICATOR MAY SUBMIT AN ALTERNATE BOLTED CONNECTION DETAIL FOR APPROVAL.

6. MOMENT CONNECTIONS ARE DENOTED THUS (▶——) ON PLAN. SEE TYPICAL DETAILS ON SHEET S- .

7. WHERE STEEL BEAMS BEAR ACROSS BUILDING EXPANSION JOINTS OR AT WALL CONTROL JOINTS, PROVIDE A "SLIP" CONNECTION PER TYPICAL DETAIL ON SHEET S- .

8. HOLES IN STEEL SHALL BE DRILLED OR PUNCHED. ALL SLOTTED HOLES SHALL BE PROVIDED WITH SMOOTH EDGES. BURNING OF HOLES AND TORCH CUTTING AT THE SITE IS NOT PERMITTED.

9. UNLESS OTHERWISE NOTED, ALL STRUCTURAL STEEL PERMANENTLY EXPOSED TO VIEW SHALL BE SHOP PAINTED WITH ONE COAT OF SSPC 15-68, TYPE I (RED OXIDE) PAINT.

10. THE STRUCTURAL STEEL ERECTOR SHALL PROVIDE ALL TEMPORARY GUYING AND BRACING (SEE 'GENERAL STRUCTURAL NOTES').

11. COLUMNS, ANCHOR BOLTS, BASE PLATES, ETC. HAVE BEEN DESIGNED FOR THE FINAL COMPLETED CONDITION AND HAVE NOT BEEN INVESTIGATED FOR POTENTIAL LOADINGS ENCOUNTERED DURING STEEL ERECTION AND CONSTRUCTION. ANY INVESTIGATION OF THE COLUMNS, ANCHOR BOLTS, BASE PLATES, ETC. FOR ADEQUACY DURING THE STEEL ERECTION AND CONSTRUCTION PROCESS IS THE SOLE RESPONSIBILITY OF THE CONTRACTOR.

12. STEEL FABRICATORS SHALL BE AN AISC CERTIFIED SHOP FOR CATEGORY I STEEL STRUCTURES AND MAINTAIN DETAILED QUALITY CONTROL PROCEDURES AS REQUIRED TO SATISFY THE SPECIAL INSPECTION REQUIREMENTS OF BOCA 1996.

13. PROVIDE GIRTS WITH A SLIGHT DOWNWARD "BOW" AT MIDSPAN.

14. UNLESS OTHERWISE NOTED, ALL STRUCTURAL STEEL PERMANENTLY EXPOSED TO THE WEATHER, INCLUDING ALL BRICK SHELF ANGLES, SHALL BE HOT-DIPPED GALVANIZED IN ACCORDANCE WITH ASTM A153.

STRUCTURAL STEEL NOTES (CONTINUED):

15. PROTECTIVE COATINGS DAMAGED DURING THE TRANSPORTING, ERECTING AND FIELD WELDING PROCESSES SHALL BE REPAIRED IN THE FIELD TO MATCH THE SHOP APPLIED COATING.

16. THE CONTRACTOR SHALL [OWNER WILL] HIRE AN INDEPENDENT TESTING AGENCY TO PROVIDE SPECIAL INSPECTIONS OF BOLTING, WELDING AND OTHER ITEMS IN ACCORDANCE WITH BOCA 1993, SECTION 1705.

17. SPECIAL OR COMPLEX CONNECTIONS THAT ARE TO BE DESIGNED BY THE FABRICATOR ARE DENOTED AS SUCH ON PLAN.
THE FABRICATOR SHALL DESIGN THESE CONNECTIONS FOR THE FORCES SHOWN AND SUBMIT CALCULATIONS AND SHOP DRAWINGS BEARING THE SIGNED AND DATED SEAL OF A PROFESSIONAL ENGINEER REGISTERED IN THE STATE OF VIRGINIA.

> NOTE: THIS IS A "SPECIAL CONNECTION" AND SHALL BE DESIGNED BY THE FABRICATOR FOR THE FORCES SHOWN.
> (SEE THE "STRUCTURAL STEEL NOTES")

18. PROVIDE ANGLE FRAMES AT ALL ROOF OPENINGS AND MECHANICAL ROOFTOP UNITS PER TYPICAL DETAILS ON SHEET S-   .

STEEL JOIST NOTES:

1. ALL STEEL JOISTS SHALL BE DESIGNED, FABRICATED AND ERECTED IN ACCORDANCE WITH SJI STANDARD SPECIFICATIONS, 1994 EDITION.

2. JOIST BRIDGING SHALL CONFORM TO SJI SPECIFICATIONS. PROVIDE DIAGONAL BRIDGING AT ALL BEAMS AND END BAYS. FIELD WELD BRIDGING AT ENDS AND INTERSECTIONS. ALL JOISTS FORTY (40) FEET AND LONGER REQUIRE A ROW OF BOLTED CROSS BRIDGING TO BE IN PLACE BEFORE SLACKENING OF HOISTING LINES.

3. ALL JOISTS SHALL BE PROPERLY ANCHORED AT BEARINGS. SEE TYPICAL DETAILS ON SHEET S-

4. JOIST BRIDGING AND CONNECTIONS SHALL BE COMPLETELY INSTALLED PRIOR TO PLACING ANY CONSTRUCTION LOADS ON THE JOISTS. CONSTRUCTION LOADING SHALL NOT EXCEED THE JOIST DESIGN LOAD.

5. ALL ROOF JOISTS SHALL BE DESIGNED FOR A NET WIND UPLIFT AS SCHEDULED. PROVIDE AN ADDITIONAL ROW OF CONTINUOUS HORIZONTAL BOTTOM CHORD BRIDGING AT THE FIRST PANEL POINT LOCATION AT EACH END OF ALL ROOF JOISTS (TO RESIST WIND UPLIFT). UPLIFT BRIDGING SHALL TERMINATE WITH DIAGONAL BRIDGING AT ALL END BAYS. JOIST FABRICATOR MAY TAKE A 1/3 ALLOWABLE STRESS INCREASE FOR WIND LOADING DESIGN PER SJI SPECIFICATIONS.

6. ALL JOISTS SHALL BE SHOP PAINTED IN ACCORDANCE WITH SJI REQUIREMENTS.

7. THE JOIST MANUFACTURER SHALL SUBMIT CALCULATIONS FOR ALL SPECIAL JOISTS TO THE ENGINEER FOR RECORD PURPOSES PRIOR TO FABRICATION. THESE CALCULATIONS SHALL BEAR THE SIGNED AND DATED SEAL OF A PROFESSIONAL ENGINEER REGISTERED IN THE STATE WHERE MANUFACTURED.

8. THE JOIST MANUFACTURER SHALL BE A SJI CERTIFIED SHOP AND MAINTAIN APPROVED FABRICATION PROCEDURES AS REQUIRED TO SATISFY THE SPECIAL INSPECTION REQUIREMENTS OF BOCA 1996.

9. AT STANDING SEAM ROOFS THE JOIST MANUFACTURER SHALL PROVIDE ADDITIONAL BRIDGING TO ADEQUATELY BRACE THE TOP CHORDS AGAINST LATERAL MOVEMENT UNDER A FULL LOADING CONDITION.

10. JOISTS ON COLUMN CENTERLINES SHALL HAVE EXTENDED BOTTOM CHORD CONNECTIONS PER TYPICAL DETAILS ON SHEET S- . DO NOT CONNECT BOTTOM CHORD EXTENSION UNTIL ALL GRAVITY DEAD LOADS ARE IN PLACE. THESE JOISTS ARE INDICATED AS "SP" ON PLAN AND SHALL BE DESIGNED BY THE JOIST MANUFACTURER FOR LIVE LOAD END MOMENTS ($0.1WL^2$) BASED ON THE INDICATED LIVE LOADS.

ROOF/NONCOMPOSITE FLOOR DECK NOTES:

1. ALL METAL DECK SHALL BE MANUFACTURED AND ERECTED IN ACCORDANCE WITH THE LATEST EDITION OF THE "DESIGN MANUAL FOR COMPOSITE DECKS, FORM DECKS AND ROOF DECKS" BY THE STEEL DECK INSTITUTE (SDI).

2. ALL ROOF DECKING SHALL BE 1 1/2" DEEP, 22 GAUGE, WIDE RIB DECK (MIN. I_x = 0.16 IN^4 / FT AND S_p= 0.180 IN^3 / FT) SPANNING PERPENDICULAR TO SUPPORTS. CONNECT WITH 5/8" DIA. PUDDLE WELDS AND MECHANICALLY FASTENED SIDELAPS PER THE "TYPICAL ROOF DECK ATTACHMENT DETAIL".

3. ALL FLOOR DECKING SHALL BE 9/16" x 28 GAUGE FORM DECK (MIN. I_x = 0.011 IN^4/ FT AND S_x = 0.035 IN^3 / FT) WITH A 3" NORMAL WEIGHT CONCRETE SLAB REINFORCED WITH 6x6-W1.4xW1.4 WELDED WIRE FABRIC (W.W.F.). FASTEN METAL FLOOR DECK TO STEEL SUPPORTS WITH 5/8" DIA. PUDDLE WELDS AT 12" O/C. FASTEN FLOOR DECK SIDELAPS IN ACCORDANCE WITH THE SUPPLIER'S RECOMMENDATIONS.

4. ALL WELDED WIRE FABRIC SHALL BE IN ACCORDANCE WITH ASTM A185, CENTERED IN SLAB. CONCRETE SHALL HAVE A MINIMUM 28 DAY COMPRESSIVE STRENGTH OF 3000 PSI. CONCRETE SHALL HAVE A MAX. SLUMP OF 4" AND MAX. AGGREGATE SIZE OF 1/2" UNLESS OTHERWISE NOTED.

5. ALL METAL DECK WELDING SHALL BE IN ACCORDANCE WITH AMERICAN WELDING SOCIETY SPECIFICATION D1.3. PROVIDE WELDING WASHERS FOR ALL FLOOR DECK WELDS.

6. SUSPENDED CEILINGS, LIGHT FIXTURES, DUCTS AND OTHER PERMANENT SUSPENDED LOADS SHALL NOT BE SUPPORTED BY THE METAL DECKING.

7. ALL ROOF DECKING SHALL BE PAINTED [GALVANIZED]. FLOOR DECKING SHALL BE GALVANIZED. ALL DECK WELDS SHALL BE TOUCHED UP WITH PAINT (GALVANIZING REPAIR PAINT FOR GALVANIZED DECKS).

8. SUBMIT DETAILED SHOP DRAWINGS PRIOR TO FABRICATION SHOWING LAYOUT, TYPES OF METAL DECK UNITS, CONNECTION DETAILS, ACCESSORIES AND OTHER RELATED ITEMS.

9. ROOF DECK SIDELAPS SHALL BE ATTACHED AT ENDS OF CANTILEVERS AND AT A MAXIMUM SPACING OF 12" O/C FROM CANTILEVERED ROOF DECK ENDS. THE ROOF DECK MUST BE COMPLETELY ATTACHED TO THE SUPPORTS AND AT THE SIDELAPS BEFORE ANY LOAD IS APPLIED TO THE CANTILEVER.

STEEL JOIST GIRDER NOTES:

1. ALL JOIST GIRDERS SHALL BE DESIGNED, FABRICATED AND ERECTED IN ACCORDANCE WITH SJI STANDARD SPECIFICATIONS, LATEST EDITION.

2. ALL ROOF GIRDERS SHALL BE DESIGNED FOR A NET WIND UPLIFT OF 30 PSF. JOIST GIRDER MANUFACTURER SHALL PROVIDE BOTTOM CHORD ANGLE BRACING AS REQUIRED TO RESIST WIND UPLIFT.

3. GIRDERS ON COLUMN CENTERLINES SHALL HAVE EXTENDED BOTTOM CHORD CONNECTIONS AS DETAILED. DO NOT CONNECT THESE BOTTOM CHORD EXTENSION UNTIL ALL GRAVITY (DEAD) LOADS ARE IN PLACE. THESE STEEL JOIST GIRDERS ARE INDICATED AS "SP" ON PLAN AND SHALL BE DESIGNED BY THE GIRDER MANUFACTURER FOR END FORCES AS SHOWN ON PLAN, IN COMBINATION WITH THE DESIGN PANEL POINT LOADING.

4. THE MANUFACTURER SHALL SUBMIT CALCULATIONS FOR ALL JOIST GIRDERS TO THE ENGINEER FOR RECORD PURPOSES PRIOR TO FABRICATION. THESE CALCULATIONS SHALL BEAR THE SIGNED AND DATED SEAL OF A PROFESSIONAL ENGINEER REGISTERED IN THE STATE WHERE MANUFACTURED.

5. THE GIRDER MANUFACTURER SHALL BE A SJI CERTIFIED SHOP AND MAINTAIN APPROVED FABRICATION PROCEDURES AS REQUIRED TO SATISFY THE SPECIAL INSPECTION REQUIREMENTS OF BOCA 1996.

6. ALL GIRDERS SHALL BE SHOP PAINTED IN ACCORDANCE WITH SJI REQUIREMENTS.

7. GIRDER MANUFACTURER MAY TAKE A 1/3 ALLOWABLE STRESS INCREASE FOR WIND LOADING IN ACCORDANCE WITH SJI SPECIFICATIONS.

COMPOSITE DECKING SYSTEM NOTES:

1. THE FLOOR DECKING OF THIS BUILDING HAS BEEN DESIGNED IN ACCORDANCE WITH THE AISC ALLOWABLE STRESS DESIGN SPECIFICATIONS FOR UNSHORED, PARTIAL COMPOSITE CONSTRUCTION.

2. ALL COMPOSITE STEEL DECKING SHALL BE:
   A. TYPE 1.5 VL, 20 GAUGE, AS MANUFACTURED BY VULCRAFT, OR APPROVED EQUAL.
   B. DESIGNED IN ACCORDANCE WITH THE COMPOSITE STEEL FLOOR DECK SPECIFICATIONS OF THE STEEL DECK INSTITUTE (SDI).
   C. FABRICATED FROM STEEL SHEET CONFORMING TO ASTM A611 OR A446 (OR EQUAL) HAVING A MINIMUM YIELD STRENGTH OF 33 KSI.
   D. GALVANIZED BY THE HOT-DIP PROCESS CONFORMING TO ASTM A525 CLASS G60 OR G90.
   E. ERECTED IN ACCORDANCE WITH SDI SPECIFICATIONS. ANCHOR DECKING AS INDICATED. MINIMUM BEARING OF THE DECK SHALL BE 1 1/2" UNLESS OTHERWISE NOTED.
   F. WELDED IN ACCORDANCE WITH AWS D1.4.

3. PROVIDE STEEL DECKING ACCESSORIES, SUCH AS CLOSURES, POUR STOPS AND FILLERS AS INDICATED. ACCESSORIES SHALL BE 20 GAUGE UNLESS OTHERWISE NOTED, AND CONFORM TO ITEMS 2.B THRU 2.F ABOVE.

4. ALL CONCRETE DECKING SHALL BE:
   A. NORMALWEIGHT WITH A 28 DAY COMPRESSIVE STRENGTH, f'c EQUAL TO 3,000 PSI (UNIT WEIGHT = 145 PCF). TOTAL SLAB THICKNESS FROM UNDERSIDE OF METAL DECK TO TOP OF FINISHED SLAB SURFACE SHALL BE 4".
   B. REINFORCED WITH ONE LAYER OF 6x6-W1.4xW1.4 WELDED WIRE FABRIC LOCATED 1" CLEAR FROM THE TOP OF THE SLAB.
   C. IN ACCORDANCE WITH ACI 318 AND 301 AND HAVE A MAXIMUM SLUMP OF 4 INCHES. ADMIXTURES CONTAINING CHLORIDES ARE PROHIBITED.

5. ALL SHEAR STUD CONNECTORS SHALL BE:
   A. 3/4" DIAMETER X 3" LONG (AFTER WELDING) HEADED STUDS WITH A MINIMUM CAPACITY OF 9.9 KIPS PER CONNECTOR.
   B. AS MANUFACTURED BY NELSON STUD WELDING COMPANY, OR APPROVED EQUAL.
   C. FABRICATED FROM COLD-FINISHED CARBON STEEL PER ASTM A108 GRADES C1015 THROUGH C1020.
   D. SPACED AS INDICATED ON THE TYPICAL DETAILS. THE TOTAL NUMBER OF STUDS FOR EACH BEAM IS INDICATED ON THE "COMPOSITE BEAM SCHEDULE".
   E. WELDED IN ACCORDANCE WITH AWS D1.1, SECTION F. STUDS MAY BE WELDED THROUGH THE DECK OR DIRECTLY TO THE STEEL MEMBER.

6. ALL STEEL BEAM CONNECTIONS SHALL BE IN ACCORDANCE WITH THE "STRUCTURAL STEEL NOTES", DESIGNED FOR THE REACTIONS INDICATED IN THE "COMPOSITE BEAM SCHEDULE".

COLD FORMED STEEL FRAMING NOTES:

1. ALL COLD FORMED STEEL FRAMING MEMBERS, THEIR DESIGN, FABRICATION, AND ERECTION SHALL CONFORM TO THE "SPECIFICATION FOR THE DESIGN OF COLD-FORMED STEEL STRUCTURAL MEMBERS" OF THE A.I.S.I. (1986 ED.).

2. ALL FRAMING MEMBERS SHALL BE FORMED FROM STEEL CONFORMING TO ASTM A446 WITH A MINIMUM YIELD STRENGTH AS FOLLOWS:

   12, 14 & 16 GAUGE MEMBERS: FY=50 KSI (GRADE D)
   18 & 20 GAUGE MEMBERS: FY=33 KSI (GRADE A)

3. ALL FRAMING MEMBERS SHALL BE GALVANIZED WITH A G-60 COATING MEETING THE REQUIREMENTS OF ASTM A525.

4. MEMBERS SHALL BE THE MANUFACTURER'S STANDARD "C" SHAPED STUDS/JOISTS OF THE SIZE, FLANGE WIDTH, AND GAUGE INDICATED. ALL MEMBERS SHALL HAVE A MINIMUM FLANGE LIP RETURN OF 1/2" AND SATISFY THE MINIMUM PROPERTIES AS PER "DIETRICH INDUSTRIES", OR APPROVED EQUAL.

5. THE GAUGE OF ALL TRACKS SHALL BE NO LIGHTER THAN THE FRAMING BEING CONNECTED. UNLESS OTHERWISE INDICATED, CONNECT TRACKS TO CONCRETE WITH 0.205" DIA. POWER DRIVEN FASTENERS (WITH 1.25" EMBEDMENT) AT 16" ON CENTER.

6. ALL WELDING SHALL BE IN CONFORMANCE WITH AMERICAN WELDING SOCIETY SPECIFICATION D1.3. ALL WELDS SHALL BE TOUCHED UP WITH ZINC RICH PAINT.

7. ALL STRUCTURAL MEMBERS SHALL BE PROPERLY CONNECTED TO EACH OTHER AND TO THE SUPPORTING BACK-UP FRAMING. FASTENINGS SHALL BE MADE WITH SELF TAPPING SCREWS OR WELDS OF SUFFICIENT SIZE TO INSURE THE CONNECTION STRENGTH. UNLESS OTHERWISE NOTED, CONNECT ALL MEMBERS BASED ON THE FOLLOWING LOADINGS:

   JOISTS/RAFTERS - DEAD LOAD AND LIVE LOAD PER THE "DESIGN CRITERIA NOTES"
   RAFTERS - NET WIND UPLIFT OF 35 PSF
   STUDS LESS THAN 8 FEET LONG- 45 PSF EXTERIOR (5 PSF INTERIOR)
   STUDS BETWEEN 8 AND 10 FEET LONG- 40 PSF EXTERIOR (5 PSF INTERIOR)
   STUDS BETWEEN 10.5 AND 16 FEET LONG- 35 PSF EXTERIOR (5 PSF INTERIOR)

8. PROVIDE BRIDGING FOR STUDS, JOISTS AND RAFTERS AT MIDSPAN AND AT A MAXIMUM SPACING NOT TO EXCEED 6'-0". ALL BRIDGING SHALL BE INSTALLED PRIOR TO THE ADDITION OF ANY LOADING. CONNECT BRIDGING TO EACH MEMBER BY WELDING, CLIP ANGLES OR OTHER APPROVED METHOD PER THE MANUFACTUER'S REQUIREMENTS.

9. PROVIDE WEB STIFFENERS AT JOIST AND RAFTER BEARINGS IN ACCORDANCE WITH THE MANUFACTURER'S REQUIREMENTS.

10. ALL AXIALLY LOADED STUDS SHALL HAVE FULL BEARING AGAINST THE INSIDE TRACK WEB, PRIOR TO STUD AND TRACK ALIGNMENT. SPLICES IN AXIALLY LOADED STUDS ARE NOT PERMITTED.

11. PROVIDE THE MANUFACTURER'S STANDARD TRACK, CLIP ANGLES, BRACING, REINFORCEMENTS, FASTENERS AND ACCESSORIES AS RECOMMENDED BY THE MANUFACTURER FOR THE APPLICATION INDICATED AND AS NEEDED TO PROVIDE A COMPLETE FRAMING SYSTEM. UNLESS OTHERWISE NOTED, INSTALL THE METAL FRAMING SYSTEM IN ACCORDANCE WITH THE MANUFACTURER'S WRITTEN INSTRUCTIONS AND RECOMMENDATIONS.

12. THE CONTRACTOR SHALL SUBMIT THE FOLLOWING FOR APPROVAL:

   A. MANUFACTURER'S PRODUCT DATA AND LATEST TECHNICAL DATA.
   B. ERECTION DRAWINGS SHOWING THE NUMBER, TYPE, LOCATION AND SPACING OF ALL MEMBERS. ALL CONNECTIONS AND ATTACHMENTS SHALL BE CLEARLY SHOWN.
   C. THE PROPERTIES OF ALL FRAMING MEMBERS THAT ARE USED IN LOADBEARING APPLICATIONS, DEMONSTRATING CONFORMANCE WITH THE MINIMUM ACCEPTABLE PROPERTIES NOTED HEREIN.
   D. STRUCTURAL CALCULATIONS FOR ALL CONNECTIONS NOT OTHERWISE DETAILED ON THE DRAWINGS.

13. UNLESS OTHERWISE NOTED, PROVIDE DOUBLE JACK STUDS AT ALL BEAM BEARINGS.

PRE-ENGINEERED COLD-FORMED METAL TRUSS NOTES:

1. THE CONTRACTOR SHALL EMPLOY A STRUCTURAL ENGINEER LICENSED IN VIRGINIA WHO SHALL BE RESPONSIBLE FOR THE ACTUAL DESIGN OF ALL ASPECTS OF THE TRUSSES, INCLUDING MEMBER SIZES, GAUGES, CONNECTIONS, BRACINGS, WEB STIFFENERS, ETC.

2. METAL TRUSSES SHALL BE DESIGNED BY THE MANUFACTURER TO SUPPORT THE FOLLOWING LOADS:

   A. GRAVITY LOADING CASE
   - TOP CHORD LOADING
     LIVE LOAD - 20 PSF (ON THE HORIZONTAL PROJECTION)
     DEAD LOAD - 10 PSF (ON THE SURFACE AREA)
          - ADDITIONAL 5 PSF AT BUILT-UP FRAMING AREAS

   - BOTTOM CHORD LOADING
     ATTIC LIVE LOAD - XX PSF (PER BOCA 96, SECTION 1606.2.3)
     DEAD LOAD - 10 PSF

   B. WIND LOADING CASE (PER BOCA 96, SECTION 1609.8)
   SEE 'DESIGN CRITERIA NOTES' FOR WIND COMPONENT CRITERIA

   - TOP CHORD LOADING, ON THE SURFACE AREA
     NET UPLIFT = P - (TOP CHORD DL × .67)
     P = Pv × I × Kh × (GCp)   WITH Gp PER BOCA FIG. 1609.8.1(2)
   - BOTTOM CHORD LOADING, ON THE SURFACE AREA
     NET UPLIFT = P - (BOTTOM CHORD DL × .67)
     P = Pv × I × Kh × (GCpi)   WITH Gpi AS NOTED HEREIN

3. METAL TRUSSES SHALL BE DESIGNED BY THE CONTRACTOR IN ACCORDANCE WITH THE APPLICABLE PROVISIONS OF THE LATEST EDITION OF A.I.S.I.

4. FRAMING MATERIALS SHALL BE IN ACCORDANCE WITH THE "COLD-FORMED STEEL NOTES" THIS SHEET AND MIN. REQUIREMENTS SHOWN ON SHEET S-

5. SUBMIT COMPLETE SHOP DRAWINGS FOR ALL METAL TRUSSES SHOWING CALCULATIONS, MEMBER SIZES, GAUGE, YIELD STRENGTH, CONNECTIONS, SPAN, CAMBER, DIMENSIONS, CHORD PITCH AND DESIGN LOADINGS. SHOP DRAWINGS AND CALCULATIONS SHALL BE SUBMITTED TO THE ENGINEER AND SHALL BEAR THE SEAL OF A PROFESSIONAL ENGINEER REGISTERED IN VIRGINIA.

6. METAL TRUSSES SHALL BE ERECTED IN ACCORDANCE WITH THE MAN-UFACTURER'S REQUIREMENTS. THE CONTRACTOR SHALL PROVIDE ALL TEMPORARY AND PERMANENT BRACING AS REQUIRED FOR SAFE ERECTION AND PERFORMANCE OF THE TRUSSES.

7. METAL TRUSSES SHALL BE DESIGNED SO THAT THE MAXIMUM LIVE LOAD DEFLECTION IS LIMITED TO L/360.

PRE-ENGINEERED WOOD TRUSS NOTES:

1. WOOD TRUSSES SHALL BE DESIGNED BY THE MANUFACTURER TO SUPPORT THE FOLLOWING LOADS:

   A. GRAVITY LOADING CASE
   - TOP CHORD LOADING
     LIVE LOAD - 20 PSF (ON THE HORIZONTAL PROJECTION)
     DEAD LOAD - 10 PSF (ON THE SURFACE AREA)
       - ADDITIONAL 5 PSF AT BUILT-UP FRAMING AREAS
   - BOTTOM CHORD LOADING
     ATTIC LIVE LOAD - 20PSF (PER BOCA 1996, SECTION 1606.2.3)
     DEAD LOAD - 10 PSF

   B. WIND LOADING CASE (PER BOCA 1996, SECTION 1609.8)
   SEE 'DESIGN CRITERIA NOTES' FOR WIND COMPONENT CRITERIA
   - TOP CHORD LOADING, ON THE SURFACE AREA
     NET UPLIFT = P - (TOP CHORD DL x .67)
     P = Pv x I x Kh x (GCp)   WITH Gp PER BOCA FIG. 1609.8.1(2)
   - BOTTOM CHORD LOADING, ON THE SURFACE AREA
     NET UPLIFT = P - (BOTTOM CHORD DL x .67)
     P = Pv x I x Kh x (GCpi)   WITH Gpi AS NOTED HEREIN

2. WOOD TRUSSES SHALL BE DESIGNED BY THE MANUFACTURER IN ACCORDANCE WITH THE APPLICABLE PROVISIONS OF THE LATEST EDITION OF THE NATIONAL DESIGN SPECIFICATION OF THE NATIONAL FOREST PRODUCTS ASSOCIATION, THE DESIGN SPECIFICATION FOR METAL PLATE CONNECTED WOOD TRUSSES OF THE TRUSS PLATE INSTITUTE AND BOCA 1996.

3. WOOD MATERIALS SHALL BE SOUTHERN PINE, DOUGLAS FIR OR LARCH AND SHALL BE KILN DRIED AND USED AT 19% MAXIMUM MOISTURE CONTENT. PROVIDE GRADE NO. 2 OR AS REQUIRED TO SATISFY STRESS REQUIREMENTS.

4. CONNECTOR PLATES SHALL BE NOT LESS THAN 0.036 INCHES (20 GAUGE) IN COATED THICKNESS, SHALL MEET OR EXCEED ASTM GRADE A OR HIGHER AND SHALL BE HOT DIPPED GALVANIZED ACCORDING TO ASTM A-525 (COATING G60). MINIMUM STEEL YIELD STRESS SHALL BE 33,000 PSI.

5. TRUSSES SHALL BE FABRICATED IN A PROPERLY EQUIPPED MANUFACTURING FACILITY OF A PERMANENT NATURE. TRUSSES SHALL BE MANUFACTURED BY EXPERIENCED WORKMEN, USING PRECISION CUTTING, JIGGING AND PRESSING EQUIPMENT UNDER THE REQUIREMENTS IN QUALITY CONTROL STANDARD QST-88 OF THE TRUSS PLATE INSTITUTE.

6. SECONDARY BENDING STRESSES IN TRUSS TOP AND BOTTOM CHORDS DUE TO DEAD, LIVE AND WIND LOADS SHALL BE CONSIDERED IN THE DESIGN. LOAD DURATION FACTORS SHALL BE PER THE "NATIONAL DESIGN SPECIFICATION FOR WOOD CONSTRUCTION".

7. WOOD TRUSSES SHALL BE ERECTED IN ACCORDANCE WITH THE TRUSS MANUFACTURER'S REQUIREMENTS. THIS WORK SHALL BE DONE BY A QUALIFIED AND EXPERIENCED CONTRACTOR. TRUSS ERECTION BY AN INEXPERIENCED OR NONQUALIFIED CONTRACTOR CAN RESULT IN CONSTRUCTION COLLAPSE AND/OR SERIOUS INJURY AND DAMAGE.

8. THE CONTRACTOR SHALL PROVIDE ALL TEMPORARY AND PERMANENT BRACING AS REQUIRED FOR SAFE ERECTION AND PERFORMANCE OF THE TRUSSES. THE GUIDELINES SET FORTH BY THE TRUSS PLATE INSTITUTE PUBLICATION "HIB-91, COMMENTARY AND RECOMMENDATIONS FOR HANDLING, INSTALLING AND BRACING METAL PLATE CONNECTED WOOD TRUSSES" SHALL BE A MINIMUM REQUIREMENT.

9. TRUSS MEMBERS AND COMPONENTS SHALL NOT BE CUT, NOTCHED, DRILLED NOR OTHERWISE ALTERED IN ANY WAY WITHOUT THE WRITTEN APPROVAL OF THE ENGINEER.

10. SUBMIT COMPLETE SHOP DRAWINGS FOR ALL WOOD TRUSSES SHOWING MEMBER SIZES, SPECIES, GRADE, MOISTURE CONTENT, SPAN, CAMBER, DIMENSIONS, CHORD PITCH, BRACING REQUIREMENTS AND LOADINGS. SHOP DRAWINGS SHALL BE SUBMITTED TO THE ENGINEER AND SHALL BEAR THE SEAL OF A PROFESSIONAL ENGINEER REGISTERED IN VIRGINIA.

11. SEE THE "SCISSOR TRUSS NOTES" FOR ADDITIONAL REQUIREMENTS.

PLYWOOD/GYPBOARD SHEATHING NOTES:

1. ALL PLYWOOD CONSTRUCTION SHALL BE IN ACCORDANCE WITH THE AMERICAN PLYWOOD ASSOCIATION (APA) SPECIFICATIONS.

2. ALL ROOF PANEL SHEATHING SHALL BE 5/8" (NOM.) TYPE CDX, EXP. I APA RATED SHEATHING. SUITABLE EDGE SUPPORT SHALL BE PROVIDED BY USE OF PANEL CLIPS OR BLOCKING BETWEEN FRAMING. UNLESS OTHERWISE NOTED CONNECT ROOF SHEATHING WITH 8d COMMON NAILS AT 6" O/C AT SUPPORTED PANEL EDGES AND 6" O/C AT INTERMEDIATE SUPPORTS.

3. ALL FLOOR SHEATHING SHALL BE 3/4" (NOM.) APA RATED STURD-I-FLOOR, EXP. I, WITH TONGUE AND GROOVE EDGE. UNLESS OTHERWISE NOTED CONNECT FLOOR SHEATHING WITH 10d COMMON NAILS SPACED 6" O/C AT SUPPORTED EDGES AND 12" O/C AT INTERMEDIATE SUPPORTS. FIELD-GLUE USING ADHESIVES MEETING APA SPECIFICATION AFG-01, APPLIED IN ACCORDANCE WITH THE MANUFACTURER'S RECOMMENDATIONS.

4. ALL WALL PANEL SHEATHING SHALL BE 1/2" (NOM.) TYPE CDX, EXP. I APA RATED SHEATHING. UNLESS OTHERWISE INDICATED, CONNECT WALL SHEATHING WITH 10d COMMON NAILS SPACED 6" O/C AT SUP-PORTED PANEL EDGES AND 12" O/C AT INTERMEDIATE SUPPORTS.

5. INSTALL ALL PLYWOOD SHEATHING WITH THE LONG DIMENSION OF THE PANEL ACROSS SUPPORTS AND WITH PANEL CONTINUOUS OVER TWO OR MORE SPANS. STAGGER PANEL END JOINTS. ALLOW 1/8" SPACING AT PANEL ENDS AND EDGES UNLESS OTHERWISE RECOMMENDED BY THE SHEATHING MANUFACTURER.

6. ALL NAILING SHALL BE CAREFULLY DRIVEN AND NOT OVERDRIVEN. THE USE OF STAPLES AND PNUEMATIC NAIL GUNS ARE PROHIBITED FROM USE.

7. ALL EXT. WALLS SHALL BE SHEATHED ON BOTH FACES WITH GYP-BOARD SHEATHING (SEE ARCH. DWGS. FOR THICKNESSES) AND CONNECTED WITH 5d COOLER NAILS SPACED 7" O/C AT SUPPORTED PANEL EDGES AND INTERMEDIATE SUPPORTS.

8. PROVIDE 2x BLOCKING AT UNSUPPORTED PANEL EDGES AS FOLLOWS:
   ROOFS AND FLOORS- ONLY WHERE INDICATED ON PLAN
   WALLS- PER THE SHEARWALL SCHEDULE ON SHEET S-

WOOD FRAMING NOTES:

1. ALL WOOD FRAMING MATERIAL SHALL BE SURFACED DRY AND USED AT 19% MAXIMUM MOISTURE CONTENT. ALLOWABLE STRESS REQUIREMENTS OF ALL MATERIAL SHALL BE IN ACCORDANCE WITH THE "SCHEDULE OF REQUIRED STRESS VALUES", ON SHEET S- .

2. ALL STUD AND WALL FRAMING SHALL BE EITHER OF THE FOLLOWING:
   A. NO. 2 GRADE SOUTHERN YELLOW PINE (SYP)
   B. NO. 2 GRADE SPRUCE-PINE-FIR (SPF)

   "STUD" GRADE MATERIAL IS STRICTLY PROHIBITED FROM USE.

3. ALL JOIST, RAFTER & MISC. FRAMING SHALL BE NO. 2 GRADE, SOUTHERN PINE. PROVIDE FULL-DEPTH (OR METAL) BRIDGING AT MIDSPAN AND AT A MAXIMUM SPACING OF 8'-0" O/C IN BETWEEN.

4. ALL FRAMING EXPOSED TO THE WEATHER OR IN CONTACT WITH MASONRY OR CONCRETE SHALL BE PRESSURE-TREATED IN ACCORDANCE WITH THE AMERICAN WOOD PRESERVERS ASSOCIATION SPECIFICATIONS.

   WHERE POSSIBLE, ALL CUTS AND HOLES SHOULD BE COMPLETED BEFORE TREATMENT. CUTS AND HOLES DUE TO ON-SITE FABRICATION SHALL BE BRUSHED WITH 2 COATS OF COPPER NAPHTHENATE SOLUTION CONTAINING A MINIMUM OF 2% METALLIC COPPER IN SOLUTION (PER AWPA STD. M4).

5. THE CONTRACTOR SHALL CAREFULLY SELECT LUMBER TO BE USED IN LOADBEARING APPLICATIONS. THE LENGTH OF SPLIT ON THE WIDE FACE OF 2" NOMINAL LOADBEARING FRAMING SHALL BE LIMITED TO LESS THAN 1/2 OF THE WIDE FACE DIMENSION. THE LENGTH OF SPLIT ON THE WIDE FACE OF 3" (NOMINAL) AND THICKER LUMBER SHALL BE LIMITED TO 1/2 OF THE NARROW FACE DIMENSION.

6. ALL NAILING NOT OTHERWISE INDICATED SHALL BE IN ACCORDANCE WITH THE "NAILING SCHEDULE" ON SHEET S- . NAILING SHALL NOT BE OVERDRIVEN.

7. PROVIDE DOUBLE JOISTS UNDER ALL PARTITIONS WHICH RUN PARALLEL WITH JOISTS AND UNDER ALL CONCENTRATED LOADS FROM FRAMING ABOVE.

8. PROVIDE HEADER BEAMS OF THE SAME SIZE AS JOISTS OR RAFTERS TO FRAME AROUND OPENINGS IN THE PLYWOOD DECK UNLESS OTHERWISE INDICATED.

9. STRUCTURAL STEEL PLATE CONNECTORS SHALL CONFORM TO ASTM A-36 SPECIFICATIONS AND BE 1/4" THICK UNLESS OTHERWISE INDICATED. BOLTS CONNECTING WOOD MEMBERS SHALL BE PER ASTM A-307 AND BE 3/4" DIAMETER UNLESS OTHERWISE INDICATED. PROVIDE WASHERS FOR ALL BOLT HEADS AND NUTS IN CONTACT WITH WOOD SURFACES.

10. BOLT HOLES SHALL BE CAREFULLY CENTERED AND DRILLED NOT MORE THAN 1/16" LARGER THAN THE BOLT DIAMETER. BOLTED CONNECTIONS SHALL BE SNUGGED TIGHT BUT NOT TO THE EXTENT OF CRUSHING WOOD UNDER WASHERS.

11. PREFABRICATED "MICRO-LAM" LUMBER HEADERS AND BEAMS SHALL BE AS MANUFACTURED BY "TRUS JOIST MacMILLAN CORP.", BOISE, IDAHO OR APPROVED EQUAL. MICRO-LAM MATERIAL SHALL BE 2.OE, SOUTHERN PINE. DO NOT CUT OR NOTCH MICRO-LAM MATERIAL WITHOUT THE MANUFACTURER'S APPROVAL.

12. PREFABRICATED METAL JOIST HANGERS, HURRICANE CLIPS, HOLD-DOWN ANCHORS AND OTHER ACCESSORIES SHALL BE AS MANUFACTURED BY "SIMPSON STRONG-TIE COMPANY", TEL. 800-999-5099), OR APPROVED EQUAL. INSTALL ALL ACCESSORIES PER THE MANUFACTURER'S REQUIREMENTS. ALL STEEL SHALL HAVE A MINIMUM THICKNESS OF 0.04 INCHES (PER ASTM A446, GRADE A) AND BE GALVANIZED (COATING G60).

13. HOLES AND NOTCHES DRILLED OR CUT INTO WOOD FRAMING SHALL NOT EXCEED THE REQUIREMENTS OF BOCA 1996, SECTION 2305.3.

14. ALL PLATES, ANCHORS, NAILS, BOLTS, NUTS, WASHERS, AND OTHER MISCELLANEOUS HARDWARE SHALL BE HOT DIP GALVANIZED.

PREFABRICATED WOOD JOIST NOTES:

1. PREFABRICATED WOOD I-JOISTS SHALL BE DESIGNED AND FURNISHED IN ACCORDANCE WITH A CURRENT CODE-ACCEPTED EVALUATION REPORT. STRUCTURAL CAPACITIES AND DESIGN PROVISIONS SHALL BE ESTABLISHED AND MONITORED IN ACCORDANCE WITH ASTM D 5055.

2. WOOD JOISTS SHALL BE DESIGNED BY THE MANUFACTURER TO SUPPORT THE LOADS INDICATED PER THE JOIST LOADING DIAGRAMS. UNLESS OTHERWISE INDICATED, JOISTS SHALL BE DESIGNED FOR THE FOLLOWING:

   TOP CHORD LOADING
     LIVE LOAD - 40 PSF
     DEAD LOAD - 15 PSF

   BOTTOM CHORD LOADING
     DEAD LOAD - 5 PSF

   WIND LOAD - 35 PSF NET UPLIFT PRESSURE (ROOF JOISTS ONLY)

3. WOOD I-JOISTS SHALL BE ERECTED IN ACCORDANCE WITH THE MANUFACTURER'S REQUIREMENTS. THE CONTRACTOR SHALL PROVIDE ALL TEMPORARY AND PERMANENT BRACING AS REQUIRED FOR SAFE ERECTION AND PERFORMANCE OF THE JOISTS.

4. WOOD I-JOISTS SHALL NOT BE CUT, NOTCHED, COPED, DRILLED NOR OTHERWISE ALTERED IN ANY WAY UNLESS SPECIFICALLY CONDUCTED IN ACCORDANCE WITH THE MANUFACTURER'S WRITTEN REQUIREMENTS. DO NOT CUT FLANGES.

5. WOOD I-JOISTS SHALL BE PRODUCED BY A CODE ACCEPTED FABRICATOR WITH A MINIMUM OF FIVE (5) YEARS EXPERIENCE PRODUCING PREFABRICATED WOOD I-JOISTS. QUALITY CONTROL SHALL BE AUDITED BY AN AGENCY ACCEPTED BY THE "BUILDING OFFICIALS & CODE ADMINISTRATORS, INC.".

6. WEB PANELS MUST BE JOINED WITH A MACHINED AND GLUED VEE JOINT TO FORM A CONTINUOUS MEMBER. ALL JOINTS SHALL BE GLUED USING AN EXTERIOR TYPE ADHESIVE PER ASTM D 2559.

7. WOOD I-JOISTS SHALL BE STORED IN BUNDLES IN AN UPRIGHT POSITION AND AWAY FROM GROUND CONTACT. ANY DAMAGE TO JOISTS SHALL BE BROUGHT TO THE IMMEDIATE ATTENTION OF THE JOIST SUPPLIER. FIELD REPAIR OR MOD-IFICATION OF JOISTS MUST NOT BE MADE WITHOUT THE WRITTEN APPROVAL BY THE SUPPLIER, EXCEPT FOR TRIMMING TO CORRECT LENGTH.

8. WOOD I-JOISTS SHALL BE CAREFULLY HANDLED TO PREVENT DAMAGE AND DIS-TORTION. EACH JOIST SHALL BE ANCHORED AND BRACED, AS IT IS ERECTED, USING BLOCKING PANELS AND ANCHORAGE INDICATED (AND PER THE SUPPLIER'S REQUIREMENTS). ERECTOR SHALL PROVIDE SUPPLEMENTAL LATERAL BRACING OF THE TOP FLANGE UNTIL SHEATHING IS PROPERLY NAILED.

9. ALL ROOF JOISTS SHALL BE POSITIONED WITH THE NATURAL CAMBER TURNED UP.

SCISSOR TRUSS NOTES:

1. SEE THE ARCHITECTURAL DRAWINGS FOR THE BOTTOM CHORD PITCH.

2. TRUSS INTERIOR CONFIGURATION IS SCHEMATIC AND MAY BE MODIFIED BY THE TRUSS MANUFACTURER TO SUIT THE ACTUAL TRUSS DESIGN.

3. TRUSS MANUFACTURER SHALL DESIGN THE SCISSOR TRUSS WITH ONE END "FIXED" AND THE OTHER END FREE TO "MOVE". THE SUPPORTING STRUCTURE IS NOT DESIGNED TO PROVIDE HORIZONATAL THRUST FORCE RESISTANCE.

4. CONNECT TRUSS AT "FIXED" END AS INDICATED, USING HURRICANE CLIPS. CONNECT TRUSS AT "SLIP" END WITH SIMPSON TYPE TC26 TRUSS CONNECTOR, INSTALLING NAILS TO ALLOW TRUSS MOVEMENT (DURING CONSTRUCTION).

5. DURING CONSTRUCTION, BRACE SUPPORTING WALLS AT BOTH ENDS AND APPLY ALL DEAD LOADS (EXCEPT OMIT STRIP OF CEILING AT "SLIP" END).

6. AFTER ALL DEAD LOADS ARE IN PLACE, PROVIDE NAILS AT THE "SLIP" END SO THAT IT IS NOW "FIXED". NO ADDITIONAL HORIZONTAL SLIPPING WILL OCCUR.

7. SOME FUTURE LONG-TERM MOVEMENT WILL OCCUR, DUE TO ROOF LIVE LOADS (SNOW) AND LONG-TERM CREEP (DUE TO SUSTAINED DEAD LOADS). ANY FUTURE MOVEMENT WILL NOW CONSIST OF MOVEMENT OF THE WALL SUPPORTS.

8. TRUSS MANUFACTURER SHALL DESIGN TRUSS TO LIMIT THE LONG-TERM HORIZONTAL DEFLECTION TO BE LESS THAN 0.xx" TOTAL. LONG-TERM HORIZONTAL DEFLECTION SHALL BE CALCULATED USING 1.5 x DEAD LOAD DEFLECTION PLUS THE DEFLECTION DUE TO SNOW LOADS.

Appendix B

Structural Schedules

CONCRETE GRADE BEAM SCHEDULE

MARK	'W'	'D'	BOTT. REINF.		TOP REINF.		CLOSED TIES		
			CONT.	ADDL.	CONT.	ADDL.	SIZE	TYPE	SPACING FROM EACH END
GB-1	16"	16"	(2)-#5	—	(2)-#5	—	#3	☐	4 @ 6", REM @ 12"
GB-2	18"	24"	(2)-#7	—	(2)-#7	—	#3	☐	7 @ 6", REM @ 10"
GB-3	20"	28"	(2)-#7	(2)-#7	(2)-#5	(2)-#5	#3	☐	7 @ 6", REM @ 12"

CONCRETE STRAP BEAM SCHEDULE

MARK	'W'	'D'	CONT. BOTT.	CONT. TOP	CLOSED TIES		
					SIZE	TYPE	SPACING
SB-1	16"	16"	(2)-#6	(2)-#6	#3	☐	12"
SB-2	18"	18"	(2)-#7	(2)-#7	#3	☐	12"

ISOLATED COLUMN FOOTING SCHEDULE

MARK	SIZE	THICK.	REINFORCING	
			TRANSVERSE	LONGITUDINAL
F3-0	3'-0" × 3'-0"	12"	4-#4 × 2'-6" BOTT.	4-#4 × 2'-6" BOTT.
F3-6	3'-6" × 3'-6"	12"	3-#5 × 3'-0" BOTT.	3-#5 × 3'-0" BOTT.
F4-0	4'-0" × 4'-0"	12"	4-#5 × 3'-6" BOTT.	4-#5 × 3'-6" BOTT.
F4-6	4'-6" × 4'-6"	12"	4-#5 × 4'-0" BOTT.	4-#5 × 4'-0" BOTT.
F5-0	5'-0" × 5'-0"	12"	5-#5 × 4'-6" BOTT.	5-#5 × 4'-6" BOTT.
F5-6	5'-6" × 5'-6"	12"	5-#5 × 5'-0" BOTT.	5-#5 × 5'-0" BOTT.
F6-0	6'-0" × 6'-0"	12"	5-#5 × 5'-6" BOTT.	5-#5 × 5'-6" BOTT.
F6-6	6'-6" × 6'-6"	13"	7-#5 × 6'-0" BOTT.	7-#5 × 6'-0" BOTT.
F4×3	4'-0" × 3'-0"	12"	4-#5 × 2'-6" BOTT.	3-#5 × 3'-6" BOTT.
F5×3	5'-0" × 3'-0"	12"	5-#5 × 2'-6" BOTT.	3-#5 × 4'-6" BOTT.

WALL FOOTING SCHEDULE

MARK	WIDTH 'W'	THICK. 'T'	REINFORCING	
			CONTINUOUS	TIE BARS
WF1-8	1'-8"	12"	(3)-#4	#4 AT 6'-0" O/C
WF2-0	2'-0"	12"	(3)-#4	#4 AT 6'-0" O/C
WF2-6	2'-6"	12"	(4)-#4	#4 AT 6'-0" O/C
WF3-0	3'-0"	12"	(4)-#4	#4 AT 6'-0" O/C
WF3-6	3'-6"	12"	(5)-#4	#4 AT 6'-0" O/C

CONCRETE PEDESTAL SCHEDULE

MARK	SIZE	REINFORCING		PLAN VIEW
		CONT.	TIES	
P-1	14"x14"	(4)-#5	#3@12"	
P-2	14"x24"	(6)-#5	#3@12"	

COLUMN BASE PLATE AND ANCHOR BOLT SCHEDULE

COLUMN	PL SIZE 'B' x 'N'	PL THICK. 'T'	ANCHOR BOLTS		
			NO. & SIZE	BOLT HOLE	EMBEDMENT
TS4x4	10" x 10"	5/8"	(4) - 3/4" DIA.	1 1/16"	9"
TS8x8	18" x 18"	1 3/8"	(4) - 1" DIA.	1 1/2"	12"
TS10x10	20" x 20"	1 5/8"	(4) - 1 1/8" DIA.	1 5/8"	14"

THICKENED SLAB EDGE SCHEDULE

MARK	'W'	'D'	BOTT. REINF.	TOP REINF.	TRANSVERSE
TS-1	12"	18"	(2) - #4	(2) - #5	#3 TIES AT 24"
TS-2	16"	24"	(2) - #6	(2) - #6	#3 TIES AT 24"

CONCRETE BEAM SCHEDULE

MARK	'W'	'D'	BOTT. REINF.		TOP REINF.		CLOSED TIES		
			CONT.	ADDL.	CONT.	ADDL.	SIZE	TYPE	SPACING FROM EACH END
B-1	16"	18"	(2)-#5	—	(2)-#5	—	#3	▢	4 @ 6", REM @ 12"
B-2	16"	24"	(2)-#7	—	(2)-#7	—	#3	▢	7 @ 6", REM @ 10"
B-3	18"	28"	(2)-#7	(2)-#7	(2)-#5	(2)-#5	#3	▢	7 @ 6", REM @ 12"

ONE-WAY SLAB SCHEDULE

MARK	'T'	BOTT. BARS	TOP BARS		TEMP. BARS	REMARKS
			ENDS	INTERIOR		
S-1	8"	#4 @ 14"	#4 @ 14"	#4 @ 14"	#4 @ 14"	
S-2	8"	#4 @ 12"	#4 @ 14"	#4 @ 8"	#4 @ 14"	
S-3	—	#5 @ 12"	#4 @ 14"	#5 @ 8"	#4 @ 14"	'T' VARIES, FROM 7" MAX. TO 6" MIN.

TWO-WAY SLAB REINFORCING SCHEDULE

| BOTTOM BARS (#4) | | | | | TOP BARS (#5) | | | | | |
| COLUMN STRIP | | MIDDLE STRIP | | | COLUMN STRIP | | | MIDDLE STRIP | | |
MARK	REINF.	MARK	REINF.		MARK	CONTINUOUS	ADDED	MARK	REINF.	
CB-1	(10)-#4	MB-1	(7)-#4		CT-1	(5)-#5	(8)-#5	MT-1	(5)-#5	
CB-2	(8)-#4	MB-2	(6)-#4		CT-2	(5)-#5	(5)-#5	MT-2	(6)-#5	
CB-3	(9)-#4				CT-3	(5)-#5	(7)-#5			
					CT-4	(5)-#5	(4)-#5			

MASONRY WALL REINFORCING SCHEDULE

MARK	VERTICAL REINFORCING	HORIZONTAL BOND BEAMS		
		SIZE	REINF.	LOCATIONS
MW-1	#6 AT 40" O/C	10"W × 8"D	2-#5	EL. (0'-0") EL. (+11'-4") EL. (+23'-4")
MW-2	#6 AT 80" O/C	10"W × 8"D	2-#5	EL. (0'-0") EL. (+11'-4") EL. (+23'-4")
MW-3	#4 AT 80" O/C	8"W × 8"D	2-#4	EL. (+11'-4") HIGH J.B.E.

BRICK SHELF ANGLE SCHEDULE

CLEAR SPAN	ANGLE SIZE
UP TO 3'-4"	L4x4x1/4
3'-5" TO 5'-4"	L4X4X3/8
5'-5" TO 6'-8"	L6x4x5/16 (LLV)
6'-9" TO 8'-0"	L7x4x3/8 (LLV)
OVER 8'-0"	L5x5x3/8 BOLTED TO HDR.

LINTEL SCHEDULE

MARK	SECTION	LINTEL DESCRIPTION	BRICK SUPPORT	REMARKS
L-1		10"W × 8"D BOND BEAM REINF. W/ 2-#4 CONT.	SEE SHELF ANGLE SCHEDULE	8" MIN. BEARING EACH END
L-2		10"W × 16"D BOND BEAM REINF. W/ 2-#5 CONT. TOP & BOTTOM	SEE SHELF ANGLE SCHEDULE	8" MIN. BEARING EACH END
L-3		8"W × 8"D BOND BEAM REINF. W/ 2-#4 CONT.	—	8" MIN. BEARING EACH END
L-4		W8×18 STEEL BEAM	PL 3/8" × 15" (WHERE OCCURS)	BRG PL BP-1 EACH END

470

STEEL BEAM BEARING PLATE SCHEDULE

MARK	PLATE SIZE B x N x T	DIST'D TO CL PL	ANCHORS	EMBED. E	MAX. REACTION
BP-1	9 x 6 x 3/4	CL WALL	(2)-5/8"	10"	27 KIPS
BP-2	6 x 6 x 5/8	4"	(2)-5/8"	10"	16 KIPS
BP-3	12 x 6 x 5/8	SEE A/S-5	SEE A/S-5	SEE A/S-5	25 KIPS
BP-4	6 x 6 x 5/8	SEE B/S-5	SEE B/S-5	SEE B/S-5	5 KIPS
BP-5	6 x 6 x 5/8	SEE B/S-5	SEE B/S-5	SEE B/S-5	10 KIPS
BP-6	6 x 6 x 5/8	SEE C/S-5	SEE C/S-5	SEE C/S-5	10 KIPS

COMPOSITE BEAM SCHEDULE

MARK	BEAM	MAX. END REACTION	MIDSPAN CAMBER	NUMBER OF STUDS	REMARKS
CB-1	W10x15	27 KIPS	—	8	INTERIOR STRINGER
CB-2	W10x22	30 KIPS	—	10	INTERIOR STRINGER
CB-3	W12x19	40 KIPS	—	12	INTERIOR STRINGER
CB-4	W16x31	50 KIPS	1/2 IN.	20	INTERIOR GIRDER
CB-5	W18x35	60 KIPS	1/2 IN.	25	INTERIOR GIRDER

BRACED BAY MEMBER SCHEDULE			
LINES	THREADED ROD	TURNBUCKLE	WIND FORCE
A & E	7/8" DIA.	1 1/4" DIA.	12k (SERVICE)
4	1 1/4" DIA.	1 7/8" DIA.	31k (SERVICE)

NOTES:

1. TURNBUCKLE MANUFACTURER TO VERIFY THAT TURNBUCKLE CAPACITY EXCEEDS WIND FORCE REQUIREMENT.

2. PAINT TURNBUCKLE THREADS AFTER TIGHTENING.

COLD FORMED CONNECTION SCHEDULE	
MARK	CONNECTION DESCRIPTION
CN-1	(3) NO. 10-16 SCREWS BETWEEN STUDS (STUDS ARE BACK TO BACK)
CN-2	0.145" DIA. POWER DRIVEN FASTENER FROM CONT. TRACK TO STRUCTURAL STEEL AT EACH STUD (24" MAX.)
CN-3	0.145" DIA. POWER DRIVEN FASTENER FROM CONT. TRACK TO MASONRY WALL AT EACH STUD (24" MAX.)
CN-4	(2) NO. 7 (x 7/16" LONG) FRAMING SCREWS FROM CONT. TRACK TO FLANGE OF STUDS (AT EACH STUD)

COLD FORMED STEEL FRAMING SCHEDULE

MARK	SIZE	GAUGE	SPACING	REMARKS
W-1	6" (x1 5/8")	16 GA.	24" O/C	TYP. WALL
W-2	6" (x1 5/8")	16 GA.	16" O/C	HIGH WALL
J-1	6" (x1 5/8")	16 GA.	24" O/C	JOISTS
R-1	6" (x1 5/8")	16 GA.	24" O/C	RAFTERS

SCHEDULE OF COLD FORMED STEEL MINIMUM STRUCTURAL PROPERTIES

MEMBER SIZE (DEPTH xFLANGE)	GAUGE	AREA (IN.x2)	Ix (IN.x4)	Sx (IN.x3)
3 5/8" (x1 5/8")	18 GA.	0.283	0.721	0.398
4" (x2")	16 GA.	0.433	1.32	0.66
6" (x2")	16 GA.	0.553	3.40	1.13
8" (x2")	16 GA.	0.672	6.73	1.68
10" (x2")	16 GA.	0.792	11.57	2.31
12" (x2")	16 GA.	0.912	18.15	3.02

METAL STRAP X-BRACING SCHEDULE

MARK	DESCRIPTION	REMARKS
XB-1	3", 14 GA. STRAP EACH SIDE	FULL WALL HT.
XB-2	2", 16 GA. STRAP EACH SIDE	FULL WALL HT.

STEEL COLUMN SCHEDULE

MARK	COLUMN	BASE PLATE	CAP PLATE
C-1	TS10x10x5/16	16x16x7/8	3/4"
C-2	W8x31	14x14x3/4	3/4"

STEEL GIRT SCHEDULE

MARK	SECTION	CONNECTION SHEAR FORCE
G-1	MC6x13.1	10 KIPS (SERVICE)
G-2	MC6x18	10 KIPS (SERVICE)

MECHANICAL ROOFTOP UNIT SCHEDULE

MARK	MECHANICAL CAPACITY	TOTAL WEIGHT	CORNER WEIGHT
M-1	3 TONS	600 LBS.	255 LBS.
M-2	4 TONS	700 LBS.	275 LBS.
M-3	5 TONS	800 LBS.	300 LBS.

STEEL JOIST WIND UPLIFT SCHEDULE

SPAN	"NET" WIND UPLIFT PRESSURE	
	WITHIN 10' OF BLDG. PERIMETER	ALL OTHER AREAS
LESS THAN 27 FEET	55 PSF	45 PSF
FROM 27 TO 46 FEET	45 PSF	35 PSF
OVER 46 FEET	35 PSF	25 PSF

STEEL JOIST END MOMENT SCHEDULE

MARK	DUE TO LIVE LOAD	DUE TO WIND LOAD
JM-1	2.5 FT-KIPS	2.1 FT-KIPS
JM-2	3.2 FT-KIPS	4.5 FT-KIPS

SHEARWALL SCHEDULE

WALL LEVEL	SHEATHING	NAILS		CHORD STUDS	BLOCKED EDGES	HOLD DOWN ANCHOR		CHORD STUD THRU-BOLTS
		EDGES	FIELD			BOLT TO FDN.	EMBED.	
SW-1 1ST TO 2ND	NOM. 1/2" PWD ON EXTERIOR / 1/2" GYP. ON INTERIOR	8d@6" / 5d@7"	8d@12" / 5d@7"	DOUBLE	NOT REQD.	3/4" AT HD5A	16"	SEE HOLD-DOWN ANCHOR SCHED.
SW-1 2ND TO ROOF	NOM. 1/2" PWD ON EXTERIOR / 1/2" GYP. ON INTERIOR	8d@6" / 5d@7"	8d@12" / 5d@7"	DOUBLE	NOT REQD.	—	—	SEE HOLD-DOWN ANCHOR SCHED.
SW-2 1ST TO 2ND	NOM. 1/2" PWD ON EXTERIOR / 1/2" GYP. ON INTERIOR	8d@4" / 5d@7"	8d@12" / 5d@7"	DOUBLE	REQUIRED	7/8" AT HD10A	32"	SEE HOLD-DOWN ANCHOR SCHED.
SW-2 2ND TO ROOF	NOM. 1/2" PWD ON EXTERIOR / 1/2" GYP. ON INTERIOR	8d@6" / 5d@7"	8d@12" / 5d@7"	DOUBLE	REQUIRED	—	—	SEE HOLD-DOWN ANCHOR SCHED.

NOTES:

1. PROVIDE "COOLER NAILS" AT GYPSUM WALL BOARD SHEATHING.

2. COORDINATE HOLD DOWN ANCHOR BOLTS, THRU-BOLTS AND OTHER REQUIREMENTS WITH THE HARDWARE SELECTED.

SCHEDULE OF REQUIRED STRESS VALUES
VISUALLY GRADED, NO. 2 MATERIAL (@100% STRESS RATING)

WOOD FRAMING SIZES	REQUIRED VALUES, IN PSI					
	Fb	Ft	Fv	Fc⊥	Fc	E
SOUTHERN YELLOW PINE						
2×2, 2×3, 2×4, 4×4	1500	825	90	565	1650	1,600,000
2×6, 3×6, & 4×6	1250	725	90	565	1600	1,600,000
2×8, 3×8, & 4×8	1200	650	90	565	1550	1,600,000
2×10, 3×10, & 4×10	1050	575	90	565	1500	1,600,000
2×12, 3×12, & 4×12	975	550	90	565	1450	1,600,000
SPRUCE PINE FIR						
2×2 THRU 4×12	875	425	70	425	1100	1,400,000

NAILING SCHEDULE

BUILDING ELEMENT	NAILS	NUMBER & LOCATION
WALL FRAMING:		
STUD TO SILL PL	8d COMMON	4 TOE NAIL
	OR 16d COMMON	2 DIRECT NAIL
STUD TO CAP PL	16d COMMON	2 TOE NAIL OR DIRECT
DOUBLE STUDS	10d COMMON	12" O/C DIRECT
CORNER STUDS	16d COMMON	24" O/C DIRECT
SILL PL TO JOIST/BLOCKING	16d COMMON	16" O/C
DOUBLE TOP PL	10d COMMON	16" O/C DIRECT
TOP PL LAPS	10d COMMON	2 DIRECT
RAFTER FRAMING:		
RAFTER TO PL	8d COMMON	3 TOE NAIL
RAFTER TO RIDGE	16d COMMON	2 TOE NAIL OR DIRECT
JACK RAFTER TO HIP	10d COMMON	3 TOE NAIL
	OR 16d COMMON	2 DIRECT NAIL
JOIST FRAMING:		
JOIST TO PL OR GIRDER	8d COMMON	3 TOE NAIL
CEILING JOIST TO PL	16d COMMON	3 TOE NAIL
CEILING JOIST LAPS	10d COMMON	3 DIRECT
CEILING JOIST TO RAFTER	10d COMMON	3 DIRECT
MISCELLANEOUS:		
BRIDGING	8d COMMON	2 DIRECT EACH END
BLOCKING	8d COMMON	2 TOE NAIL, ALL 3 SIDES
BUILT-UP GIRDER	20d COMMON	32" O/C DIRECT
CONT. HEADER TO STUD	8d COMMON	4 TOE NAIL
CONT. HEADER, 2 PIECES	16d COMMON	16" O/C DIRECT, 2 ROWS

NOTE: THIS SCHEDULE IS BASED UPON THE "FASTENING SCHEDULE" IN BOCA 1996. ANY QUESTIONS OR ITEMS NOT INCLUDED ON THE ABOVE SCHEDULE SHALL BE BROUGHT TO THE ENGINEER'S IMMEDIATE ATTENTION.

WOOD GIRDER SCHEDULE

MARK	SIZE	REMARKS
G-1	(2) - 2×12	BOTH ONE SIDE
G-2	(3) - 2×12	ALL ONE SIDE
G-3	(4) - 2×12	2 EA. SIDE

HOLD-DOWN ANCHOR SCHEDULE

MARK	VERTICAL BOLT	CONCRETE EMBED.	CHORD THRU-BOLTS	ALLOWABLE UPLIFT **
HD-2A	5/8"	16"	(2) - 5/8"	2775 #
HD-5A	3/4"	16"	(2) - 3/4"	3705 #
HD-6A	7/8"	16"	(2) - 7/8"	4405 #
HD-8A	7/8"	16"	(3) - 7/8"	6465 #

** - UPLIFT CAPACITIES ARE BASED ON THRU-BOLTS IN DOUBLE STUDS.

WOOD I-JOIST SCHEDULE

MARK	JOIST DEPTH	JOIST SPACING	MINIMUM REQUIREMENTS		
			ALLOWABLE END REACTION	ALLOWABLE MOMENT	EI (x 10^6 LBS-IN^2)
FJ-1	14"	24"	1040 LBS	5200 FT-LBS	560
FJ-2	14"	16"	1040 LBS	5200 FT-LBS	560
FJ-3	14"	24"	624 LBS	2000 FT-LBS	200

Index

About the Author

David R. Williams, P.E., is the founder of Williams Engineering Associates, P.C., a Virginia-based engineering firm which specializes in the design and renovation of building-related projects. Previously, he was a project manager and structural department head at Hayes, Seay, Mattern & Mattern, Inc., a full-service A/E firm in Virginia. A member of the National Society of Professional Engineers and American Society of Civil Engineers, among several other professional groups, Mr. Williams publishes a quarterly newsletter *Building Design and Construction News*.